Selected Titles in This Series

7 **Dmitri Fomin, Sergey Genkin, and Ilia Itenberg,** Mathematical circles (Russian experience), 1996

6 **David W. Farmer and Theodore B. Stanford,** Knots and surfaces: A guide to discovering mathematics, 1996

5 **David W. Farmer,** Groups and symmetry: A guide to discovering mathematics, 1996

4 **V. V. Prasolov,** Intuitive topology, 1995

3 **L. E. Sadovskiĭ and A. L. Sadovskiĭ,** Mathematics and sports, 1993

2 **Yu. A. Shashkin,** Fixed points, 1991

1 **V. M. Tikhomirov,** Stories about maxima and minima, 1990

Mathematical World • Volume 7

Mathematical Circles
(Russian Experience)

Dmitri Fomin
Sergey Genkin
Ilia Itenberg

Translated from the Russian by
Mark Saul

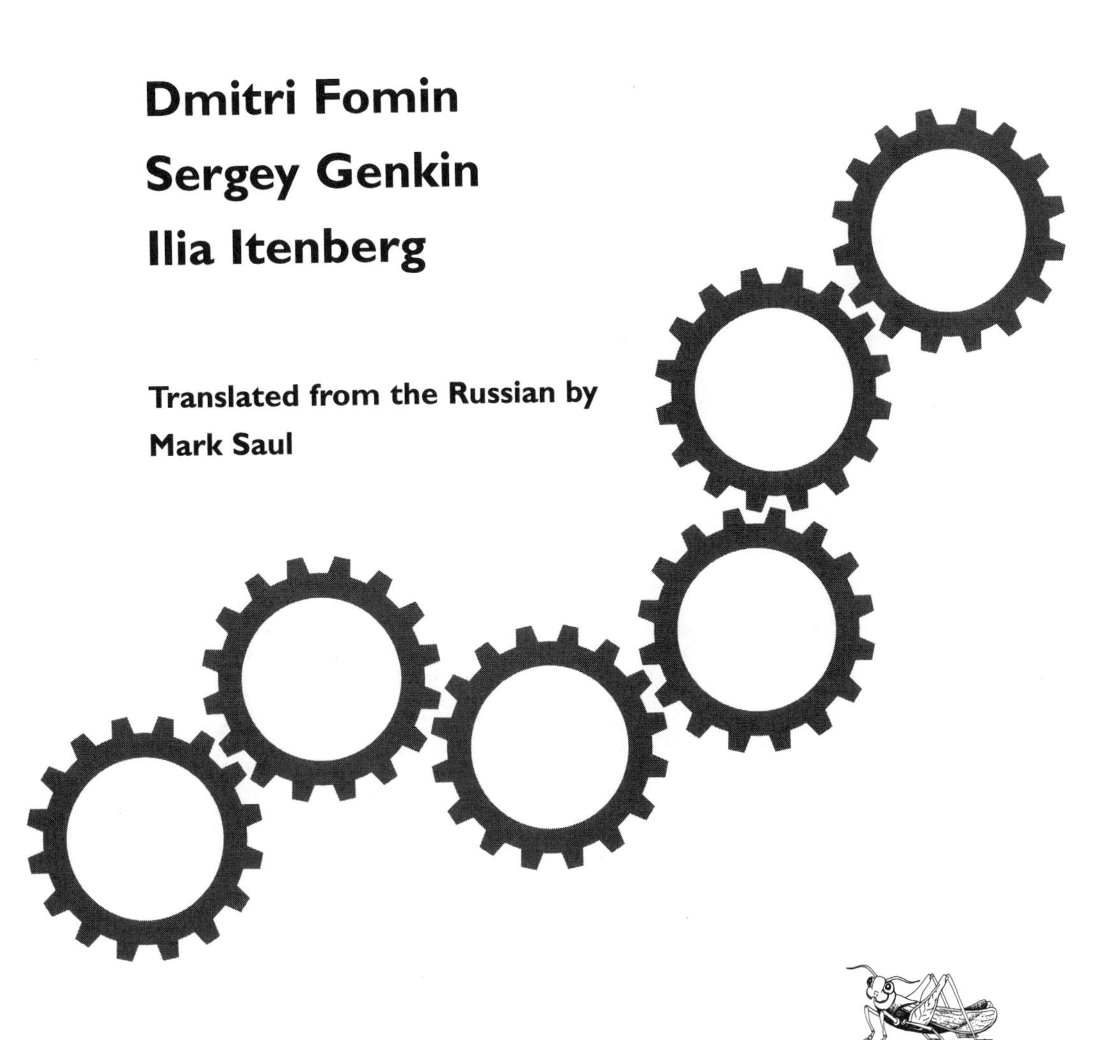

American Mathematical Society

С. А. ГЕНКИН, И. В. ИТЕНБЕРГ, Д. В. ФОМИН

МАТЕМАТИЧЕСКИЙ КРУЖОК

САНКТ-ПЕТЕРБУРГ 1992, 1993

Translated from the Russian by Mark Saul

1991 *Mathematics Subject Classification.* Primary 00A08; Secondary 00A07.

ABSTRACT. This book is intended for students and teachers who love mathematics and want to study its various branches beyond the limits of school curriculum. It is also a book of mathematical recreations, and at the same time a book containing vast theoretical and problem material in some areas of what authors consider to be "extracurricular mathematics". The book is based on an experience gained by several generations of Russian educators and scholars.

Library of Congress Cataloging-in-Publication Data
Genkin, S. A. (Sergeĭ Aleksandrovich)
[Matematicheskiĭ kruzhok. English]
Mathematical circles : (Russian experience) / Dmitri Fomin, Sergey Genkin, Ilia Itenberg; translated from the Russian by Mark Saul.
p. cm. — (Mathematical world, ISSN 1055-9426; v. 7)
On Russian ed., Genkin's name appears first on t.p.
Includes bibliographical references (p. –).
ISBN 0-8218-0430-8 (alk. paper)
1. Mathematical recreations. I. Fomin, D. V. (Dmitriĭ Vladimirovich) II. Itenberg, I. V. (Il′i͡a Vladimirovich) III. Title. IV. Series.
QA95.G3813 1996
510′.76—dc20
96-17683
CIP

∞ The paper used in this book is acid-free and falls within the guidelines
established to ensure permanence and durability.

10 9 8 7 6 5 4 3 2 01 00 99 98 97

Contents

Foreword

This is not a textbook. It is not a contest booklet. It is not a set of lessons for classroom instruction. It does not give a series of projects for students, nor does it offer a development of parts of mathematics for self-instruction.

So what kind of book is this? It is a book produced by a remarkable cultural circumstance, which fostered the creation of groups of students, teachers, and mathematicians, called mathematical circles, in the former Soviet Union. It is predicated on the idea that studying mathematics can generate the same enthusiasm as playing a team sport, without necessarily being competitive.

Thus it is more like a book of mathematical recreations—except that it is more serious. Written by research mathematicians holding university appointments, it is the result of these same mathematicians' years of experience with groups of high school students. The sequences of problems are structured so that virtually any student can tackle the first few examples. Yet the same principles of problem solving developed in the early stages make possible the solution of extremely challenging problems later on. In between, there are problems for every level of interest or ability.

The mathematical circles of the former Soviet Union, and particularly of Leningrad (now St. Petersburg, where these problems were developed) are quite different from most math clubs in the United States. Typically, they were run not by teachers, but by graduate students or faculty members at a university, who considered it part of their professional duty to show younger students the joys of mathematics. Students often met far into the night, and went on weekend trips or summer retreats together, achieving a closeness and mutual support usually reserved in our country for members of athletic teams.

We are fortunate to be living in a time when Russians and Americans can easily communicate and share their cultures. The development of mathematics education is an aspect of Russian culture from which we have much to learn. It is still very rare to find research mathematicians in America willing to devote time, energy, and thought to the development of materials for high school students.

So we must borrow from our Russian colleagues. The present book is the result of such borrowings. Some chapters, such as the one on the triangle inequality, can be used directly in American classrooms, to supplement the development in the usual textbooks. Others, such as the discussion of graph theory, stretch the curriculum with gems of mathematics which are not usually touched on in the classroom. Still others, such as the chapter on games, offer a rich source of extra-curricular materials with more structure and meaning than many.

Each chapter gives examples of mathematical methods in some of their barest forms. A game of nim, which can be enjoyed and even analyzed by a third grader, turns out to be the same as a game played with a single pawn on a chessboard. This becomes a lesson for seventh graders in restating problems, then offers an introduction to the nature of isomorphism for the high school student. The Pigeon Hole Principle, among the simplest yet most profound mathematics has to offer, becomes a tool for proof in number theory and geometry.

Yet the tone of the work remains light. The chapter on combinatorics does not require an understanding of generating functions or mathematical induction. The problems in graph theory, too, remain on the surface of this important branch of mathematics. The approach to each topic lends itself to mind play, not weighty reflection. And yet the work manages to strike some deep notes.

It is this quality of the work which the mathematicians of the former Soviet Union developed to a high art. The exposition of mathematics, and not just its development, became a part of the Russian mathematician's work. This book is thus part of a literary genre which remains largely undeveloped in the English language.

Mark Saul, Ph.D.
Bronxville Schools
Bronxville, New York

Preface to the Russian Edition

§1. Introduction

This book was originally written to help people in the former Soviet Union who dealt with extracurricular mathematical education: school teachers, university professors participating in mathematical education programs, various enthusiasts running mathematical circles, or people who just wanted to read something both mathematical and recreational. And, certainly, students can also use this book independently.

Another reason for writing this book was that we considered it necessary to record the role played by the traditions of mathematical education in Leningrad (now St. Petersburg) over the last 60 years. Though our city was, indeed, the cradle of the olympiad movement in the USSR (having seen the very first mathematical seminars for students in 1931–32, and the first city olympiad in 1934), and still remains one of the leaders in this particular area, its huge educational experience has not been adequately recorded for the interested readers.

* * *

In spite of the stylistic variety of this book's material, it is methodologically homogeneous. Here we have, we believe, all the basic topics for sessions of a mathematical circle for the first two years of extracurricular education (approximately, for students of age 12–14). Our main objective was to make the preparation of sessions and the gathering of problems easier for the teacher (or any enthusiast willing to spend time with children, teaching them non-standard mathematics). We wanted to talk about mathematical ideas which are important for students, and about how to draw the students' attention to these ideas.

We must emphasize that the work of preparing and leading a session is itself a creative process. Therefore, it would be unwise to follow our recommendations blindly. However, we hope that your work with this book will provide you with material for most of your sessions. The following use of this book seems to be natural: while working on a specific topic the teacher reads and analyzes a chapter from the book, and after that begins to construct a sketch of the session. Certainly, some adjustments will have to be made because of the level of a given group of students. As supplementary sources of problems we recommend [**13, 16, 24, 31, 33**], and [**40**].

* * *

We would like to mention two significant points of the Leningrad tradition of extracurricular mathematical educational activity:

(1) Sessions feature vivid, spontaneous communication between students and teachers, in which each student is treated individually, if possible.

(2) The process begins at a rather early age: usually during the 6th grade (age 11–12), and sometimes even earlier.

This book was written as a guide especially for secondary school students and for their teachers. The age of the students will undoubtedly influence the style of the sessions. Thus, a few suggestions:

A) We consider it wrong to hold a long session for younger students devoted to only one topic. We believe that it is helpful to change the direction of the activity even within one session.

B) It is necessary to keep going back to material already covered. One can do this by using problems from olympiads and other mathematical contests (see Appendix A).

C) In discussing a topic, try to emphasize a few of the most basic landmarks and obtain a complete understanding (not just memorization!) of these facts and ideas.

D) We recommend constant use of non-standard and "gamelike" activities in the sessions, with complete discussion of solutions and proofs. It is important also to use recreational problems and mathematical jokes. These can be found in [**5–7, 16–18, 26–30**].

We must mention here our predecessors—those who have tried earlier to create a sort of anthology for Leningrad mathematical circles. Their books [**32**] and [**43**], unfortunately, did not reach a large number of readers interested in mathematics education in secondary school.

In 1990–91 the original version of the first part of our book was published by the Academy of Pedagogical Sciences of USSR as a collection of articles [**21**] written by a number of authors. We would like to thank all our colleagues whose materials we used when working on the preparation of the present book: Denis G. Benua, Igor B. Zhukov, Oleg A. Ivanov, Alexey L. Kirichenko, Konstantin P. Kokhas, Nikita Yu. Netsvetaev, and Anna G. Frolova.

We also express our sincere gratitude to Igor S. Rubanov, whose paper on induction written especially for the second part of the book [**21**] (but never published, unfortunately) is included here as the chapter "Induction".

Our special thanks go to Alexey Kirichenko whose help in the early stages of writing this book cannot be overestimated. We would also like to thank Anna Nikolaeva for drawing the figures.

§2. Structure of the book

The book consists of this preface, two main parts, Appendix A "Mathematical Contests", Appendix B "Answers, Hints, Solutions", and Appendix C "References".

The first part ("The First Year of Education") begins with Chapter Zero, consisting of test questions intended mostly for students of ages 10–11. The problems of this chapter have virtually no mathematical content, and their main objective is to reveal the abilities of the students in mathematics and logic. The rest of the first part is divided into 8 chapters. The first seven of these are devoted to particular topics, and the eighth ("Problems for the first year") is simply a compilation of problems on a variety of themes.

The second part ("The Second Year of Education") consists of 9 chapters, some of which just continue the discussion in the first part (for example, the chapters "Graphs–2" and "Combinatorics–2"). Other chapters are comprised of material considered to be too complicated for the first year: "Invariants", "Induction", "Inequalities".

Appendix A tells about five main types of mathematical contests popular in the former Soviet Union. These contests can be held at sessions of mathematical circles or used to organize contests between different circles or even schools.

Advice to the teacher is usually given under the remark labelled **"For teachers"**. Rare occasions of **"Methodological remarks"** contain mostly recommendations about the methodology of problem solving: they draw attention to the basic patterns of proofs or methods of recognizing and classifying problems.

§3. Technicalities and legend

(1) The most difficult problems are marked with an asterisk (*).

(2) Almost all of the problems are commented on in Appendix B: either a full solution or at least a hint and answer. If a problem is computational, then we usually provide only an answer. We do not give the solutions to problems for independent solution (this, in particular, goes for all the problems from Chapters 8 and 17).

(3) All the references can be found at the end of the book in the list of references. The books we recommend most are marked with an asterisk.

CHAPTER 0

Chapter Zero

In this chapter we have gathered 25 simple problems. To solve them you do not need anything but common sense and the simplest calculational skills. These problems can be used at sessions of a mathematical circle to probe the logical and mathematical abilities of students, or as recreational questions.

*   *   *

Problem 1. A number of bacteria are placed in a glass. One second later each bacterium divides in two, the next second each of the resulting bacteria divides in two again, et cetera. After one minute the glass is full. When was the glass half-full?

Problem 2. Ann, John, and Alex took a bus tour of Disneyland. Each of them must pay 5 plastic chips for the ride, but they have only plastic coins of values 10, 15, and 20 chips (each has an unlimited number of each type of coin). How can they pay for the ride?

Problem 3. Jack tore out several successive pages from a book. The number of the first page he tore out was 183, and it is known that the number of the last page is written with the same digits in some order. How many pages did Jack tear out of the book?

Problem 4. There are 24 pounds of nails in a sack. Can you measure out 9 pounds of nails using only a balance with two pans? (See Figure 1.)

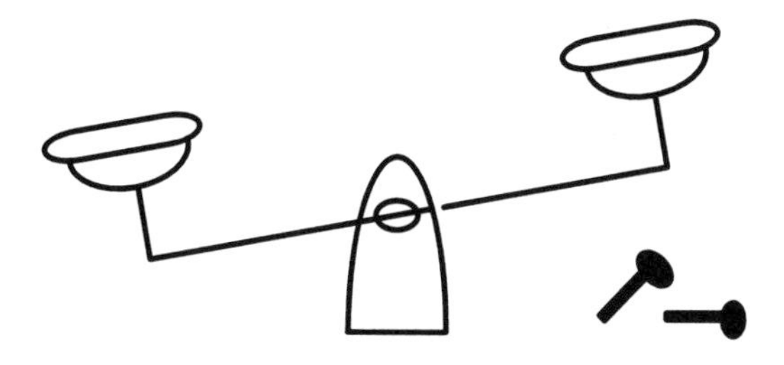

FIGURE 1

Problem 5. A caterpillar crawls up a pole 75 inches high, starting from the ground. Each day it crawls up 5 inches, and each night it slides down 4 inches. When will it first reach the top of the pole?

Problem 6. In a certain year there were exactly four Fridays and exactly four Mondays in January. On what day of the week did the 20th of January fall that year?

Problem 7. How many boxes are crossed by a diagonal in a rectangular table formed by 199×991 small squares?

Problem 8. Cross out 10 digits from the number 1234512345123451234512345 so that the remaining number is as large as possible.

* * *

Problem 9. Peter said: "The day before yesterday I was 10, but I will turn 13 in the next year." Is this possible?

Problem 10. Pete's cat always sneezes before it rains. She sneezed today. "This means it will be raining," Pete thinks. Is he right?

Problem 11. A teacher drew several circles on a sheet of paper. Then he asked a student "How many circles are there?" "Seven," was the answer. "Correct! So, how many circles are there?" the teacher asked another student. "Five," answered the student. "Absolutely right!" replied the teacher. How many circles were really drawn on the sheet?

Problem 12. The son of a professor's father is talking to the father of the professor's son, and the professor does not take part in the conversation. Is this possible?

Problem 13. Three turtles are crawling along a straight road heading in the same direction. "Two other turtles are behind me," says the first turtle. "One turtle is behind me and one other is ahead," says the second. "Two turtles are ahead of me and one other is behind," says the third turtle. How can this be possible?

Problem 14. Three scholars are riding in a railway car. The train passes through a tunnel for several minutes, and they are plunged into darkness. When they emerge, each of them sees that the faces of his colleagues are black with the soot that flew in through the open window. They start laughing at each other, but, all of a sudden, the smartest of them realizes that his face must be soiled too. How does he arrive at this conclusion?

Problem 15. Three tablespoons of milk from a glass of milk are poured into a glass of tea, and the liquid is thoroughly mixed. Then three tablespoons of this mixture are poured back into the glass of milk. Which is greater now: the percentage of milk in the tea or the percentage of tea in the milk?

* * *

Problem 16. Form a magic square with the digits 1, 2, 3, 4, 5, 6, 7, 8, and 9; that is, place them in the boxes of a 3×3 table so that all the sums of the numbers along the rows, columns, and two diagonals are equal.

Problem 17. In an arithmetic addition problem the digits were replaced with letters (equal digits by same letters, and different digits by different letters). The result is: LOVES + LIVE = THERE. How many "loves" are "there"? The answer is the maximum possible value of the word THERE.

Problem 18. The secret service of The Federation intercepted a coded message from The Dominion which read: BLASE + LBSA = BASES. It is known that equal digits are coded with equal letters, and different digits with different letters. Two giant computers came up with two different answers to the riddle. Is this possible or does one of them need repair?

Problem 19. Distribute 127 one dollar bills among 7 wallets so that any integer sum from 1 through 127 dollars can be paid without opening the wallets.

* * *

Problem 20. Cut the figure shown in Figure 2 into four figures, each similar to the original with dimensions twice as small.

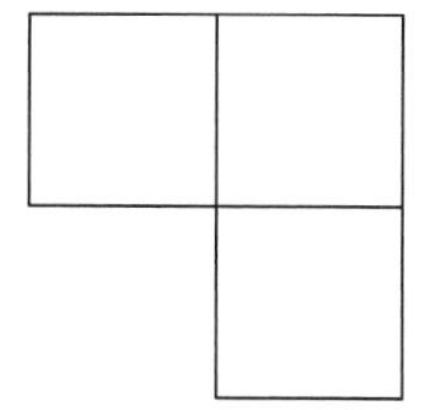

FIGURE 2

Problem 21. Matches are arranged to form the figure shown in Figure 3. Move two matches to change this figure into four squares with sides equal in length to one match.

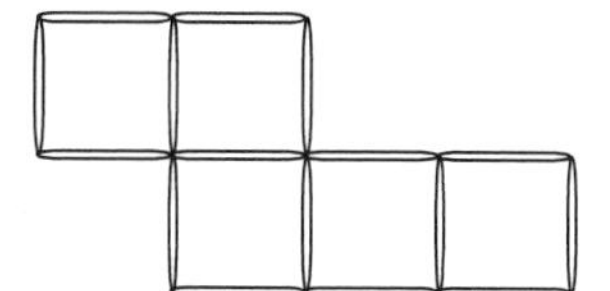

FIGURE 3

Problem 22. A river 4 meters wide makes a $90°$ turn (see Figure 4). Is it possible to cross the river by bridging it with only two planks, each 3.9 meters long?

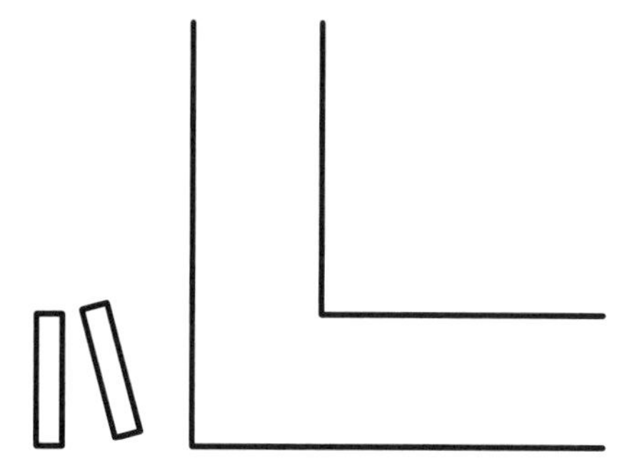

FIGURE 4

Problem 23. Is it possible to arrange six long round pencils so that each of them touches all the others?

Problem 24. Using scissors, cut a hole in a sheet of ordinary paper (say, the size of this page) through which an elephant can pass.

Problem 25. Ten coins are arranged as shown in Figure 5. What is the minimum number of coins we must remove so that no three of the remaining coins lie on the vertices of an equilateral triangle?

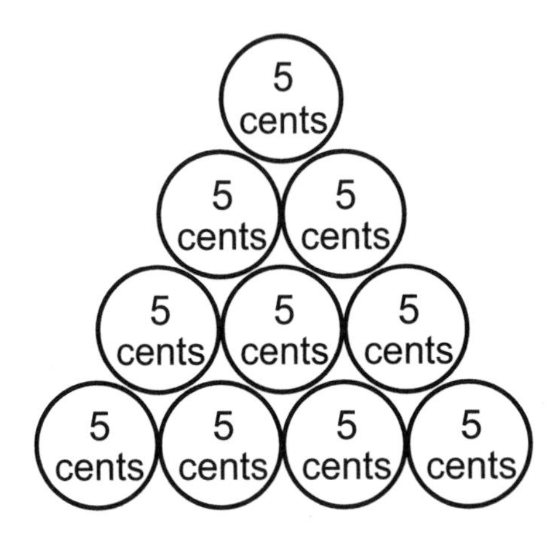

FIGURE 5

CHAPTER 1

Parity

An even number is said to have even parity, and an odd number, odd parity. This concept, despite its utmost simplicity, appears in the solution of the most varied sorts of questions. It turns out to be useful in the solution of many problems, including some which are quite difficult.

The very simplicity of this theme makes it possible to pose interesting problems for students with almost no background. The same simplicity makes it even more important than usual to point out the common theme in all such problems.

§1. Alternations

Problem 1. Eleven gears are placed on a plane, arranged in a chain as shown (see Figure 6). Can all the gears rotate simultaneously?

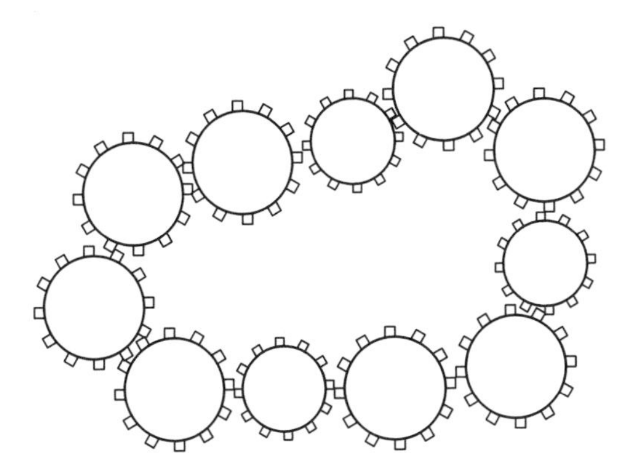

FIGURE 6

Solution. The answer is no. Suppose that the first gear rotates clockwise. Then the second gear must rotate counter-clockwise, the third clockwise again, the fourth counter-clockwise, and so on. It is clear that the "odd" gears must rotate clockwise, while the "even" gears must rotate counter-clockwise. But then the first and eleventh gears must rotate in the same direction. This is a contradiction.

The main idea in the solution to this problem is that the gears rotating clockwise and counter-clockwise alternate. Finding objects that alternate is the basic idea in the solution of the following problems as well.

Problem 2. On a chessboard, a knight starts from square $a1$, and returns there after making several moves. Show that the knight makes an even number of moves.

Problem 3. Can a knight start at square $a1$ of a chessboard, and go to square $h8$, visiting each of the remaining squares exactly once on the way?

Solution. No, he cannot. At each move, a knight jumps from a square of one color to a square of the opposite color. Since the knight must make 63 moves, the last (odd) move must bring him to a square of the opposite color from the square on which he started. However, squares $a1$ and $h8$ are of the same color.

Like Problem 3, many of the problems in this section deal with proofs that certain situations are impossible. Indeed, when a question asks whether some situation is possible, the answer in this section is invariably "no". This poses some difficulty for mathematically naive students. Their first reaction is either frustration that they cannot find the "correct" situation (fulfilling the impossible conditions) or a declaration that the situation is impossible, without a clear conception of what it might take to prove this. Here is a simple problem, related to the "odd and even" problems later in this section, which might clear up this point:

Can you find five odd numbers whose sum is 100?

A discussion can ensue, through which students are made aware that it is not just their own human failing that prevents them from finding this set of numbers, but a contradiction in the nature of the set itself. It is proof by contradiction that is at the basis of the students' confusion, as well as the notion of proof of impossibility. Problems in parity are a simple yet effective way to introduce both these concepts.

Problem 4. A closed path is made up of 11 line segments. Can one line, not containing a vertex of the path, intersect each of its segments?

Problem 5. Three hockey pucks, A, B, and C, lie on a playing field. A hockey player hits one of them in such a way that it passes between the other two. He does this 25 times. Can he return the three pucks to their starting points?

Problem 6. Katya and her friends stand in a circle. It turns out that both neighbors of each child are of the same gender. If there are five boys in the circle, how many girls are there?

Let us note an additional principle, which comes up in the solution of the previous problem: in a closed alternating chain of objects, there are as many objects of one type (boys) as there are of the other (girls).

§2. Partitioning into pairs

Problem 7. Can we draw a closed path made up of 9 line segments, each of which intersects exactly one of the other segments?

Solution. If such a closed path were possible, then all the line segments could be partitioned into pairs of intersecting segments. But then the number of segments would have to be even.

Let us single out the central point in this solution: if a set of objects can be partitioned into pairs, then there are evenly many of them. Here are some similar problems:

Problem 8. Can a 5×5 square checkerboard be covered by 1×2 dominoes?

Problem 9. Given a convex 101-gon which has an axis of symmetry, prove that the axis of symmetry passes through one of its vertices. What can you say about a 10-gon with the same properties?

Problems 10 and 11 concern a set of dominoes consisting of 2×1 rectangles with 0 to 6 spots on each square. All 28 possible pairs of numbers of spots (including doubles) are represented. The game is played by forming a chain in which squares of adjacent dominoes have equal numbers of spots.

Problem 10. All the dominoes in a set are laid out in a chain (so that the number of spots on the ends of adjacent dominoes match). If one end of the chain is a 5, what is at the other end?

Comment. A set of dominoes consists of 2×1 rectangles with 0 to 6 spots on each square. All 28 possible pairs of numbers of spots (including doubles) are represented.

Problem 11. In a set of dominoes, all those in which one square has no spots are discarded. Can the remaining dominoes be arranged in a chain?

Problem 12. Can a convex 13-gon be divided into parallelograms?

Problem 13. Twenty-five checkers are placed on a 25×25 checkerboard in such a way that their positions are symmetric with respect to one of its diagonals. Prove that at least one of the checkers is positioned on that diagonal.

Solution. If no checker occurred on the diagonal, then the checkers could be partitioned into pairs, placed symmetrically with respect to the diagonal. Therefore, there must be one (and in fact an odd number) of checkers on the diagonal.

In solving this problem, students often have trouble understanding that there may be not just one, but any odd number of checkers on the diagonal. For this problem, we may formulate our assertion about partitions into pairs thus: if we form a number of pairs from a set of oddly many objects, then at least one object will remain unpaired.

Problem 14. Let us now assume that the positions of the checkers in Problem 13 are symmetric with respect to both diagonals of the checkerboard. Prove that one of the checkers is placed in the center square.

Problem 15. In each box of a 15×15 square table one of the numbers 1, 2, 3, ..., 15 is written. Boxes which are symmetric to one of the main diagonals contain equal numbers, and no row or column contains two copies of the same number. Show that no two of the numbers along the main diagonal are the same.

§3. Odd and even

Problem 16. Can one make change of a 25-ruble bill, using in all ten bills each having a value of 1, 3, or 5 rubles?

Solution. It is not possible. This conclusion is based on a simple observation: the sum of evenly many odd numbers is even. A generalization of this fact is this: the parity of the sum of several numbers depends only on the parity of the number of its odd addends. If there are oddly (evenly) many odd addends, then the sum is odd (even).

Problem 17. Pete bought a notebook containing 96 pages, and numbered them from 1 through 192. Victor tore out 25 pages of Pete's notebook, and added the 50 numbers he found on the pages. Could Victor have gotten 1990 as the sum?

Problem 18. The product of 22 integers is equal to 1. Show that their sum cannot be zero.

Problem 19. Can one form a "magic square" out of the first 36 prime numbers?

A "magic square" here means a 6×6 array of boxes, with a number in each box, and such that the sum of the numbers along any row, column, or diagonal is constant.

Problem 20. The numbers 1 through 10 are written in a row. Can the signs "+" and "−" be placed between them, so that the value of the resulting expression is 0?

Note that negative numbers can also be odd or even.

Problem 21. A grasshopper jumps along a line. His first jump takes him 1 cm, his second 2 cm, and so on. Each jump can take him to the right or to the left. Show that after 1985 jumps the grasshopper cannot return to the point at which he started.

Problem 22. The numbers 1, 2, 3, ... , 1984, 1985 are written on a blackboard. We decide to erase from the blackboard any two numbers, and replace them with their positive difference. After this is done several times, a single number remains on the blackboard. Can this number equal 0?

§4. Assorted problems

Some more difficult problems are collected in this section. Their solutions use the ideas of parity, but also additional considerations.

Problem 23. Can an ordinary 8×8 chessboard be covered with 1×2 dominoes so that only squares $a1$ and $h8$ remain uncovered?

Problem 24. A 17-digit number is chosen, and its digits are reversed, forming a new number. These two numbers are added together. Show that their sum contains at least one even digit.

Problem 25. There are 100 soldiers in a detachment, and every evening three of them are on duty. Can it happen that after a certain period of time each soldier has shared duty with every other soldier exactly once?

Problem 26. Forty-five points are chosen along line AB, all lying outside of segment AB. Prove that the sum of the distances from these points to point A is not equal to the sum of the distances of these points to point B.

Problem 27. Nine numbers are placed around a circle: four 1's and five 0's. The following operation is performed on the numbers: between each adjacent pair of numbers is placed a 0 if the numbers are different, and a 1 if the numbers are the same. The "old" numbers are then erased. After several of these operations, can all the remaining numbers be equal?

Problem 28. Twenty-five boys and 25 girls are seated at a round table. Show that both neighbors of at least one student are boys.

Problem 29. A snail crawls along a plane with constant velocity, turning through a right angle every 15 minutes. Show that the snail can return to its starting point only after a whole number of hours.

Problem 30. Three grasshoppers play leapfrog along a line. At each turn, one grasshopper leaps over another, but not over two others. Can the grasshoppers return to their initial positions after 1991 leaps? (See Figure 7.)

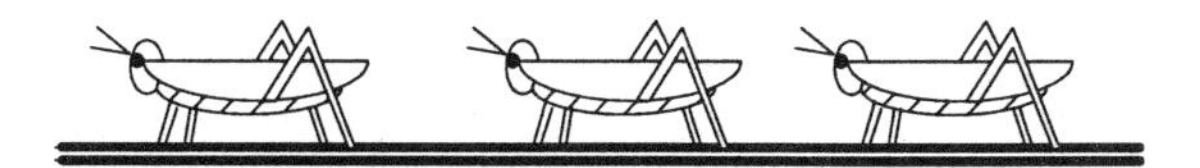

FIGURE 7

Problem 31. Of 101 coins, 50 are counterfeit, and differ from the genuine coins in weight by 1 gram. Peter has a scale in the form of a balance which shows the difference in weights between the objects placed in each pan. He chooses one coin, and wants to find out in one weighing whether it is counterfeit. Can he do this?

Problem 32. Is it possible to arrange the numbers from 1 through 9 in a sequence so that there are oddly many numbers between 1 and 2, between 2 and 3, ... , and between 8 and 9?

CHAPTER 2

Combinatorics–1

How many ways are there to drive from A to B? How many words does the Hermetian language contain? How many "lucky" six-digit numbers are there? How many ... ? These and many other similar questions will be discussed in this chapter.

We will start with a few simple problems.

Problem 1. There are five different teacups and three different tea saucers in the "Tea Party" store. How many ways are there to buy a cup and a saucer?

Solution. First, let us choose a cup. Then, to complete the set, we can choose any of three saucers. Thus we have 3 different sets containing the chosen cup. Since there are five cups, we have 15 different sets ($15 = 5 \cdot 3$).

Problem 2. There are also four different teaspoons in the "Tea Party" store. How many ways are there to buy a set consisting of a cup, a saucer, and a spoon?

Solution. Let us start with any of the 15 sets from the previous problem. There are four different ways to complete it by choosing a spoon. Therefore, the number of all possible sets is 60 (since $60 = 15 \cdot 4 = 5 \cdot 3 \cdot 4$).

In just the same way we can solve the following problem.

Problem 3. There are three towns A, B, and C, in Wonderland. Six roads go from A to B, and four roads go from B to C (see Figure 8). In how many ways can one drive from A to C?

Answer. $24 = 6 \cdot 4$.

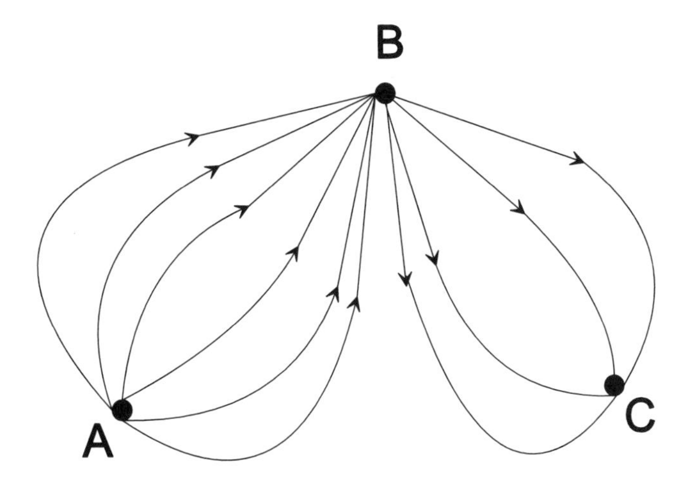

FIGURE 8

In the solution to Problem 4 we use a new idea.

Problem 4. A new town called D and several new roads were built in Wonderland (see Figure 9). How many ways are there to drive from A to C now?

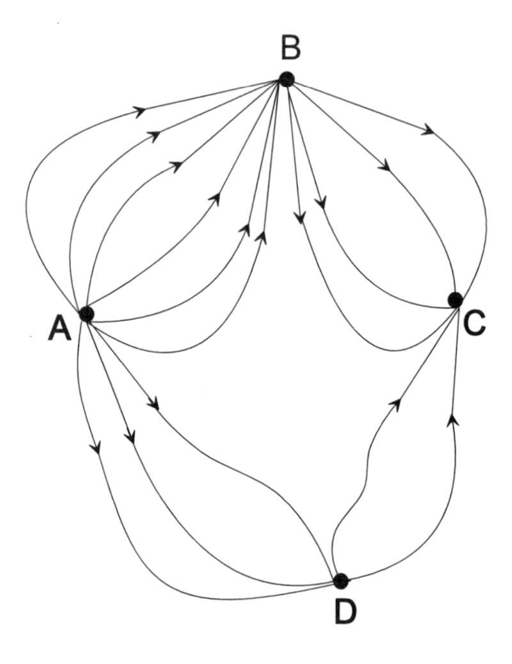

FIGURE 9

Solution. Consider two cases: our route passes either through B or through D. In each case it is quite easy to calculate the number of routes—if we drive through B then we have 24 ways to drive from A to C; otherwise we have 6 ways. To obtain the answer we must add up these two numbers. Thus we have 30 possible routes.

Dividing the problem into several cases is a very useful idea. It also helps in solving Problem 5.

Problem 5. There are five different teacups, three saucers, and four teaspoons in the "Tea Party" store. How many ways are there to buy two items with different names?

Solution. Three cases are possible: we buy a cup and a saucer, or we buy a cup and a spoon, or we buy a saucer and a spoon. It is not difficult to calculate the number of ways each of these cases can occur: 15, 20, and 12 ways respectively. Adding, we have the answer: 47.

For teachers. The main goal which the teacher must pursue during a discussion of these problems is making the students understand when we must add the numbers of ways and when we must multiply them. Of course, many problems should be presented (some can be found at the end of this chapter (Problems 28–32), in [**49**], or created by the teacher). Some possible subjects are shopping, traffic maps, arrangement of objects, etc.

Problem 6. We call a natural number "odd-looking" if all of its digits are odd. How many four-digit odd-looking numbers are there?

Solution. It is obvious that there are 5 one-digit odd-looking numbers. We can add another odd digit to the right of any odd-looking one-digit number in five ways. Thus, we have $5 \cdot 5 = 25$ two-digit odd-looking numbers. Similarly, we get $5 \cdot 5 \cdot 5 = 125$ three-digit odd-looking numbers, and $5 \cdot 5 \cdot 5 \cdot 5 = 5^4 = 625$ four-digit odd-looking numbers.

__For teachers.__ In the last problem the answer has the form m^n. Usually, an answer of this type results from problems where we can place an element of some given m-element set in each of n given places. In such problems the students may encounter difficulty distinguishing the two numbers m and n, therefore confusing the base and the exponent.

Here are four more similar problems.

Problem 7. We toss a coin three times. How many different sequences of heads and tails can we obtain?
Answer. 2^3.

Problem 8. Each box in a 2×2 table can be colored black or white. How many different colorings of the table are there?
Answer. 2^4.

Problem 9. How many ways are there to fill in a Special Sport Lotto card? In this lotto you must predict the results of 13 hockey games, indicating either a victory for one of two teams, or a draw.
Answer. 3^{13}.

Problem 10. The Hermetian alphabet consists of only three letters: A, B, and C. A word in this language is an arbitrary sequence of no more than four letters. How many words does the Hermetian language contain?

Hint. Calculate separately the numbers of one-letter, two-letter, three-letter, and four-letter words.
Answer. $3 + 3^2 + 3^3 + 3^4 = 120$.

Let us continue with another set of problems.

Problem 11. A captain and a deputy captain must be elected in a soccer team with 11 players. How many ways are there to do this?

Solution. Any of 11 players can be elected as captain. After that, any of the 10 remaining players can be chosen for deputy. Therefore, we have $11 \cdot 10 = 110$ different outcomes of elections.

This problem differs from the previous ones in that the choice of captain influences the set of candidates for deputy position, since the captain cannot be his or her own deputy. Thus, the choices of captain and deputy are not independent (as the choices of a cup and a saucer were in Problem 1, for example).

Below we have four more problems on the same theme.

Problem 12. How many ways are there to sew one three-colored flag with three horizontal strips of equal height if we have pieces of fabric of six colors? We can distinguish the top of the flag from the bottom.

Solution. There are six possible choices of a color for the bottom strip. After that we have only five colors to use for the middle strip, and then only four colors for the top strip. Therefore, we have $6 \cdot 5 \cdot 4 = 120$ ways to sew the flag.

Problem 13. How many ways are there to put one white and one black rook on a chessboard so that they do not attack each other?

Solution. The white rook can be placed on any of the 64 squares. No matter where it stands, it attacks exactly 15 squares (including the square it stands on). Thus we are left with 49 squares where the black rook can be placed. Hence there are $64 \cdot 49 = 3136$ different ways.

Problem 14. How many ways are there to put one white and one black king on a chessboard so that they do not attack each other?

Solution. The white king can be placed on any of the 64 squares. However, the number of squares it attacks depends on its position. Therefore, we have three cases:

a) If the white king stands in one of the corners then it attacks 4 squares (including the square it stands on). We have 60 squares left, and we can place the black king on any of them.

b) If the white king stands on the edge of the chessboard but not in the corner (there are 24 squares of this type) then it attacks 6 squares, and we have 58 squares to place the black king on.

c) If the white king does not stand on the edge of the chessboard (we have 36 squares of this type) then it attacks 9 squares, and only 55 squares are left for the black king.

Finally, we have $4 \cdot 60 + 24 \cdot 58 + 36 \cdot 55 = 3612$ ways to put both kings on the chessboard.

* * *

Let us now calculate the number of ways to arrange n objects in a row. Such arrangements are called *permutations*, and they play a significant role in combinatorics and in algebra. But before this we must digress a little bit.

If n is a natural number, then $n!$ (pronounced n *factorial*) is the product $1 \cdot 2 \cdot 3 \cdot \ldots \cdot n$. Therefore, $2! = 2$, $3! = 6$, $4! = 24$, and $5! = 120$. For convenience of calculations and for consistency, $0!$ is defined to be equal to 1.

<u>**Methodological remark.**</u> Before working with permutations one must know the definition of factorial and learn how to deal with this function. The following exercises may be useful.

Exercise 1. Simplify the expressions a) $10! \cdot 11$; b) $n! \cdot (n+1)$.

Exercise 2. a) Calculate $100!/98!$; b) Simplify $n!/(n-1)!$.

Exercise 3. Prove that if p is a prime number, then $(p-1)!$ is not divisible by p.

Now let us go back to the permutations.

Problem 15. How many three-digit numbers can be written using the digits 1, 2, and 3 (without repetitions) in some order?

Solution. Let us reason just the same way we did in solving Problem 12. The first digit can be any of the three given, the second can be any of the two remaining digits, and the third must be the one remaining digit. Thus we have $3 \cdot 2 \cdot 1 = 3!$ numbers.

Problem 16. How many ways are there to lay four balls, colored red, black, blue, and green, in a row?

Solution. The first place in the row can be occupied by any of the given balls. The second can be occupied by any of the three remaining balls, et cetera. Finally, we have the answer (similar to that of Problem 15): $4 \cdot 3 \cdot 2 \cdot 1 = 4!$.

Analogously we can prove that n different objects can be laid out in a row in $n \cdot (n-1) \cdot (n-2) \cdot \ldots \cdot 2 \cdot 1$ ways; that is

the number of permutations of n objects is $n!$.

For convenience of notation we introduce the following convention. Any finite sequence of English letters will be called "a word" (whether or not it can be found in a dictionary). For example, we can form six words using the letters A, B, and C each exactly once: ABC, ACB, BAC, BCA, CAB, and CBA. In the following five problems you must calculate the number of different words that can be obtained by rearranging the letters of a particular word.

Problem 17. "VECTOR".

Solution. Since all the letters in this word are different, the answer is $6!$ words.

Problem 18. "TRUST".

Solution. This word contains two letters T, and all the other letters are different. Let us temporarily think of these letters T as two different letters T_1 and T_2. Under this assumption we have $5! = 120$ different words. However, any two words which can be obtained from each other just by transposing the letters T_1 and T_2 are, in fact, identical. Thus, our 120 words split into pairs of identical words. This means that the answer is $120/2 = 60$.

Problem 19. "CARAVAN".

Solution. Thinking of the three letters A in this word as different letters A_1, A_2, and A_3, we get $8!$ different words. However, any words which can be obtained from each other just by transposing the letters A_i are identical. Since the letters A_i can be rearranged in their places in $3! = 6$ ways, all $8!$ words split into groups of $3!$ identical words. Therefore the answer is $8!/3!$.

Problem 20. "CLOSENESS".

Solution. We have three letters S and two letters E in this word. Temporarily thinking of all of them as different letters, we have $9!$ words. When we remember that the letters E are identical the number of different words reduces to $9!/2!$. Then, recalling that the letters S are identical, we come to the final answer: $9!/(2! \cdot 3!)$.

Problem 21. "MATHEMATICAL".
Answer. $12!/(3! \cdot 2! \cdot 2!)$.

This set of problems about words demonstrates one very interesting and important idea—the idea of multiple counting. That is, instead of counting the number of objects we are interested in, it may be easier to count some other objects whose number is some known multiple of the number of objects.

Here are four more problems using this method.

Problem 22. There are 20 towns in a certain country, and every pair of them is connected by an air route. How many air routes are there?

Solution. Every route connects two towns. We can choose any of the 20 towns in the country (say, town A) as the beginning of a route, and we have 19 remaining towns to choose the end of a route (say, town B) from. Multiplying, we have $20 \cdot 19 = 380$. However, this calculation counted every route AB twice: when A was chosen as the beginning of the route, and when B was chosen as the beginning. Hence, the number of routes is $380/2 = 190$.

A similar problem is discussed in the chapter "Graphs–1" where we count the number of edges of a graph.

Problem 23. How many diagonals are there in a convex n-gon?

Solution. Any of the n vertices can be chosen as the first endpoint of a diagonal, and we have $n-3$ vertices to choose from for the second end (any vertex, except the chosen one and its two neighbors). Counting the diagonals this way, we have counted every diagonal exactly twice. Hence, the answer is $n(n-3)/2$. (See Figure 10.)

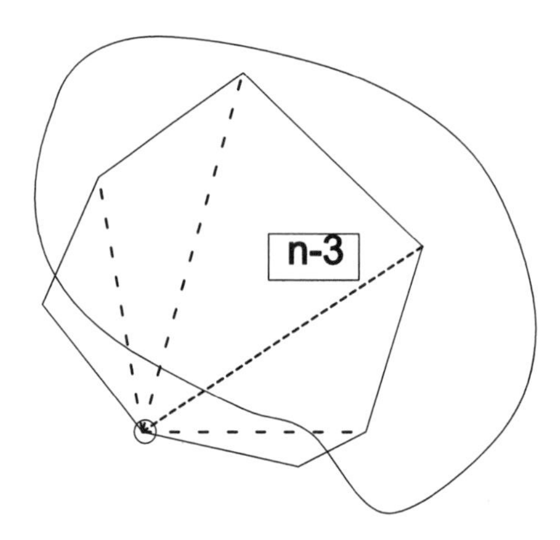

FIGURE 10

Problem 24. A "necklace" is a circular string with several beads on it. It is allowed to rotate a necklace but not to turn it over. How many different necklaces can be made using 13 different beads?

Solution. Let us first assume that it is prohibited to rotate the necklace. Then it is clear that we have 13! different necklaces. However, any arrangement of beads must be considered identical to those 12 that can be obtained from it by rotation. (See Figure 11.)
Answer: $13!/13 = 12!$.

Problem 25. Assume now that it is allowed to turn a necklace over. How many necklaces can be made using 13 different beads?

Solution. Turning the necklace over divides the number of necklaces by 2.
Answer: $12!/2$.

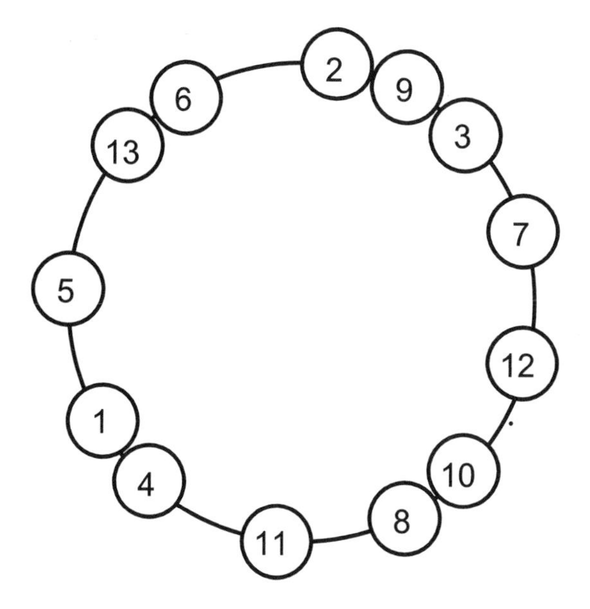

FIGURE 11

The following problem illustrates another important combinatorial idea.

Problem 26. How many six-digit numbers have at least one even digit?

Solution. Instead of counting the numbers with at least one even digit, let us find the number of six-digit numbers that do not possess this property. Since these are exactly the numbers with all their digits odd, there are $5^6 = 15625$ of them (see Problem 6). Since there are 900000 six-digit numbers in all, we conclude that the number of six-digit numbers with at least one even digit is $900000 - 15625 = 884375$.

The main idea in this solution was to use the method of complements; that is, counting (or, considering) the "unrequested" objects instead of those "requested". Here is another problem which can be solved using this method.

Problem 27. There are six letters in the Hermetian language. A word is any sequence of six letters, some pair of which are the same. How many words are there in the Hermetian language?

Answer. $6^6 - 6!$.

<u>**For teachers.**</u> In conclusion we would like to note that it is reasonable to devote a separate session to any idea which ties together the problems of each set in this chapter (and, perhaps, with other themes more distant from combinatorics). We also recommend reviewing the material already covered in previous sessions. For this reason we present here a list of problems for independent solution and for homework. In addition, you can take problems from [**49**] or create them yourself.

Problems for independent solution

Problem 28. There are five types of envelopes and four types of stamps in a post office. How many ways are there to buy an envelope and a stamp?

Problem 29. How many ways are there to choose a vowel and a consonant from the word "RINGER"?

Problem 30. Seven nouns, five verbs, and two adjectives are written on a blackboard. We can form a sentence by choosing one word of each type, and we do not care about how much sense the sentence makes. How many ways are there to do this?

Problem 31. Each of two novice collectors has 20 stamps and 10 postcards. We call an exchange fair if they exchange a stamp for a stamp or a postcard for a postcard. How many ways are there to carry out one fair exchange between these two collectors?

Problem 32. How many six-digit numbers have all their digits of equal parity (all odd or all even)?

Problem 33. In how many ways can we send six urgent letters if we can use three messengers and each letter can be given to any of them?

Problem 34. How many ways are there to choose four cards of different suits and different values from a deck of 52 cards?

Problem 35. There are five books on a shelf. How many ways are there to arrange some (or all) of them in a stack? The stack may consist of a single book.

Problem 36. How many ways are there to put eight rooks on a chessboard so that they do not attack each other?

Problem 37. There are N boys and N girls in a dance class. How many ways are there to arrange them in pairs for a dance?

Problem 38. The rules of a chess tournament say that each contestant must play every other contestant exactly once. How many games will be played if there are 18 participants?

Problem 39. How many ways are there to place a) two bishops; b) two knights; c) two queens on a chessboard so that they do not attack each other?

Problem 40. Mother has two apples, three pears, and four oranges. Every morning, for nine days, she gives one fruit to her son for breakfast. How many ways are there to do this?

Problem 41. There are three rooms in a dormitory: one single, one double, and one for four students. How many ways are there to house seven students in these rooms?

Problem 42. How many ways are there to place a set of chess pieces on the first row of a chessboard? The set consists of a king, a queen, two identical rooks, two identical knights, and two identical bishops.

Problem 43. How many "words" can be written using exactly five letters A and no more than three letters B (and no other letters)?

Problem 44. How many ten-digit numbers have at least two equal digits?

Problem 45*. Do seven-digit numbers with no digits 1 in their decimal representations constitute more than 50% of all seven-digit numbers?

Problem 46. We toss a die three times. Among all possible outcomes, how many have at least one occurrence of six?

Problem 47. How many ways are there to split 14 people into seven pairs?

Problem 48*. How many nine-digit numbers have an even sum of their digits?

CHAPTER 3

Divisibility and Remainders

For teachers. This theme is not so recreational as some others, yet it contains large amounts of important theoretical material. Try to introduce elements of play in your sessions. Even very routine problems like the factoring of integers can be turned into a contest by asking "Who can factor this huge number first?" or "Who can find the greatest prime divisor of this number first?" Thus, sessions devoted to this topic must be prepared more carefully than others. Since divisibility also enters into the school curriculum, you can use the knowledge acquired by students there.

§1. Prime and composite numbers

Among natural numbers we can distinguish prime and composite numbers. A number is *composite* if it is equal to the product of two smaller natural numbers. For example, $6 = 2 \cdot 3$. Otherwise, and if the number is not equal to 1, it is called *prime*. The number 1 is neither prime nor composite.

Prime numbers are like "bricks", which you can use to construct all natural numbers. How can this be done? Let us consider the number 420. It is certainly composite. It can be represented, for instance, as $42 \cdot 10$. But each of the numbers 42 and 10 is composite, too. Indeed, $42 = 6 \cdot 7$, and $10 = 2 \cdot 5$. Since $6 = 2 \cdot 3$, we have $420 = 42 \cdot 10 = 6 \cdot 7 \cdot 2 \cdot 5 = 2 \cdot 3 \cdot 7 \cdot 2 \cdot 5 = 2 \cdot 2 \cdot 3 \cdot 5 \cdot 7$ (see Figure 12). This is the complete "decomposition" of our number (its representation as a product of prime numbers).

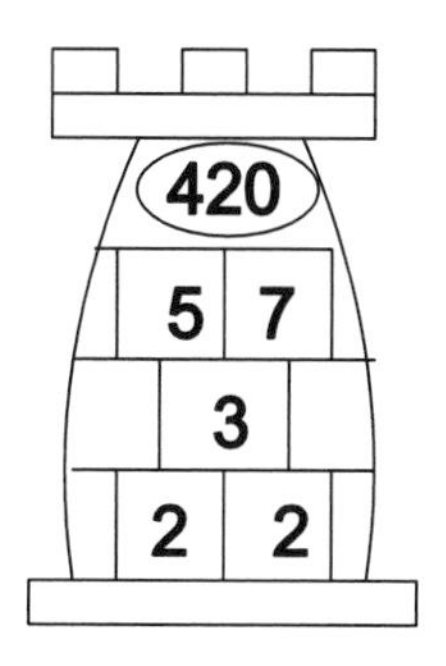

FIGURE 12

It is clear that we can factor any natural number greater than 1 in the same way. We just keep factoring the numbers we have into pairs of smaller numbers as long as we can (and if any one of the factors cannot be represented as such a product, then it is a prime factor).

But what if we try to factor the number 420 in some other way? For example, we can start with $420 = 15 \cdot 28$. It may surprise you that we will always end up with the same representation (products which differ only in the order of their factors are considered identical—we usually arrange the factors in increasing order).

This may seem evident, but it is not easy to prove. It is called the **Fundamental Theorem of Arithmetic**: any natural number different from 1 can be uniquely represented as a product of prime numbers in increasing order.

For teachers. Most of the contents of this section are connected with the Fundamental Theorem of Arithmetic.

Students should understand that the properties of divisibility are almost completely determined by the representation of a natural number as the product of prime numbers. The following exercises will help.

1. Is $2^9 \cdot 3$ divisible by 2?
Answer. Yes, since 2 is one of the factors in the decomposition of the given number.

2. Is $2^9 \cdot 3$ divisible by 5?
Answer. No, since the decomposition of this number does not contain the prime number 5.

3. Is $2^9 \cdot 3$ divisible by 8?
Answer. Yes, since $8 = 2^3$, and there are nine 2's in the decomposition of the given number.

4. Is $2^9 \cdot 3$ divisible by 9?
Answer. No, since $9 = 3 \cdot 3$, and there is only one 3 in the decomposition of the given number (see Figure 13).

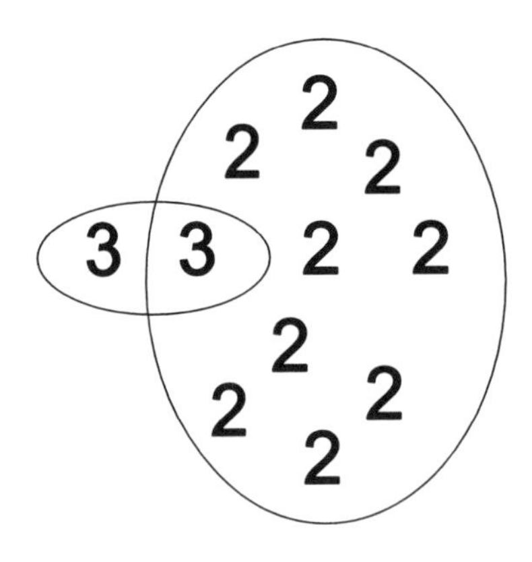

FIGURE 13

5. Is $2^9 \cdot 3$ divisible by 6?
Answer. Yes, since $6 = 2 \cdot 3$, and the decomposition of the given number contains both the prime numbers 2 and 3 (see Figure 14).

6. Is it true that if a natural number is divisible by 4 and by 3, then it must be divisible by $4 \cdot 3 = 12$?

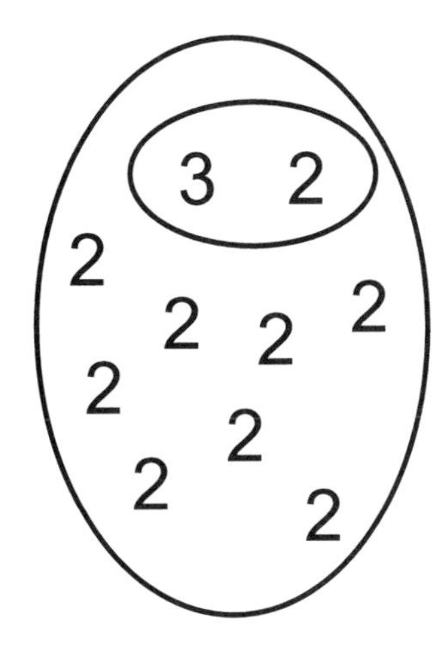

FIGURE 14

Answer. Yes. Indeed, the decomposition of a natural number which is divisible by 4 must contain at least two 2's. Since this number is also divisible by 3, there is also at least one 3. Therefore, our number is divisible by $3 \cdot 2 \cdot 2 = 12$.

7. Is it true that if a natural number is divisible by 4 and by 6, then it must be divisible by $4 \cdot 6 = 24$?
Answer. No. For example, the number 12 can serve as a counterexample. The reason is that if a number is divisible by 4, then its decomposition contains at least two 2's; if the same number is divisible by 6, then it means that its decomposition contains 2 and 3. Therefore, we can be sure that the decomposition has two 2's (but not necessarily three!) and 3, so we can only claim divisibility by 12.

8. The number A is not divisible by 3. Is it possible that the number $2A$ is divisible by 3?
Answer. No, since 3 does not belong to the decomposition of A, and, therefore, does not belong to the decomposition of $2A$.

9. The number A is even. Is it true that $3A$ must be divisible by 6?
Answer. Yes, since both 2 and 3 belong to the decomposition of the number $3A$.

10. The number $5A$ is divisible by 3. Is it true that A must be divisible by 3?
Answer. Yes, since the decomposition of $5A$ contains 3, while the decomposition of 5 does not.

11. The number $15A$ is divisible by 6. Is it true that A must be divisible by 6?
Answer. No. For example, A might be 2. The reason is that the number 3, which is one of the prime factors of the number 6, also belongs to the decomposition of the number 15. Thus we can only be sure that A is even.

IMPORTANT DEFINITION

Two natural numbers are called *relatively prime*, or *coprime*, if they have no common divisors greater than 1.

For example, two different prime numbers are, of course, relatively prime. Also, the number 1 is relatively prime to any other natural number.

Using reasoning similar to that used in exercises 6 and 10, we can prove the following two facts.

a) If some natural number is divisible by two relatively prime numbers p and q, then it is divisible by their product pq.

b) If the number pA is divisible by q, where p and q are relatively prime, then A is also divisible by q.

For teachers. Students should discuss and solve a few examples. Problems using relatively prime numbers can be found at the end of the section.

TWO MORE IMPORTANT DEFINITIONS

1. The *Greatest Common Divisor* (G.C.D. or $\gcd(x, y)$) of two natural numbers is ... what do you think? ... the greatest natural number which divides them both.

2. The *Least Common Multiple* (L.C.M. or $\operatorname{lcm}(x, y)$) of two natural numbers is ... guess again ... the least natural number which is divisible by both of them.

For example, $\gcd(18, 24) = 6$, $\operatorname{lcm}(18, 24) = 72$.

These definitions allow us to state a few more exercises.

12. Given the numbers $A = 2^3 \cdot 3^{10} \cdot 5 \cdot 7^2$ and $B = 2^5 \cdot 3 \cdot 11$ find $\gcd(A, B)$.
Answer. $\gcd(A, B) = 24 = 2^3 \cdot 3$. This is the common part ("intersection") of the decompositions of the numbers.

13. Given the numbers $A = 2^8 \cdot 5^3 \cdot 7$ and $B = 2^5 \cdot 3 \cdot 5^7$ find $\operatorname{lcm}(A, B)$.
Answer. $\operatorname{lcm}(A, B) = 420000000 = 2^8 \cdot 3 \cdot 5^7 \cdot 7$. This, as you can see, is the "union" of the numbers' decompositions.

For teachers. We ask you to think of the material of this section as just a sketch of a scenario for an actual session. As a teacher, you will want to create a more elaborate version. In some places you will probably give a series of very similar problems or exercises one after another, yet try to keep things varied. Encourage students to form their own conjectures regarding the problems and theorems you discuss.

However, this topic will be learned best if, in further sessions, there are problems using the ideas explained above.

We give here a list of some such problems. Methods and ideas introduced in this section will be used for the solution of problems in other sections of this chapter as well as in other chapters of the present book.

Problem 1. Given two different prime numbers p and q, find the number of different divisors of the number a) pq; b) p^2q; c) p^2q^2; d) p^nq^m.

Problem 2. Prove that the product of any three consecutive natural numbers is divisible by 6.

Hint. There is at least one even number, and at least one number divisible by 3, among any three consecutive numbers.

Solution. Any number divisible by 2 and by 3 is divisible by 6, so the result follows directly from the hint.

Problem 3. Prove that the product of any five consecutive natural numbers is a) divisible by 30; b) divisible by 120.

Problem 4. Given a prime number p, find the number of natural numbers which are a) less than p and relatively prime to it; b) less than p^2 and relatively prime to it.

Problem 5. Find the smallest natural number n such that $n!$ is divisible by 990.

Problem 6. How many zeros are there at the end of the decimal representation of the number 100! ?

Problem 7. For some number n, can the number $n!$ have exactly five zeros at the end of its decimal representation?

Problem 8. Prove that if a number has an odd number of divisors, then it is a perfect square.

Problem 9. Tom multiplied two two-digit numbers on the blackboard. Then he changed all the digits to letters (different digits were changed to different letters, and equal digits were changed to the same letter). He obtained $\mathrm{AB} \cdot \mathrm{CD} = \mathrm{EEFF}$. Prove that Tom made a mistake somewhere.

Problem 10. Can a number written with one hundred 0's, one hundred 1's, and one hundred 2's be a perfect square?

Hint. This number is divisible by 3, but not by 9.

Solution. The sum of the digits of any number such as described in the problem is $100(0+1+2) = 300$, which is divisible by 3 but not by 9. This, then, must be true of the number in the problem, regardless of the order in which its digits appear.

For teachers. You should draw the students' attention to the idea of the solution to the last problem. This could be done, for example, by asking what if the number described had two hundred 0's, 1's, and 2's? Three hundred 0's, 1's, and 2's?

Problem 11*. The numbers a and b satisfy the equation $56a = 65b$. Prove that $a + b$ is composite.

Problem 12. Find all solutions in natural numbers of the equations a) $x^2 - y^2 = 31$; b) $x^2 - y^2 = 303$.

Hint. $x^2 - y^2 = (x - y)(x + y)$.

Problem 13. Find the integer roots of the equation $x^3 + x^2 + x - 3 = 0$.

Hint. Add 3 to both sides of the equation, then factor the left-hand side.

Problem 14. Prove that any two natural numbers a and b satisfy the equation $\gcd(a, b)\,\mathrm{lcm}(a, b) = ab$.

§2. Remainders

Assume that you are in a country where coins of certain values are in circulation, and you want to buy a stick of gum for 3 cents from a vending machine. You have a 15-cent coin in your pocket but you do not have any 3-cent coins, which you need to buy the gum. Fortunately, you see a change machine which can give you any number of 3-cent coins. Obviously, you get five 3-cent coins for your 15-cent coin. What if you had a 20-cent coin? Then, of course, you get six 3-cent coins plus two cents change. So we have $20 = 6 \cdot 3 + 2$ (see Figure 15). This is a representation of the operation of division of 20 by 3 with a remainder.

How does our change machine work? It gives out 3-cent coins until the remainder is less than 3. After that it gives you coins for this remainder, which is equal to

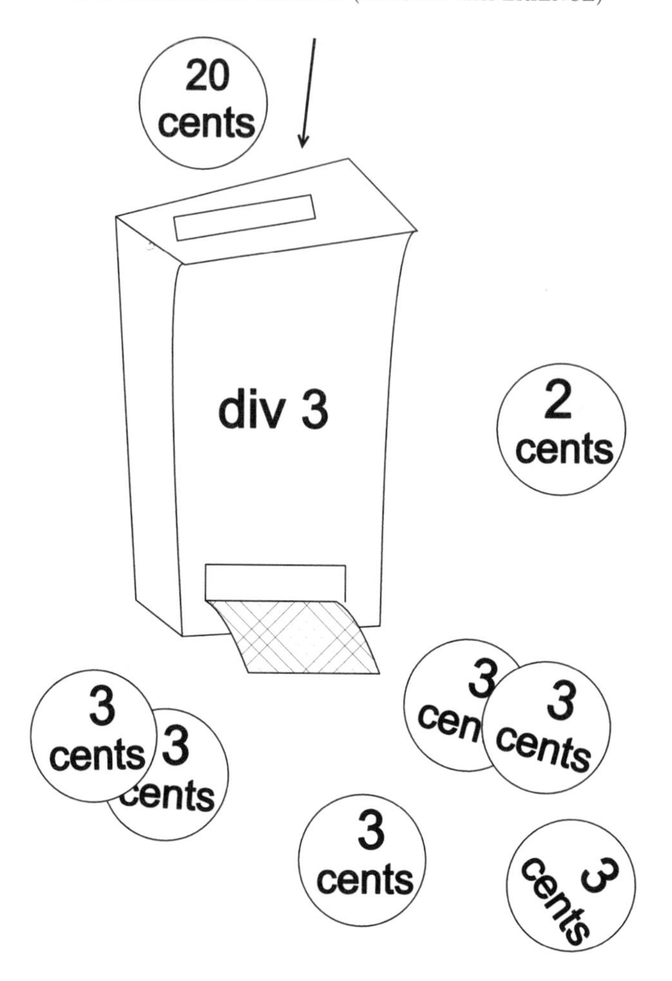

FIGURE 15

0, 1, or 2. It is clear that the remainder is zero if and only if the original number (the value of the coin you put into the machine) is divisible by 3.

Analogously, we can imagine a machine which gives m-cent coins and change which varies from 0 through $m-1$ cents. This machine would represent the operation of division by m with a remainder.

Now we give a more accurate definition:

To divide a natural number N by the natural number m with a remainder means to represent N as $N = km + r$, where $0 \leq r < m$. We call the number r the *remainder* when the number N is divided by m.

Now we can discuss the following problem: a person put twenty-two 50-cent coins and forty-four 10-cent coins into the changing machine. What is the change after he or she receives the 3-cent coins?

This is easy. It suffices to find the remainder when the number $x = 22{\cdot}50 + 44{\cdot}10$ is divided by 3. What is remarkable is that we do not have to calculate the sum of all the products. Suppose we replace each of the numbers with its remainder when divided by 3. The number x will become $1 \cdot 2 + 2 \cdot 1$. This is the number 4, which has remainder 1 when divided by 3. We claim that the remainder of the original expression (that is, of the number x) is also 1. The reason is that the following statement is always true:

Lemma On Remainders. The $\begin{smallmatrix}\text{sum}\\\text{product}\end{smallmatrix}$ of any two natural numbers has the same remainder, when divided by 3, as the $\begin{smallmatrix}\text{sum}\\\text{product}\end{smallmatrix}$ of their remainders.

Methodological remark. A formal proof of this fact is not very difficult, though for beginners it may seem full of technicalities.

For example, let us prove the second proposition. Let

$$\begin{aligned} N_1 &= k_1 \cdot 3 + r_1, \\ N_2 &= k_2 \cdot 3 + r_2 . \end{aligned}$$

Then

$$\begin{aligned} N_1 N_2 &= (k_1 \cdot 3 + r_1)(k_2 \cdot 3 + r_2) \\ &= k_1 k_2 \cdot 3^2 + k_1 r_2 \cdot 3 + k_2 r_1 \cdot 3 + r_1 r_2 \\ &= 3(3k_1 k_2 + k_1 r_2 + k_2 r_1) + r_1 r_2 . \end{aligned}$$

Thus, in changing $N_1 N_2$ cents, the machine will give out $3k_1k_2 + k_1r_2 + k_2r_1$ 3-cent coins, and will still have $r_1 r_2$ cents. Therefore, the remainder after putting $N_1 N_2$ cents into the machine is the same as the remainder for $r_1 r_2$ cents.

In the Lemma On Remainders the number 3, of course, can be changed to any other natural number: the same proof carries through.

For teachers. Generalizations of the Lemma On Remainders will be used throughout this section. Your students must learn how to apply these ideas when calculating remainders. We recommend solving a number of problems similar to Problem 15, drawing the students' attention to the use of these statements.

We do not think that a discussion of the proof of the Lemma On Remainders is absolutely necessary in the sessions.

Problem 15. Find the remainder which

a) the number $1989 \cdot 1990 \cdot 1991 + 1992^3$ gives when divided by 7;

b) the number 9^{100} gives when divided by 8.

The solution to the next problem includes one very important idea.

Problem 16. Prove that the number $n^3 + 2n$ is divisible by 3 for any natural number n.

Solution. The number n can give any of the following remainders when divided by 3: 0, 1, or 2. Thus we consider three cases.

If n has remainder 0, then both n^3 and $2n$ are divisible by 3, and therefore $n^3 + 2n$ is divisible by 3.

If n has remainder 1, then n^3 has remainder 1, $2n$ has remainder 2, and $1 + 2$ is divisible by 3.

If n has remainder 2, then n^2 has remainder 1, n^3 has remainder 2, $2n$ has remainder 1, and $2 + 1$ is divisible by 3.

This case-by-case analysis completes the required proof.

For teachers. The key moment in the last solution was the idea of a case-by-case analysis, used to examine all the possible remainders modulo some natural number. This method deserves to be pointed out to the students. They should understand that such an analysis indeed gives us a complete and rigorous proof.

Case-by-case analysis can also be used in many fields other than arithmetic. It would be excellent for students to learn to determine whether a case-by-case analysis can help in solving a problem. We hope that the following problems will help achieve this objective.

Problem 17. Prove that $n^5 + 4n$ is divisible by 5 for any integer n.

Problem 18. Prove that $n^2 + 1$ is not divisible by 3 for any integer n.

Problem 19. Prove that $n^3 + 2$ is not divisible by 9 for any integer n.

Problem 20. Prove that $n^3 - n$ is divisible by 24 for any odd n.

Hint. Prove that the given number is a multiple of both 3 and 8.

Problem 21. a) Prove that $p^2 - 1$ is divisible by 24 if p is a prime number greater than 3.

b) Prove that $p^2 - q^2$ is divisible by 24 if p and q are prime numbers greater than 3.

Problem 22. The natural numbers x, y, and z satisfy the equation $x^2 + y^2 = z^2$. Prove that at least one of them is divisible by 3.

Problem 23. Given natural numbers a and b such that $a^2 + b^2$ is divisible by 21, prove that the same sum of squares is also divisible by 441.

Problem 24. Given natural numbers a, b, and c such that $a + b + c$ is divisible by 6, prove that $a^3 + b^3 + c^3$ is also divisible by 6.

Problem 25. Three prime numbers p, q, and r, all greater than 3, form an arithmetic progression: $p = p$, $q = p + d$, and $r = p + 2d$. Prove that d is divisible by 6.

Problem 26. Prove that if we decrease by 7 the sum of the squares of any three natural numbers, then the result cannot be divisible by 8.

Problem 27. The sum of the squares of three natural numbers is divisible by 9. Prove that we can choose two of these numbers such that their difference is divisible by 9.

Hint. If two numbers have equal remainders when divided by 9, then their difference is divisible by 9.

* * *

Let us continue with another set of problems:

Problem 28. Find the last digit of the number 1989^{1989}.

Solution. To begin let us note that the last digit of the number 1989^{1989} is the same as the last digit of the number 9^{1989}. We write down the last digits of the first few powers of 9: 9, 1, 9, 1, 9, ...

To calculate the last digit of a power of 9 it is sufficient to multiply by 9 the last digit of the previous power of 9. Hence, it is quite clear that the digit 9 is always followed by the digit 1 ($9 \cdot 9 = 81$), which in its turn is always followed by 9 ($1 \cdot 9 = 9$).

Thus, the odd powers of 9 always have their last digit equal to 9. Therefore, the last digit of 1989^{1989} is also 9.

Problem 29. Find the last digit of the number 2^{50}.

Solution. Let us write down the last digits of the first few powers of two: 2, 4, 8, 6, 2, ... We can see that 2^5 ends with 2, as does 2^1. Since the last digit of any power is determined by the last digit of the previous power of 2, we have a cycle: 2^6 ends with 4 (like 2^2), 2^7 ends with 8 (like 2^3), 2^8 ends with 6, 2^9 ends with 2, et cetera. Since the length of the cycle is 4, the last digit of the number 2^{50} can be found using the remainder of the number 50 when divided by 4. This remainder is 2, and the last digit of 2^{50} coincides with the last digit of 2^2, which is 4.

Problem 30. What is the last digit of 777^{777}?

Problem 31. Find the remainder of 2^{100} when divided by 3.

Hint. Write down the remainders when several powers of 2 are divided by 3. Prove that they form another cycle.

Problem 32. Find the remainder when the number 3^{1989} is divided by 7.

Problem 33. Prove that $2222^{5555} + 5555^{2222}$ is divisible by 7.

Hint. Show that the remainder when the given number is divided by 7 is zero.

Problem 34. Find the last digit of the number 7^{7^7}.

In Problems 16–27 we used the same idea of a case-by-case analysis of the remainders modulo some natural number n. Moreover, this number n could be recognized rather easily from the statement of a problem. In the next set of problems guessing the number n will not be so easy. The "art of guessing" requires certain skills and, while there are some standard tricks, can be quite difficult.

<u>**For teachers.**</u> As exercises to maintain the skills mentioned we suggest the composition of multiplication tables for remainders when divided by "the most frequently used" numbers—2, 3, 4, 5, 6, 7, 8, 9, 11, 13, and so on. You can also try to find all possible remainders given by perfect squares and cubes when divided by these numbers.

Problem 35. a) Given that p, $p + 10$, and $p + 14$ are prime numbers, find p.
b) Given that p, $2p + 1$, and $4p + 1$ are prime numbers, find p.

Hint. Find remainders when divided by 3.

Problem 36. Given the pair of prime numbers p and $8p^2 + 1$, find p.

Problem 37. Given the pair of prime numbers p and $p^2 + 2$, prove that $p^3 + 2$ is also a prime number.

Problem 38. Prove that there are no natural numbers a and b such that $a^2 - 3b^2 = 8$.

Problem 39. a) Can the sum of two perfect squares be another perfect square?
b) Can the sum of three squares of odd natural numbers be a perfect square?

Problem 40. Prove that the sum of the squares of five consecutive natural numbers cannot be a perfect square.

Problem 41. If p, $4p^2 + 1$, and $6p^2 + 1$ are prime numbers, find p.

Methodological remark. Quite often, arithmetic problems about squares (like Problems 36–40) can be solved by using remainders modulo 3 or modulo 4. The point is that when divided by 3 or 4, perfect squares can give only remainders 0 and 1.

Problem 42. Prove that the number $100\ldots 00500\ldots 001$ (100 zeros in each group) is not a perfect cube.

Problem 43. Prove that a^3+b^3+4 is not a perfect cube for any natural numbers a and b.

Problem 44*. Prove that the number $6n^3+3$ cannot be a perfect sixth power of an integer for any natural number n.

Methodological remark. When dealing with problems about cubes of integers (like Problems 42–44) it is often useful to analyze the remainders modulo 7 or modulo 9. In either case there are only three possible remainders: $\{0,\ 1,\ 6\}$ and $\{0,\ 1,\ 8\}$ respectively.

Problem 45*. Given natural numbers x, y, and z such that $x^2+y^2=z^2$, prove that xy is divisible by 12.

For teachers. The material explained in this section may be used to create at least two sessions. The first of them should be devoted to the calculation of remainders. The second can be spent in discussing the idea of case-by-case analysis in solutions of various problems.

§3. A few more problems

This section contains a series of divisibility problems which are not united by any common statement or method of solution. However, we will use ideas and methods from the previous sections.

Problem 46. a) If it is known that $a+1$ is divisible by 3, prove that $4+7a$ is also divisible by 3.

b) It is known that $2+a$ and $35-b$ are divisible by 11. Prove that $a+b$ is also divisible by 11.

Problem 47. Find the last digit of the number $1^2+2^2+\ldots+99^2$.

Problem 48. Seven natural numbers are such that the sum of any six of them is divisible by 5. Prove that each of these numbers is divisible by 5.

Problem 49. For any $n>1$ prove that the sum of any n consecutive odd natural numbers is a composite number.

Problem 50. Find the smallest natural number which has a remainder of 1 when divided by 2, a remainder of 2 when divided by 3, a remainder of 3 when divided by 4, a remainder of 4 when divided by 5, and a remainder of 5 when divided by 6.

Problem 51. Prove that if $(n-1)!+1$ is divisible by n, then n is a prime number.

We will discuss in more detail the following two problems:

Problem 52*. Prove that there exists a natural number n such that the numbers $n+1,\ n+2,\ \ldots,\ n+1989$ are all composite.

Solution. We will try to explain how one can arrive at a solution. The number $n+1$ must be composite. Let us try to keep things simple and make this number divisible by 2. Then the number $n+2$ must be composite too, but it cannot be a multiple of 2. Let us try again to be simple and make this number divisible by 3. Proceeding as above we can try to find a number n such that $n+1$ is divisible by 2, $n+2$ is divisible by 3, $n+3$ is divisible by 4, et cetera. This is equivalent to saying that $n-1$ is divisible by 2, 3, 4, ... , and 1990. Such a number is easy to find; for example, 1990! will do. Finally, we can take $1990!+1$ as the number we are looking for.

Problem 53*. Prove that there are infinitely many prime numbers.

Solution. Assume that there are only n prime numbers, and let us denote them all by $p_1, p_2, \ldots, p_n$. Then the number $p_1 p_2 \ldots p_n + 1$ is divisible by none of the prime numbers $p_1, p_2, \ldots, p_n$. Therefore, this natural number cannot be represented as the product of primes, which is absurd. This contradiction completes the proof.

For teachers. The problems of this section should not be given for solution at one session. They can be given in the course of an entire year of classes, or used for olympiads, various types of contests, et cetera.

§4. Euclid's algorithm

In the first section we discussed the concept of the Greatest Common Divisor of two natural numbers, and we showed how to calculate the G.C.D.: you must write down the decompositions of both numbers into the products of primes and then take their common part.

For large numbers, however, this procedure is virtually impossible to carry out by hand (try to do this, for example, with the numbers 1381955 and 690713). Fortunately, there is another, less painful, way to calculate the G.C.D. It is called *Euclid's algorithm.*

This method is based on the following simple reasoning: any common divisor of two numbers a and b ($a > b$) also divides the number $a-b$; also any common divisor of b and $a-b$ divides the number a as well. Hence, $\gcd(a,b) = \gcd(b, a-b)$. In a sense, this explains all of Euclid's algorithm.

We show how it works for the two numbers 451 and 287:

$$\begin{aligned}
\gcd(451,287) &= \gcd(287,164)\\
&= \gcd(164,123)\\
&= \gcd(123,41)\\
&= \gcd(82,41)\\
&= \gcd(41,41)\\
&= 41.
\end{aligned}$$

Note that Euclid's algorithm can be shortened as follows: change a not to $a-b$ but to the remainder when a is divided by b. We can demonstrate this "improved" algorithm using the pair of numbers mentioned at the beginning of this section:

$$\begin{aligned}\gcd(1381955, 690713) &= \gcd(690713, 529)\\ &= \gcd(529, 368)\\ &= \gcd(368, 161)\\ &= \gcd(161, 46)\\ &= \gcd(46, 23)\\ &= \gcd(23, 0)\\ &= 23.\end{aligned}$$

As you can see, this method leads us to the result quite quickly.

Problem 54. Find the G.C.D. of the numbers $2n + 13$ and $n + 7$.

Solution. We have $\gcd(2n + 13, n + 7) = \gcd(n + 7, n + 6) = \gcd(n + 6, 1) = 1$.

Problem 55. Prove that the fraction $\frac{12n+1}{30n+2}$ cannot be reduced for any natural number n.

Problem 56. Find $\gcd(2^{100} - 1, 2^{120} - 1)$.

Problem 57. Find $\gcd(111\ldots111, 11\ldots11)$, where there are one hundred 1's in the decimal representation of the first number and sixty 1's in the decimal representation of the second number.

For teachers. However simple it may seem, Euclid's algorithm is a very important arithmetic fact (which can be used, for example, to prove the Fundamental Theorem of Arithmetic). Therefore, we think it would be wise to devote a separate session to this remarkable method (together with a discussion of the G.C.D., L.C.M., and their properties). For more details, see [**53**].

CHAPTER 4

The Pigeon Hole Principle

§1. Introduction

Students who have never heard of the Pigeon Hole Principle may think that it is a joke:

If we must put $N+1$ or more pigeons into N pigeon holes, then some pigeon hole must contain two or more pigeons.

Notice the vagueness of the proposition "some pigeon hole must contain ... ", "two or more ... ". This is, in fact, a distinguishing feature of the Pigeon Hole Principle, which sometimes allows us to draw quite unexpected conclusions, even when we don't seem to have enough information. (See Figure 16.)

FIGURE 16

The proof of this principle is quite simple, and uses only a trivial count of the pigeons in their pigeon holes. Suppose no more than one pigeon were in each hole. Then there would be no more than N pigeons altogether, which contradicts the assumption that we have $N+1$ pigeons. This proves the Pigeon Hole Principle, using—and we must be aware of this—the method of proof by contradiction.

But, you might ask, does the following problem concern pigeons?

Problem 1. A bag contains beads of two colors: black and white. What is the smallest number of beads which must be drawn from the bag, without looking, so that among these beads there are two of the same color?

The following problem also seems to have nothing to do with pigeons and pigeon holes:

Problem 2. One million pine trees grow in a forest. It is known that no pine tree has more than 600000 pine needles on it. Show that two pine trees in the forest must have the same number of pine needles.

Solution to Problem 1. We can draw three beads from the bag. If there were no more than one bead of each color among these, then there would be no more than two beads altogether. This is obvious, and contradicts the fact that we have chosen three beads. On the other hand, it is clear that choosing two beads is not enough. Here the beads play the role of pigeons, and the colors (black and white) play the role of pigeon holes.

Solution to Problem 2. We have one million "pigeons"—the pine trees—and, unfortunately, 600001 pigeon holes, numbered 0 through 600000. We put each "pigeon" (pine tree) in the pigeon hole numbered with the number of pine needles on the tree. Since there are many more "pigeons" than pigeon holes, there must be at least two "pigeons" (pine trees) in some pigeon hole: if there were no more than one in each pigeon hole, then there would be no more than 600001 "pigeons". But if two "pigeons" are in the same pigeon hole, that means that they have the same number of pine needles.

Notice that the statements of these problems include the same vagueness as the Pigeon Hole Principle itself. It is exactly this kind of problem that can often be solved using the Pigeon Hole Principle.

For teachers. Students have trouble dealing with this vagueness. They should first solve a few simple exercises, such as Problems 1 and 2. Sometimes they will not even remember what it is they must prove. It may be necessary to explain the difference between an intuitive understanding and an actual proof.

In discussing these first few problems, it is important to emphasize the common ideas—these are typically not obvious to students—without consciously invoking any broad principle at all. This can be followed by a series of problems in conscious imitation of the arguments just given by the teacher (Problems 3–7). Finally, we can tell students directly about the Pigeon Hole Principle, and emphasize that it was actually the basis of solution for the previous problems. From that point on, in analyzing problems, we can give some of the solutions in detail, without even mentioning the words "Pigeon Hole Principle", in order to get students to re-think the situation.

Problem 3. Given twelve integers, show that two of them can be chosen whose difference is divisible by 11.

Problem 4. The city of Leningrad has five million inhabitants. Show that two of these must have the same number of hairs on their heads, if it is known that no person has more than one million hairs on his or her head.

Problem 5. Twenty-five crates of apples are delivered to a store. The apples are of three different sorts, and all the apples in each crate are of the same sort. Show that among these crates there are at least nine containing the same sort of apple.

§2. More general pigeons

If you have read the problems above carefully, and tried to solve Problem 5 in the same way as the first two, you may not have succeeded. The Pigeon Hole Principle, after all, will only tell you that there are two crates with the same sort of apples. In solving this problem, we can use the "General Pigeon Hole Principle":

If we must put $Nk + 1$ or more pigeons into N pigeon holes, then some pigeon hole must contain at least $k + 1$ pigeons.

In the case $k = 1$, the General Pigeon Hole Principle reduces to the simple Pigeon Hole Principle. We leave the proof of the General Principle as an exercise.

Solution to Problem 5. We are putting 25 "pigeons" (crates) into 3 "pigeon holes" (sorts of apples). Since $25 = 3 \cdot 8 + 1$, we can use the General Pigeon Hole Principle for $N = 3$, $k = 8$. We find that some "pigeon hole" must contain at least 9 crates.

In analyzing this solution, it is instructive to restate it without any form of the Pigeon Hole Principle, using only a trivial counting argument (of the sort with which we proved the Pigeon Hole Principle).

Most of the following problems will require use of the General Pigeon Hole Principle.

Problem 6. In the country of Courland there are M football teams, each of which has 11 players. All the players are gathered at an airport for a trip to another country for an important game, but they are traveling on "standby". There are 10 flights to their destination, and it turns out that each flight has room for exactly M players. One football player will take his own helicopter to the game, rather than traveling standby on a plane. Show that at least one whole team will be sure to get to the important game.

Problem 7. Given 8 different natural numbers, none greater than 15, show that at least three pairs of them have the same positive difference (the pairs need not be disjoint as sets.)

In solving Problem 7 we encounter a seemingly insuperable obstacle. There are 14 possible differences between the 8 given numbers (the values of the differences being 1 through 14). These are the 14 pigeon holes. But what are our pigeons? They must be the differences between pairs of the given numbers. However, there are 28 pairs, and we can fit them in our 14 pigeon holes in such a way that there are exactly two "pigeons" in each hole (and therefore no hole containing three). Here we must use an additional consideration. We cannot put more than one pigeon in the pigeon hole numbered 14, since the number 14 can be written as a difference of two natural numbers less than 15 in only one way: $14 = 15 - 1$. This means that the remaining 13 pigeon holes contain at least 27 pigeons, and the General Pigeon Hole Principle gives us our result.

* * *

The next four problems can be solved using the Pigeon Hole Principle (Ordinary or General) plus various other considerations.

Problem 8. Show that in any group of five people, there are two who have an identical number of friends within the group.

Problem 9. Several football teams enter a tournament in which each team plays every other team exactly once. Show that at any moment during the tournament there will be two teams which have played, up to that moment, an identical number of games.

Problem 10a. What is the largest number of squares on an 8×8 checkerboard which can be colored green, so that in any arrangement of three squares (a "tromino") such as in Figure 17, at least one square is not colored green? (The tromino may appear as in the figure, or it may be rotated through some multiple of 90 degrees.)

Problem 10b. What is the smallest number of squares on an 8×8 checkerboard which can be colored green, so that in any tromino such as in Figure 17, at least one square is colored green?

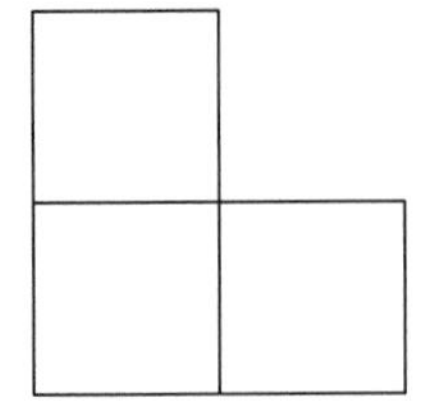

FIGURE 17

Hint for Problem 10a. Divide the checkerboard into sixteen 2×2 squares. These small squares are the pigeon holes, and the green squares will be the pigeons.

In solving some of the more complex problems (starting for example with Problem 10), it is useful to make a clear separation between the processes of identifying pigeons and pigeon holes, of introducing auxiliary considerations, and of applying the Pigeon Hole Principle itself. An important goal is to develop the skill of recognizing, from the statement of a problem, when the Pigeon Hole Principle can be applied to its solution.

Problem 11. Ten students solved a total of 35 problems in a math olympiad. Each problem was solved by exactly one student. There is at least one student who solved exactly one problem, at least one student who solved exactly two problems, and at least one student who solved exactly three problems. Prove that there is also at least one student who has solved at least five problems.

§3. Pigeons in geometry

Problem 12. What is the largest number of kings which can be placed on a chessboard so that no two of them put each other in check?

Problem 13. What is the largest number of spiders which can amicably share the spider web pictured below? A spider will tolerate a neighbor only at a distance of 1.1 meter or more, traveling along the web.

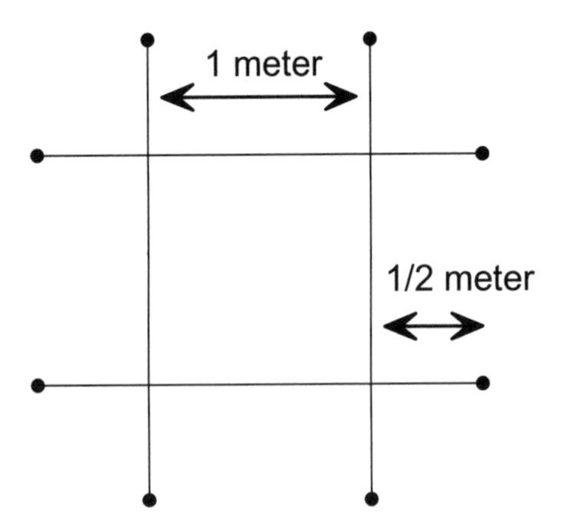

FIGURE 18

Problem 14. Show that an equilateral triangle cannot be covered completely by two smaller equilateral triangles.

Problem 15. Fifty-one points are scattered inside a square with a side of 1 meter. Prove that some set of three of these points can be covered by a square with side 20 centimeters.

Solution. If we divide the square into 25 smaller squares with sides of 20 centimeters, the General Pigeon Hole Principle assures us that one of these squares includes at least three of the 51 scattered points.

The careful reader will see a small flaw in this argument. Throughout our discussion, we have been assuming that our pigeon holes are disjoint. That is, no pigeon can belong in two different pigeon holes at the same time. However, the square "pigeon holes" in this solution have a slight overlap: points on the sides of the squares may belong to both pigeon holes.

To fix this, we must make a choice for each line segment which bounds a square, by deciding which of its two neighboring squares includes the points on the line segment. We can do this, for instance, by making the "north" and "east" borders of each square exclude their points, and the "south" and "west" border include their points (except for points on the border of the original square). With this slight adjustment, we have a set of "true pigeon holes", and the proof follows as before.

§4. Another generalization

Notice now that the proof of the Pigeon Hole Principle is based on the addition of inequalities. An important result of the process of adding inequalities, which can often be combined with the theme of the Pigeon Hole Principle, can be stated as follows:

If the sum of n or more numbers is equal to S, then among these there must be one or more numbers not greater than S/n, and also one or more numbers not less than S/n.

As with most variants of the Pigeon Hole Principle, we can prove this indirectly. If, for example, all the numbers are greater than S/n, then their sum would be bigger than S, which contradicts our assumption.

Problem 16. Five young workers received as wages 1500 rubles altogether. Each of them wants to buy a cassette player costing 320 rubles. Prove that at least one of them must wait for the next paycheck to make his purchase.

Solution. The sum S of their earnings is 1500 rubles, so the above principle guarantees that at least one worker earned no more that $1500/5 = 300$ rubles. Such a worker must wait for his cassette player.

Problem 17. In a brigade of 7 people, the sum of the ages of the members is 332 years. Prove that three members can be chosen so that the sum of their ages is no less than 142 years.

Solution. We look at all possible triples of brigade members. If we add the three ages in each group, then sum these numbers, this final sum must be $15 \cdot 332$ (since each person appears in a triple 15 times). Yet there are altogether 35 triples. This means that there is a triple of brigade members such that the sum of their ages is not less than $15 \cdot 332/35$, which is greater than 142.

Problem 18. On a certain planet in the solar system Tau Cetus, more than half the surface of the planet is dry land. Show that the Tau Cetans can dig a tunnel straight through the center of their planet, beginning and ending on dry land (assume that their technology is sufficiently developed).

§5. Number theory

Many wonderful problems touching on divisibility properties of integers can be solved using the Pigeon Hole Principle.

Problem 19. Prove that there exist two powers of two which differ by a multiple of 1987.

Problem 20. Prove that of any 52 integers, two can always be found such that the difference of their squares is divisible by 100.

Problem 21. Prove that there exists an integer whose decimal representation consists entirely of 1's, and which is divisible by 1987.

Solution to Problem 21. We look at the 1988 "pigeons" numbered 1, 11, 111, ... , 111...11 (1988 1's), and sort them into 1987 pigeon holes numbered 0, 1, 2, ... , 1986. Each number is put into the pigeon hole bearing the number equal to its remainder when divided by 1987. The Pigeon Hole Principle now assures us that there are two numbers which have the same remainder when divided by 1987. Let these numbers have m 1's and n 1's respectively, with $m > n$. Then their difference, which is divisible by 1987, is equal to 111...1100...00 ($m - n$ 1's and n zeros). We can cross out all the trailing zeros—these do not affect divisibility by 1987 since neither 2 nor 5 is a factor of 1987—to obtain a number written entirely with 1's, and which is divisible by 1987.

Problem 22. Prove that there exists a power of three which ends with the digits 001 (in decimal notation).

Problem 23. Each box in a 3×3 arrangement of boxes is filled with one of the numbers -1, 0, 1. Prove that of the eight possible sums along the rows, the columns, and the diagonals, two sums must be equal.

Problem 24. Of 100 people seated at a round table, more than half are men. Prove that there are two men who are seated diametrically opposite each other.

Problem 25. Fifteen boys gathered 100 nuts. Prove that some pair of boys gathered an identical number of nuts.

Problem 26. The digits 1, 2, ... , 9 are divided into three groups. Prove that the product of the numbers in one of the groups must exceed 71.

Problem 27. Integers are placed in each entry of a 10×10 table, with no two neighboring integers differing by more than 5 (two integers are considered neighbors if their squares share a common edge). Prove that two of the integers must be equal.

Problem 28. Prove that among any six people there are either three people, each of whom knows the other two, or three people, each of whom does not know the other two.

Problem 29. Five lattice points are chosen on an infinite square lattice. Prove that the midpoint of one of the segments joining two of these points is also a lattice point.

Problem 30. A warehouse contains 200 boots of size 41, 200 boots of size 42, and 200 boots of size 43. Of these 600 boots, there are 300 left boots and 300 right boots. Prove that one can find among these boots at least 100 usable pairs.

Problem 31. The alphabet of a certain language contains 22 consonants and 11 vowels. Any string of these letters is a word in this language, so long as no two consonants are together and no letter is used twice. The alphabet is divided into 6 (non-empty) subsets. Prove that the letters in at least one of these groups form a word in the language.

Problem 32. Prove that we can choose a subset of a set of ten given integers, such that their sum is divisible by 10.

Problem 33. Given 11 different natural numbers, none greater than 20. Prove that two of these can be chosen, one of which divides the other.

Problem 34. Eleven students have formed five study groups in a summer camp. Prove that two students can be found, say A and B, such that every study group which includes student A also includes student B.

Students should remember that even if they cannot cope with some problem right away, it always pays to go back later and try some fresh ideas. Do not jump to the solutions chapter! And do not forget that some of the problems may have alternate solutions not using the Pigeon Hole Principle.

CHAPTER 5

Graphs–1

The mathematical objects discussed in this chapter are extremely useful in solving many kinds of problems, which often bear no outward resemblance to each other. Graphs are also interesting in and of themselves. A separate subdivision of mathematics, called graph theory, is devoted to their study. We will examine several elementary ideas from this theory to show how graphs are used in solving problems.

§1. The concept of a graph

Problem 1. Cosmic liaisons are established among the nine planets of the solar system. Rockets travel along the following routes: Earth–Mercury, Pluto–Venus, Earth–Pluto, Pluto–Mercury, Mercury–Venus, Uranus–Neptune, Neptune–Saturn, Saturn–Jupiter, Jupiter–Mars, and Mars–Uranus. Can a traveler get from Earth to Mars?

Solution. We can draw a diagram, in which the planets will be represented by points, and the routes connecting them by non-intersecting line segments (see Figure 19). It is now clear that it is impossible to travel from Earth to Mars.

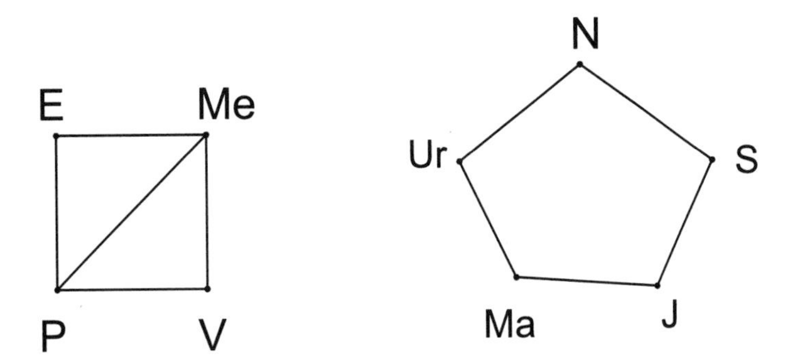

FIGURE 19

Problem 2. Several knights are situated on a 3×3 chessboard as shown in Figure 20. Can they move, using the usual chess knight's move, to the position shown in Figure 21?

Solution. The answer is no. We can show this by numbering the squares of the chessboard with the numbers 1, 2, 3, ... , 9 as shown in Figure 22. Then we can represent each square by a point. If we can get from one square to another with a

FIGURE 20

FIGURE 21

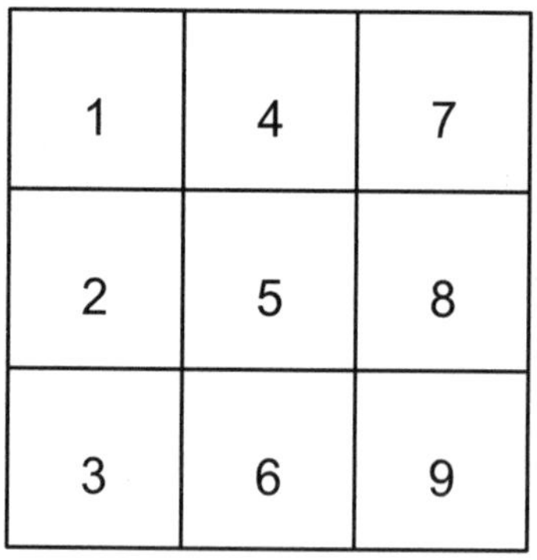

1	4	7
2	5	8
3	6	9

FIGURE 22

knight's move, we connect the corresponding points with a line (Figure 23). The starting and ending positions of the knights are shown in Figure 24.

The order in which the knights appear on the circle clearly cannot be changed. Therefore it is not possible to move the knights to the required positions.

The solution of these two problems, which do not resemble each other on the surface, have a central idea in common: the representation of the problem by a

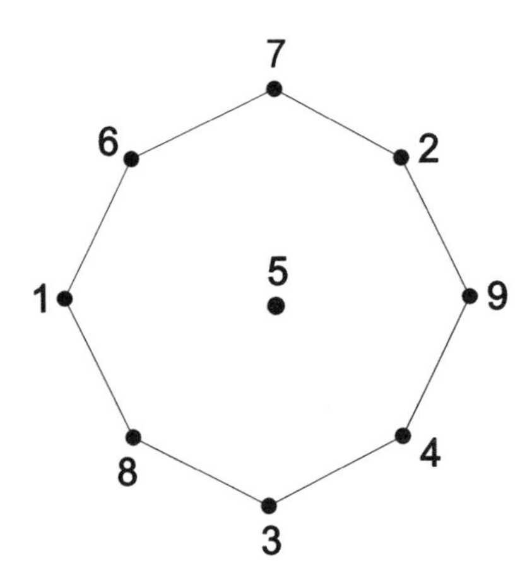

FIGURE 23

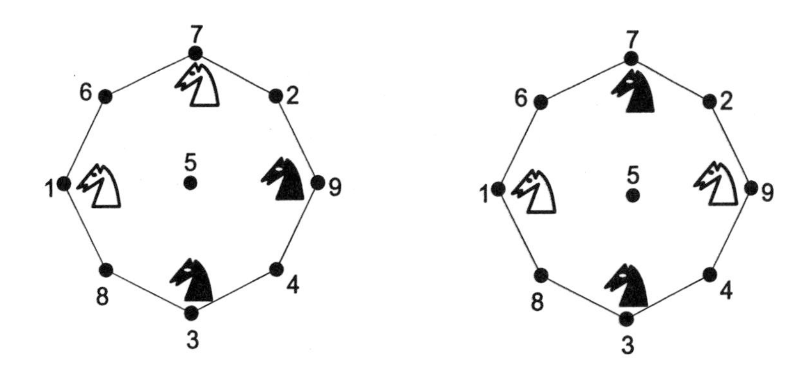

FIGURE 24

diagram. The resulting diagrams also have something in common. Each consists of a set of points, some of which are connected by line segments.

Such a diagram is called a *graph*. The points are called the *vertices* of the graph, and the lines are called its *edges*.

Methodological remark. The definition we give of a graph is actually too limiting. For example, in Problem 20 below it is rather natural to draw the edges of a graph using arcs, rather than line segments. However, an accurate definition would here be too complicated. The description above will suffice for students to get an intuitive idea of what a graph is, which they can later refine.

* * *

Here are two more problems which can be solved by drawing graphs.

Problem 3. A chessboard has the form of a cross, obtained from a 4×4 chessboard by deleting the corner squares (see Figure 25). Can a knight travel around this board, pass through each square exactly once, and end on the same square he starts on?

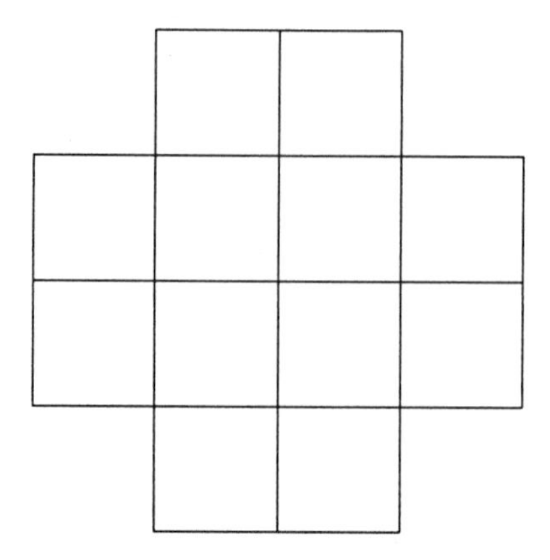

FIGURE 25

Problem 4. In the country of Figura there are nine cities, with the names 1, 2, 3, 4, 5, 6, 7, 8, 9. A traveler finds that two cities are connected by an airplane route if and only if the two-digit number formed by naming one city, then the other, is divisible by 3. Can the traveler get from City 1 to City 9?

Note that one and the same graph can be represented in different ways. For example, the graph of Problem 1 can be represented as in Figure 26.

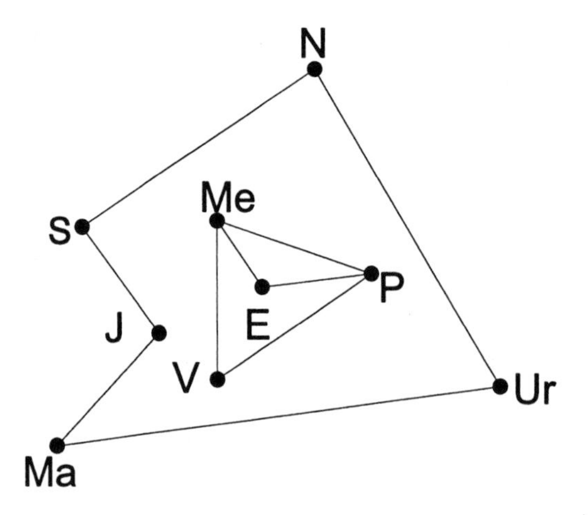

FIGURE 26

The only important thing about a graph is which vertices are connected and which are not.

Two graphs which are actually identical, but are perhaps drawn differently, are called *isomorphic.*

Problem 5. Try to find, in Figures 27, 28, and 29 a graph isomorphic to the graph of Problem 2 (see Figure 23).

Solution. The first and the third graphs are isomorphic to each other, and it is not hard to convince oneself that both of these are isomorphic to the graph of Problem

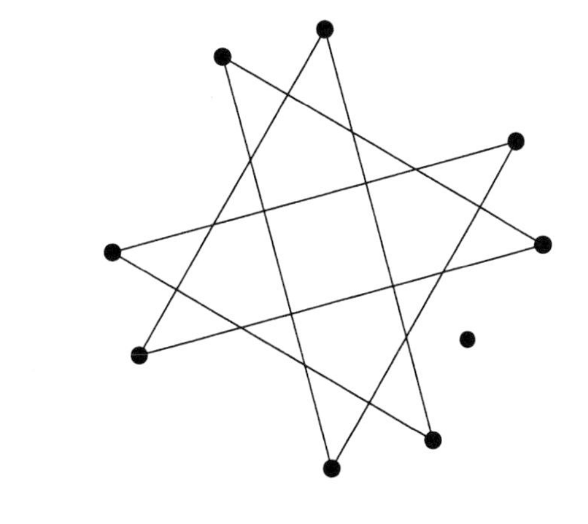

FIGURE 27

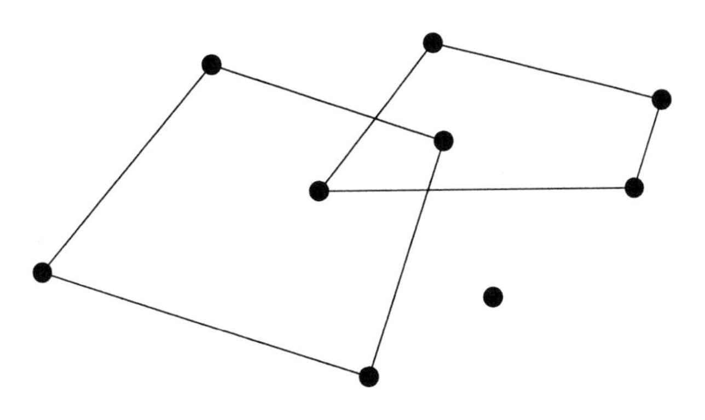

FIGURE 28

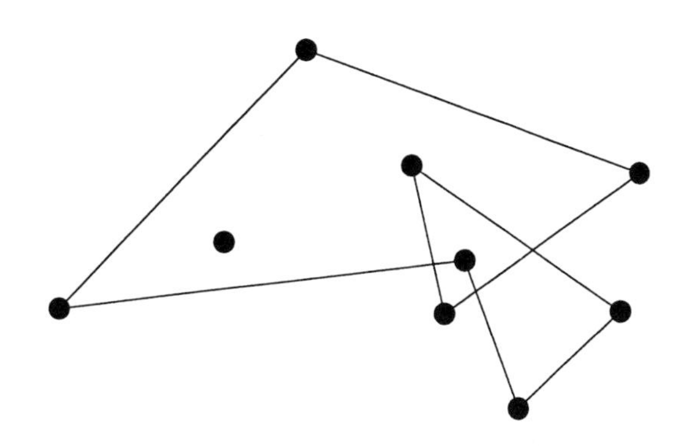

FIGURE 29

2. It suffices to renumber their vertices (see Figures 30 and 31). A proof that the graphs of Figures 28 and 23 are not isomorphic is somewhat more complicated.

For teachers. The concept of a graph should be introduced only after several problems like Problems 1 and 2 above, which involve using a graph to represent the situation of the problem. It is important that students realize right away that the

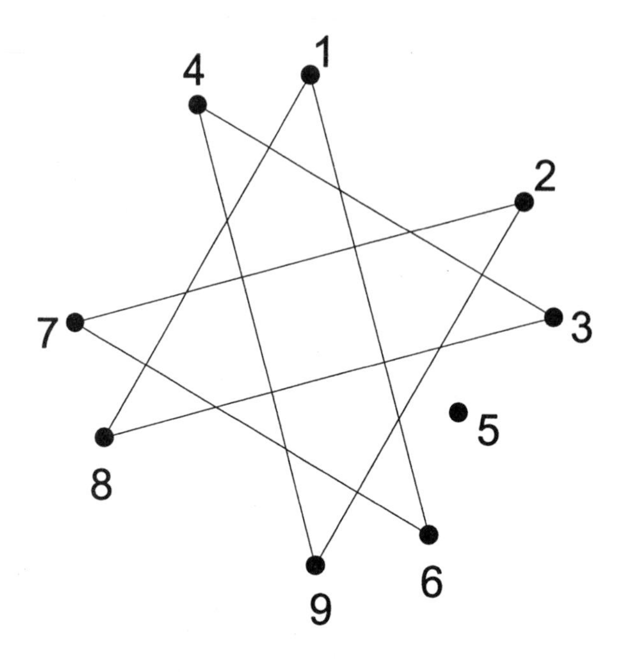

FIGURE 30

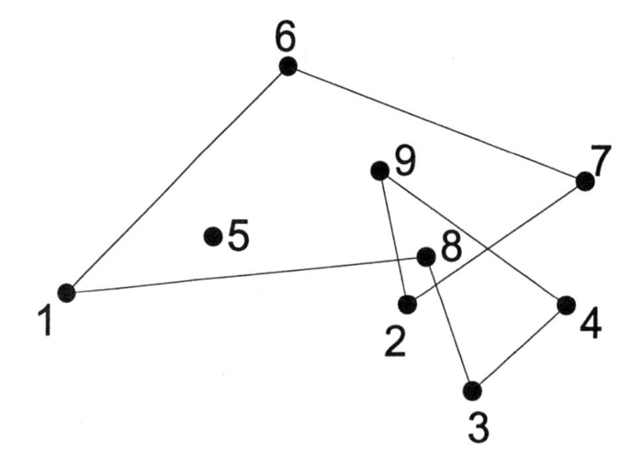

FIGURE 31

same graph can be drawn in different ways. To illustrate the idea of isomorphism, students can solve several more exercises of the type given here.

§2. The degree of a vertex: counting the edges

In the preceding section we defined a graph as a set of points (vertices), some of which are connected by lines (edges). The number of edges which start at a given

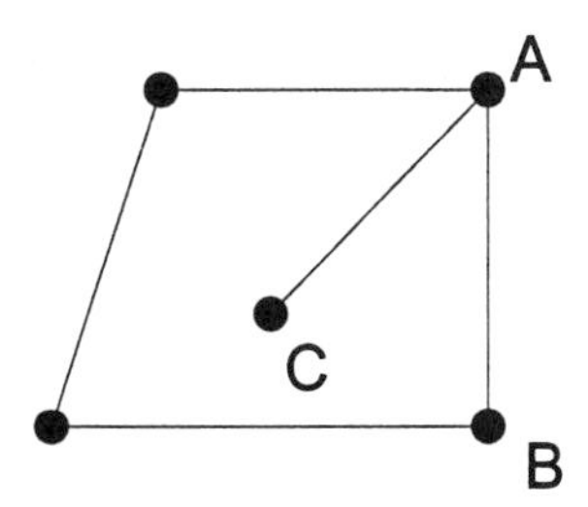

FIGURE 32

vertex is called the degree of the vertex. Thus, for example, in the graph of Figure 32, vertex A has degree 3, vertex B has degree 2, and vertex C has degree 1.

Problem 6. In Smallville there are 15 telephones. Can they be connected by wires so that each telephone is connected with exactly five others?

Solution. Suppose that this is possible. Consider the graph in which the vertices represent telephones, and the edges represent the wires. There are 15 vertices in this graph, and each has degree 5. Let us count the number of edges in this graph. To do this, we can add up the degrees of all the vertices. However, in this sum, each edge is counted twice (each edge connects two vertices). Therefore the number of edges in the graph must be equal to $15 \cdot 5/2$. But this number is not an integer. It follows that such a graph cannot exist, which means that we cannot connect the telephones as required.

In solving this problem, we have shown how to count the edges of a graph, knowing the degree of each vertex: we add the degrees of all the vertices and divide this sum by 2.

Problem 7. In a certain kingdom, there are 100 cities, and four roads lead out of each city. How many roads are there altogether in the kingdom?

Notice that our counting method for edges of a graph has the following consequence: the sum of the degrees of all the vertices in a graph must be even (otherwise, we could not divide it by 2 to get the number of edges). We can give a better formulation of this result using the following definitions:

A vertex of a graph having an odd degree is called an *odd* vertex. A vertex having an even degree is called an *even* vertex.

Theorem. The number of odd vertices in any graph must be even.

To prove this theorem it is enough to notice that the sum of several integers is even if and only if the number of odd addends is even.

<u>Methodological remark.</u> This theorem plays a central role in this chapter. It is important to keep returning to its proof, and to apply the theorem as often as possible in the solution of problems. Students should be encouraged to repeat the proof of the theorem within their solution to a problem, rather than merely quoting the theorem.

The theorem is often used to prove the existence of a certain edge of a graph, as in Problem 12. It is also used, as in Problems 8-11, to prove that a graph answering a certain description is impossible to draw. Such problems can be difficult for students to grapple with. It is essential that they first try to draw the required graph, then guess that it is not possible, and finally give a clear discussion or proof, using the theorem above, that the required graph does not exist.

Problem 8. There are 30 students in a class. Can it happen that 9 of them have 3 friends each (in the class), eleven have 4 friends each, and ten have 5 friends each?

Solution. If this were possible, then it would also be possible to draw a graph with 30 vertices (representing the students), of which 9 have degree 3, 11 have degree 4, and 10 have degree 5 (by connecting "friendly" vertices with edges). However, such a graph would have 19 odd vertices, which contradicts the theorem.

Problem 9. In Smallville there are 15 telephones. Can these be connected so that

(a) each telephone is connected with exactly 7 others;

(b) there are 4 telephones, each connected to 3 others, 8 telephones, each connected to 6 others, and 3 telephones, each connected to 5 others?

Problem 10. A king has 19 vassals. Can it happen that each vassal has either 1, 5, or 9 neighbors?

Problem 11. Can a kingdom in which 3 roads lead out of each city have exactly 100 roads?

Problem 12. John, coming home from Disneyland, said that he saw there an enchanted lake with 7 islands, to each of which there led either 1, 3, or 5 bridges. Is it true that at least one of these bridges must lead to the shore of the lake?

Problem 13. Prove that the number of people who have ever lived on earth, and who have shaken hands an odd number of times in their lives, is even.

Problem 14. Can 9 line segments be drawn in the plane, each of which intersects exactly 3 others?

§3. Some new definitions

Problem 15. In the country of Seven there are 15 towns, each of which is connected to at least 7 others. Prove that one can travel from any town to any other town, possibly passing through some towns in between.

Solution. Let us look at any 2 towns, and suppose that there is no path connecting them. This means that there is no sequence of roads such that the end of one road coincides with the beginning of the next road, connecting the 2 towns. It is given that each of the 2 towns is connected with at least 7 others. These 14 towns must be distinct: if any 2 were to coincide, there would be a path through them (or it) connecting the 2 given towns (see Figure 33). So there are at least 16 different towns, which contradicts the statement of the problem.

* * *

In light of this problem, we give two important definitions:

A graph is called *connected* if any two of its vertices can be connected by a path (a sequence of edges, each of which begins at the endpoint of the previous one).

A closed path (a path whose starting and ending vertices coincide) is called a *cycle*.

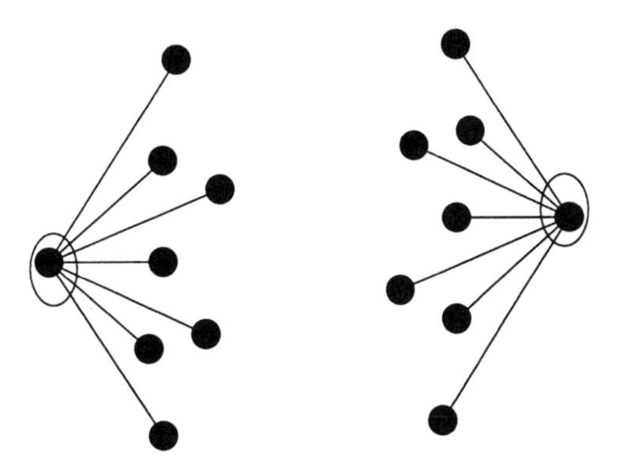

FIGURE 33

We can now reformulate the result of the previous problem: the graph of the roads of the kingdom of Seven is connected.

Problem 16. Prove that a graph with n vertices, each of which has degree at least $(n-1)/2$, is connected.

It is natural to ask how a non-connected graph looks. Such a graph is composed of several "pieces", within each of which one can travel along the edges from any vertex to any other. Thus, for example, the graph of Figure 34 consists of three "pieces", while the graph of Figure 35 consists of two.

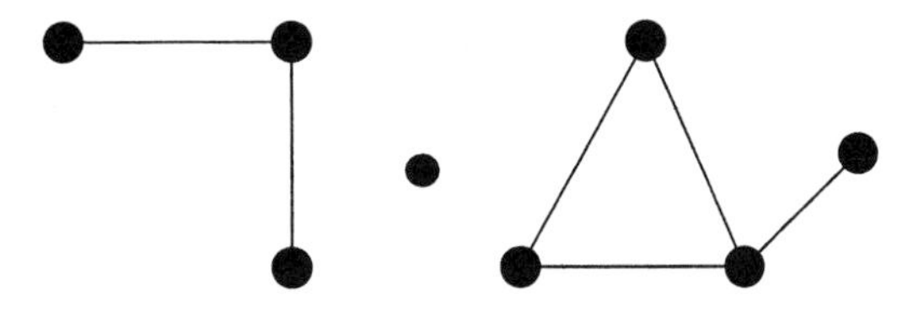

FIGURE 34

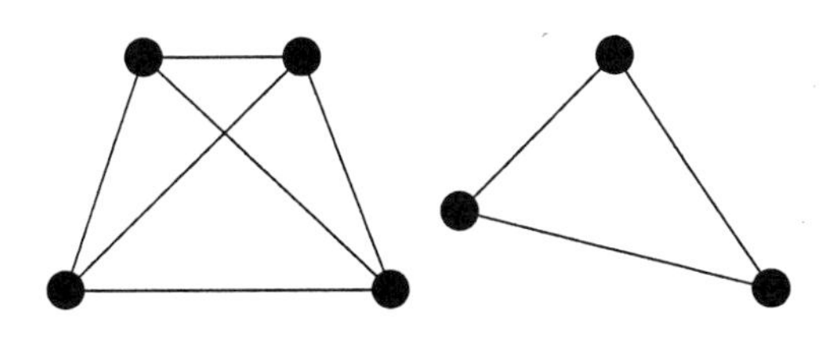

FIGURE 35

These "pieces" are called *connected components* of the graph. Each connected component is, of course, a connected graph. We note also that a connected graph consists of a single connected component.

Problem 17. In Never-Never-Land there is only one means of transportation: magic carpet. Twenty-one carpet lines serve the capital. A single line flies to Farville, and every other city is served by exactly 20 carpet lines. Show that it is possible to travel by magic carpet from the capital to Farville (perhaps by transferring from one carpet line to another).

Solution. Let us look at that connected component of the graph of carpet lines which includes the capital. We must prove that this component includes Farville. Suppose it does not. Then there are 21 edges starting at one vertex, and 20 edges starting at every other vertex. Therefore this connected component contains exactly one odd vertex. This is a contradiction.

Methodological remark. The notion of connectedness is extremely important, and is used constantly in further work in graph theory. The important point in the solution of Problem 16—consideration of a connected component—is a meaningful idea, and often turns out to be useful in solving problems.

Problem 18. In a certain country, 100 roads lead out of each city, and one can travel along those roads from any city to any other. One road is closed for repairs. Prove that one can still get from any city to any other.

§4. Eulerian graphs

Problem 19. Can one draw the graph pictured in (a) Figure 36; (b) Figure 37, without lifting the pencil from the paper, and tracing over each edge exactly once?

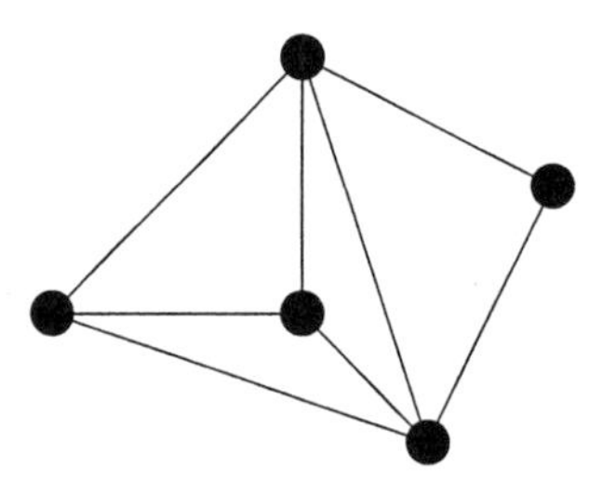

FIGURE 36

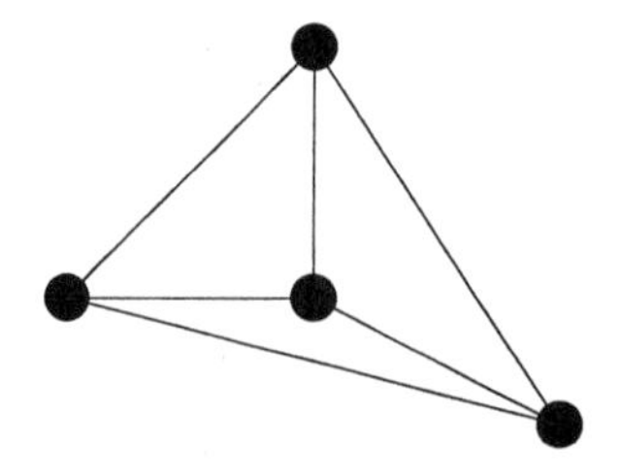

FIGURE 37

Solution. (a) Yes. One way is to start at the vertex on the extreme left, and end at the central vertex.

(b) No. Indeed, if we can trace out the graph as required in the problem, we will arrive at every vertex as many times as we leave it (with the exception of the initial and terminal vertices). Therefore the degree of each vertex, except for two, must be even. For the graph in Figure 37 this is not the case.

In solving Problem 19, we have established the following general principle:

A graph that can be traversed without lifting the pencil from the paper, while tracing each edge exactly once, can have no more than two odd vertices.

This sort of graph was first studied by the great mathematician Leonhard Euler in 1736, in connection with a famous problem about the Königsburg bridges (see also Problem 12). Graphs which can be traversed like this are called *Eulerian* graphs.

Problem 20. A map of the city of Königsburg is given in Figure 38. The city lies on both banks of a river, and there are two islands in the river. There are seven bridges connecting the various parts of the city. Can one stroll around the town, crossing each bridge exactly once?

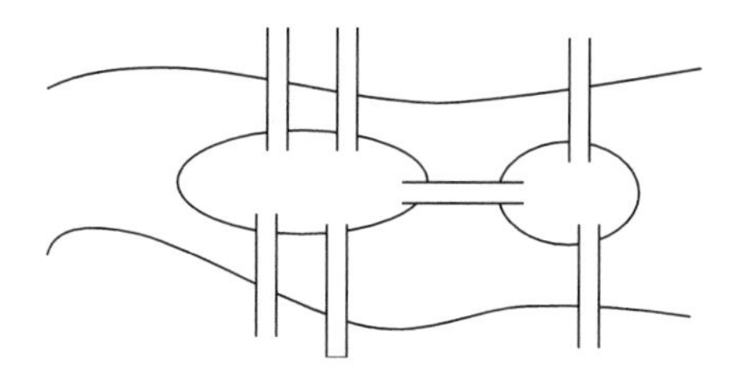

FIGURE 38

Problem 21. A group of islands are connected by bridges in such a way that one can walk from any island to any other. A tourist walked around every island, crossing each bridge exactly once. He visited the island of Thrice three times. How many bridges are there to Thrice, if

(a) the tourist neither started nor ended on Thrice;

(b) the tourist started on Thrice, but didn't end there;

(c) the tourist started and ended on Thrice?

Problem 22. (a) A piece of wire is 120 cm long. Can one use it to form the edges of a cube, each of whose edges is 10 cm?

(b) What is the smallest number of cuts one must make in the wire, so as to be able to form the required cube?

CHAPTER 6

The Triangle Inequality

§1. Introduction

The triangle inequality is easily motivated, whether or not students have had a formal introduction to geometry. But even for those students who have studied axiomatics or formal proof, there are non-trivial applications lying right beneath the surface, and problems involving the triangle inequality can be constructed which demand significant thought.

The inequality itself states that for any triangle ABC we have three inequalities

$$AB < AC + BC, \quad AC < BC + AB, \quad BC < AB + AC,$$

showing that any side of the triangle is less than the sum of two others.

Problem 1. Prove that for any three points A, B, and C we have $AC \geq |AB - BC|$.

In discussing this problem, it is important to give its geometric interpretation: the length of a side of a triangle is not less than the absolute value of the difference between the other two sides.

Problem 2. Side AC of triangle ABC has length 3.8, and side AB has length 0.6. If the length of side BC is an integer, what is this length?

Problem 3. Prove that the length of any side of a triangle is not more than half its perimeter.

Problem 4. The distance from Leningrad to Moscow is 660 kilometers. From Leningrad to the town of Likovo it is 310 kilometers, from Likovo to Klin it is 200 kilometers, and from Klin to Moscow is 150 kilometers. How far is it from Likovo to Moscow?

Hint for solution to Problem 4: Notice that the sum of the distances from Leningrad to Likovo, from Likovo to Klin, and from Klin to Moscow is equal to the distance from Leningrad to Moscow. This means that these towns are all on the same line.

Note that in solving Problem 4, we use the fact that the sum of any three sides of a quadrilateral is greater than the fourth side. This can easily be established, using the triangle inequality. In fact, for any polygon, the sum of all but one of the sides is greater than the remaining side. For many students, this fact can be established for a few cases, and then assumed intuitively for all cases. More advanced students can give a formal proof, using induction.

Problem 5. Find a point inside a convex quadrilateral such that the sum of the distances from the point to the vertices is minimal.

Solution. Since the quadrilateral is convex, its diagonals intersect at some interior point O. Suppose the vertices of the quadrilateral are A, B, C, and D (see Figure

39). Then the sum of the distances from O to the vertices is equal to $AC+BD$. But for any other point P, $PA+PC > AC$ (by the triangle inequality). Analogously, $PB+PD \geq BD$. This means that the sum of the distances from P to the vertices is not less than $AC+BD$. Clearly, this sum is equal to $AC+BD$ only if P and O coincide. Therefore O is the point we are looking for.

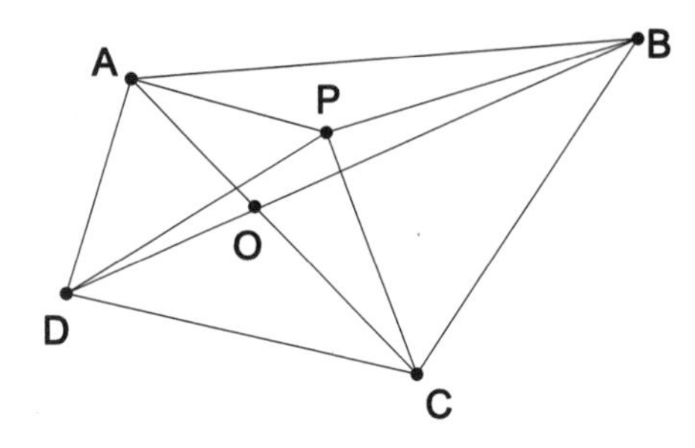

FIGURE 39

Problem 6. Point O is given on the plane of square $ABCD$. Prove that the distance from O to one of the vertices of the square is not greater than the sum of the distances from O to the other three vertices.

Problem 7. Prove that the sum of the diagonals of a convex quadrilateral is less than the perimeter but more than half the perimeter.

Problem 8. Prove that the sum of the diagonals of a convex pentagon is greater than the perimeter but less than double the perimeter.

Problem 9. Prove that the distance between any two points inside a triangle is not greater than half the perimeter of the triangle.

§2. The triangle inequality and geometric transformations

Often, the triangle to which we must apply the triangle inequality does not appear in the diagram for the problem. In these cases, a suitable choice of geometric transformation can help. The following series of problems illustrates the use of symmetry together with the triangle inequality.

Problem 10. A mushroom-gatherer leaves the woods at a given point. He must reach a highway, which follows a straight line, and go back into the woods at another given point (Figure 40). How should he do this, following the shortest path possible?

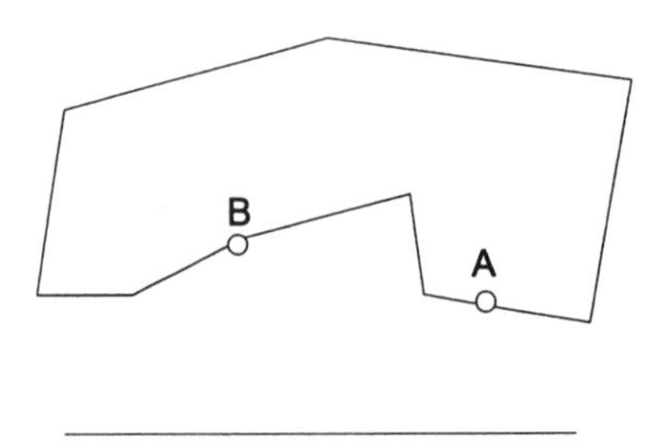

FIGURE 40

Problem 11. A woodsman's hut is in the interior of a peninsula which has the form of an acute angle. The woodsman must leave his hut, walk to one shore of the peninsula, then to the other shore, then return home. How should he choose the shortest such path?

Problem 12. Point A, inside an acute angle, is reflected in either side of the angle to obtain points B and C. Line segment BC intersects the sides of the angle at D and E (see Figure 41). Show that $BC/2 > DE$.

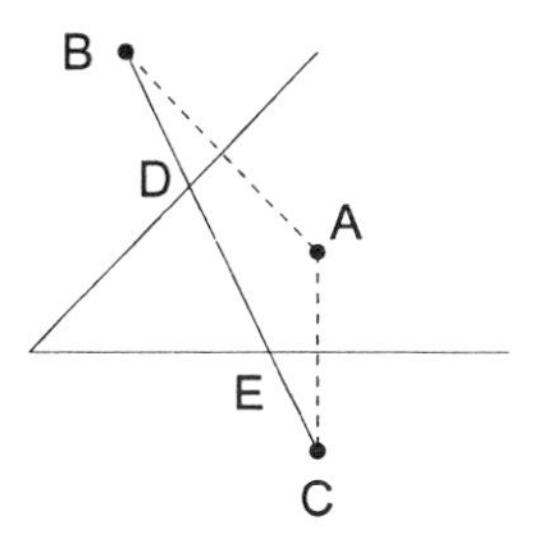

FIGURE 41

Problem 13. Point C lies inside a given right angle, and points A and B lie on its sides (see Figure 42). Prove that the perimeter of triangle ABC is not less than twice the distance OC, where O is the vertex of the given right angle.

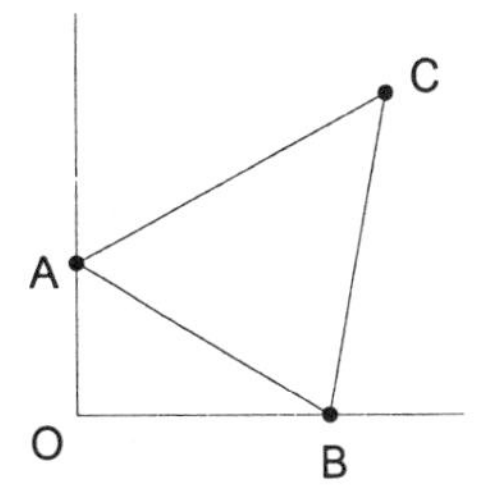

FIGURE 42

Let us analyze the solution to Problem 10. Suppose the mushroom-gatherer leaves the woods at point A, and must re-enter at point B. Reflect point A in the line of the highway (see Figure 43) to obtain point A'. If K is the point at which the mushroom-gatherer reaches the highway, then route AKB is equal in length to route $A'KB$, since we are simply reflecting segment AK in the highway. But $A'KB$ cannot be shorter than $A'B$. It follows that point K should be the point where $A'B$ intersects the highway.

Similar considerations allow us to solve the other problems in this series. For example, in Problem 13 we can reflect point C in lines OA and OB, to obtain points C' and C'' (Figure 44); it is easy to see that point O lies on straight line $C'C''$. Then we can replace the perimeter of triangle ABC with the sum of the lengths of segments $C'A$, AB, and BC''. The triangle inequality tells us that this sum is no less than the length of $C'C''$. This, in turn, is equal to $2OC$, since it is the hypotenuse of a right triangle of which OC is the median. (Students who don't know this theorem can find a more intuitive way to explain this: for example, by completing a rectangle with vertices at C', C'', and C.)

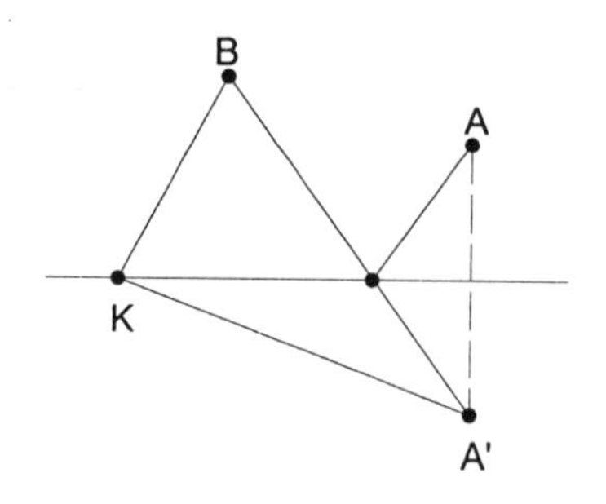

FIGURE 43

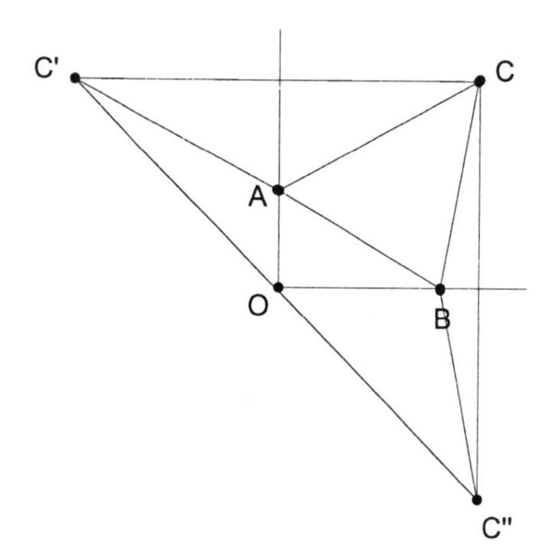

FIGURE 44

For teachers. It is important to solve these problems carefully, getting students to give a logical exposition of the solution, and not just an intuitive explanation. First we can remind the students that line reflection does not change distances. Then we can point out the common idea in these problems: to transform the required path so that its length does not change, and so that the problem becomes one of connecting two points with the shortest path possible. It is important to check that one of the transformed paths can really be a straight line so that we have an obvious answer; otherwise the solution can be much more difficult.

* * *

In many problems, the action takes place on some sort of surface in space. In such problems, the triangle inequality can be used only after we "unfold" the surface onto a plane. The following problems are typical:

Problem 14. A fly sits on one vertex of a wooden cube. What is the shortest path it can follow to the opposite vertex?

Problem 15. A fly sits on the outside surface of a cylindrical drinking glass. It must crawl to another point, situated on the inside surface of the glass. Find the shortest path possible (neglecting the thickness of the glass).

§3. Additional constructions

In many cases, the proofs of geometric inequalities require additional constructions. Such problems are often complicated, since the choice of construction requires a certain amount of practice. The following series of problems provides some such practice:

Problem 16. If point O is inside triangle ABC, prove that $AO+OC < AB+BC$.

Problem 17. Prove that the sum of the distances from point O to the vertices of a given triangle is less than the perimeter, if point O lies inside the triangle. What if point O is outside the triangle?

Problem 18. Solve Problem 11, if the peninsula has the shape of an obtuse angle.

Problem 19. Prove that the length of median AM in triangle ABC is not greater than half the sum of sides AB and AC. Prove also that the sum of the lengths of the three medians is not greater than the triangle's perimeter.

§4. Miscellaneous problems

Problem 20. A polygon is cut out of paper, and then folded in two along a straight line (see Figure 45). Prove that the perimeter of the polygon formed is not greater than the perimeter of the original polygon.

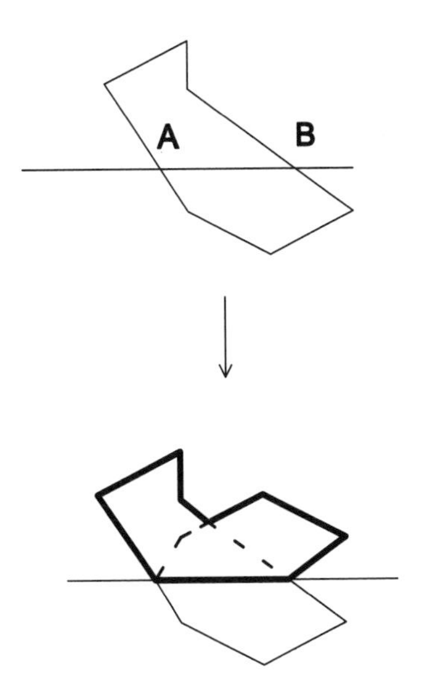

FIGURE 45

Problem 21. Prove that a convex polygon cannot have three sides, each of which is greater than the longest diagonal.

Problem 22. Prove that the perimeter of a triangle is not greater than 4/3 the sum of its medians. (For the solution of this problem, one must know the ratio into which the three medians of a triangle divide each other.)

Problem 23. Two villages lie on opposite sides of a river whose banks are parallel lines. A bridge is to be built over the river, perpendicular to the banks. Where should the bridge be built so that the path from one village to the other is as short as possible?

Problem 24. Prove that a convex pentagon (that is, a pentagon whose diagonals all lie inside the figure) has three diagonals which can form a triangle.

CHAPTER 7

Games

Students enjoy playing games. Whether the mathematics behind the game is simple or complicated, the chance for social interaction and for controlled competition will help to break up any routine patterns in school life.

At the same time, these problems hold a lot of content, and students frequently find their solution quite difficult. The chief difficulties consist first in articulating the winning strategy, and second in proving that the strategy considered always leads to a win. In surmounting these difficulties, students will learn more about accepted standards of mathematical argument, and will refine their understanding of what it means to solve a problem.

Students must understand that statements of the form: "If you do thus, I will do as follows," are usually not solutions to a game. Examples of correct solutions are given in the text.

We recommend giving no more than one or two games from this chapter in each lesson, with the exception of §4, which contains problems analyzed "backwards". The idea of symmetry (§2) and the concept of a winning position (§3) can be treated independently. This is best done after considering two or three problems on each theme.

There are many types of games considered in mathematics, and many types of game theories. This chapter considers only one type. In each of these games, there are two players who take turns making moves, and a player cannot decline to move. The problem is always the same: to find out which player (the first or the second) has a winning strategy. These notes will not be repeated for each game.

Starred problems are more difficult than the others.

§1. Pseudo-games: Games that are jokes

The first class of games we examine are games that turn out to be jokes. The outcomes of these pseudo-games do not depend on how the play proceeds. For this reason, the solution of such a pseudo-game does not consist of a winning strategy, but of a proof that one or the other of the two players will always win (regardless of how the play proceeds!).

Problem 1. Two children take turns breaking up a rectangular chocolate bar 6 squares wide by 8 squares long. They may break the bar only along the divisions between the squares. If the bar breaks into several pieces, they keep breaking the pieces up until only the individual squares remain. The player who cannot make a break loses the game. Who will win?

Solution. After each move, the number of pieces increases by one. At first, there is only one piece. At the end of the game, when no more moves are possible, the chocolate is divided into small squares, and there are 48 of these. So there must have been 47 moves, of which the last, as well as every other odd-numbered move, was made by the first player. Therefore, the first player will win, no matter how the play proceeds.

For teachers. Pseudo-games allow the students to relax and be relieved of the tension of having to solve a problem or win a game. They are very effective, for instance, if introduced right after particularly difficult material, or at the end of a lesson. It is important to let the students actually play the games before giving a solution.

Problem 2. There are three piles of stones: one with 10 stones, one with 15 stones, and one with 20 stones. At each turn, a player can choose one of the piles and divide it into two smaller piles. The loser is the player who cannot do this. Who will win, and how?

Problem 3. The numbers 1 through 20 are written in a row. Two players take turns putting plus signs and minus signs between the numbers. When all such signs have been placed, the resulting expression is evaluated (i.e., the additions and subtractions are performed). The first player wins if the sum is even, and the second wins if the sum is odd. Who will win and how?

Problem 4. Two players take turns placing rooks (castles) on a chessboard so that they cannot capture each other. The loser is the player who cannot place a castle. Who will win?

Problem 5. Ten 1's and ten 2's are written on a blackboard. In one turn, a player may erase any two figures. If the two figures erased are identical, they are replaced with a 2. If they are different, they are replaced with a 1. The first player wins if a 1 is left at the end, and the second player wins if a 2 is left.

Problem 6*. The numbers 25 and 36 are written on a blackboard. At each turn, a player writes on the blackboard the (positive) difference between two numbers already on the blackboard—if this number does not already appear on the blackboard. The loser is the player who cannot write a number.

Problem 7. Given a checkerboard with dimensions (a) 9×10; (b) 10×12; (c) 9×11. In one turn, a player is allowed to cross out one row or one column if at the beginning of the turn there is at least one square of the row or column remaining. The player who cannot make a move loses.

§2. Symmetry

Problem 8. Two players take turns putting pennies on a round table, without piling one penny on top of another. The player who cannot place a penny loses.

Solution. In this game, the first player can win, no matter how big the table may be! To do so, he must place the first penny so that its center coincides with the center of the table. After this, he replies to each move of the second player by placing a penny in a position symmetric to the penny placed by the second player, with respect to the center of the table. Notice that in such a strategy the positions of the two players are symmetric after each move of the first player. It follows that if there is a possible turn for the second player, then there is a possible response for the first player, who will therefore win.

Problem 9. Two players take turns placing bishops on the squares of a chessboard, so that they cannot capture each other (the bishops may be placed on squares of any color). The player who cannot move loses.

Solution. Since a chessboard is symmetric with respect to its center, it is natural to try a symmetric strategy. But this time, since one cannot place a bishop at the center of the chessboard, the symmetry will help the second player. It might seem, from an analogy with the previous problem, that such a strategy would allow the second player to win. However, if he follows it, he cannot even make a second move! The bishop placed by the first player can take a bishop placed in the symmetric square.

This example shows that in employing a symmetric strategy one must take into account that **a symmetric move can be blocked or prevented, but only by a move the opponent has just made**. Because of the symmetry, moves made earlier cannot affect a player's move. To solve a game using a symmetric strategy, one must find a symmetry such that the previous move does not destroy the chosen strategy.

Therefore, to solve Problem 9 we must look not to the point symmetry of the chessboard, but to its line symmetry. We can choose, for example, the line between the fourth and fifth rows as the line of symmetry. Squares which are symmetric with respect to this line will be of different colors, and therefore a bishop on one square cannot take a bishop on the symmetric square. Therefore, the second player can win this game.

The idea of a symmetric strategy need not be purely geometric. Consider the following problem.

Problem 10. There are two piles of 7 stones each. At each turn, a player may take as many stones as he chooses, but only from one of the piles. The loser is the player who cannot move.

Solution. The second player can win this game, using a symmetric strategy. At each turn, he must take as many stones as the first player has just taken, but from the other pile. Therefore the second player always has a move.

The symmetry in this problem consists in maintaining the equality of the number of stones in each pile.

Problem 11. Two players take turns placing knights on the squares of a chessboard, so that no knight can take another. The player who is unable to do this loses.

Problem 12. Two players take turns placing kings on the squares of a 9×9 chessboard, so that no king can capture another. The player who is unable to do this loses.

Problem 13. (a) Two players take turns placing bishops on the squares of a chessboard. At each turn, the bishop must threaten at least one square not threatened by another bishop. A bishop "threatens" the square it is placed on. The player who cannot move is the loser. (b)* The same game, but with rooks (castles).

Problem 14. Given a 10×10 chessboard, two players take turns covering pairs of squares with dominoes. Each domino consists of a rectangle 1 square in width and 2 squares in length (which can be held either way). The dominoes cannot overlap. The player who cannot place a domino loses.

Problem 15. A checker is placed on each square of an 11×11 checkerboard. Players take turns removing any number of checkers which lie next to each other along a row or column. The winner is the player who removes the last checker.

Problem 16. There are two piles of stones. One has 30 stones, and the other has 20 stones. Players take turns removing as many stones as they please, but from one pile only. The player removing the last stone wins.

Problem 17. Twenty points are placed around a circle. Players take turns joining two of the points with a line segment which does not cross a segment already drawn in. The player who cannot do so loses.

Problem 18. A daisy has (a) 12 petals; (b) 11 petals. Players take turns tearing off either a single petal, or two petals right next to each other. The player who cannot do so loses.

Problem 19.* Given a rectangular parallelepiped of dimensions (a) $4 \times 4 \times 4$; (b) $4 \times 4 \times 3$; (c) $4 \times 3 \times 3$, consisting of unit cubes. Players take turns skewering a row of cubes (parallel to the edges of the figure), so long as there is at least one cube which is not yet skewered in the row. The player who cannot do so loses.

Problem 20. Two players take turns breaking a piece of chocolate consisting of 5×10 small squares. At each turn, they may break along the division lines of the squares. The player who first obtains a single square of chocolate wins.

Problem 21. Two players take turns placing **x**'s and **o**'s on a 9×9 checkerboard. The first player places **x**'s, and the second player places **o**'s. At the end of the play, the first player gets a point for each row or column which contains more **x**'s than **o**'s. The second player gets a point for each row or column which contains more **o**'s than **x**'s. The player with the most points wins.

§3. Winning positions

Problem 22. On a chessboard, a rook stands on square $a1$. Players take turns moving the rook as many squares as they want, either horizontally to the right or vertically upward. The player who can place the rook on square $h8$ wins.

In this game, the second player will win. The strategy is quite simple: at each turn, place the rook on the diagonal from $a1$ to $h8$. The reason this works is that the first player is forced to move the rook off the diagonal at each turn, while the second player can always put the rook back on this diagonal. Since the winning square belongs to the diagonal, the second player will eventually be able to place the rook on it.

Let us analyze this solution a little more deeply. We have been able here to define a class of winning positions (in which the rook is on the diagonal from $a1$ to $h8$), which enjoys the following properties:

(1) The final position of the game is a winning one;

(2) A player can never move from one winning position to another in a single turn;

(3) A player can always move from a non-winning position to a winning one in a single move.

The discovery of such a class of winning positions for a given game is equivalent to solving the game. Indeed, moving to a winning position at each move constitutes a winning strategy. If the initial position of the game is a winning one, then the second player will win (as in the game described above). Otherwise, the first player will win.

<u>**For teachers.**</u> As the concept of a winning position generalizes a set of strategies, it can only be understood after solving several of the games presented in this section. As always, it is important to have students play each game before solving it.

Problem 23. A king is placed on square $a1$ of a chessboard. Players take turns moving the king either upwards, to the right, or along a diagonal going upwards and to the right. The player who places the king on square $h8$ is the winner.

Problem 24. There are two piles of candy. One contains 20 pieces, and the other 21. Players take turns eating all the candy in one pile, and separating the remaining candy into two (not necessarily equal) non-empty piles. The player who cannot move loses.

Problem 25. A checker is placed at each end of a strip of squares measuring 1×20. Players take turns moving either checker in the direction of the other, each by one or by two squares. A checker cannot jump over another checker. The player who cannot move loses.

Problem 26. A box contains 300 matches. Players take turns removing no more than half the matches in the box. The player who cannot move loses.

Problem 27. There are three piles of stones. The first contains 50 stones, the second 60 stones, and the third 70. A turn consists in dividing each of the piles containing more than one stone into two smaller piles. The player who leaves piles of individual stones is the winner.

Problem 28. The number 60 is written on a blackboard. Players take turns subtracting from the number on the blackboard any of its divisors, and replacing the original number with the result of this subtraction. The player who writes the number 0 loses.

Problem 29*. There are two piles of matches:

(a) a pile of 101 matches and a pile of 201 matches;

(b) a pile of 100 matches and a pile of 201 matches.

Players take turns removing a number of matches from one pile which is equal to one of the divisors of the number of matches in the other pile. The player removing the last match wins.

§4. Analysis from the endgame: A method of finding winning positions

Readers of the previous section may get the feeling that the discovery of a set of winning positions is based only on intuition, and is therefore not simple. We now describe a general method which will allow us to find a set of winning positions in many games.

We return to Problem 23, the problem about the single king on a chessboard. Let us try to find a set of winning positions. As always, the final position of the game, with the king in square $h8$, must be a winning one. We therefore place a plus sign in square $h8$ (see Figure 46). We will place the same sign in every other square at which the king occupies a winning position, and a minus sign in every square which is not a winning position (we will call them losing positions).

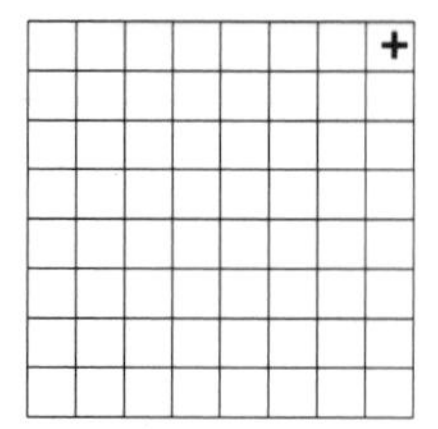

FIGURE 46

Since those squares from which the king can move to a winning square in a single move are losing squares, we arrive at Figure 47. From squares $h6$ and $f8$ we can move only to losing squares, so these must be winning positions (Figure 48). These new winning positions lead to new losing positions: $h5$, $g5$, $g6$, $f7$, $e7$, $e8$ (Figure 49). We continue in an analogous fashion (see Figures 50 and 51). After obtaining a set of minuses, we place plus signs in those squares from which any move at all leads to a losing square, then place minuses in those squares from which there is at least one move to a winning square. The pluses and minuses will finally be arranged as in Figure 52. It is not difficult to see that the squares with plus signs in them are exactly the winning squares indicated in the previous section.

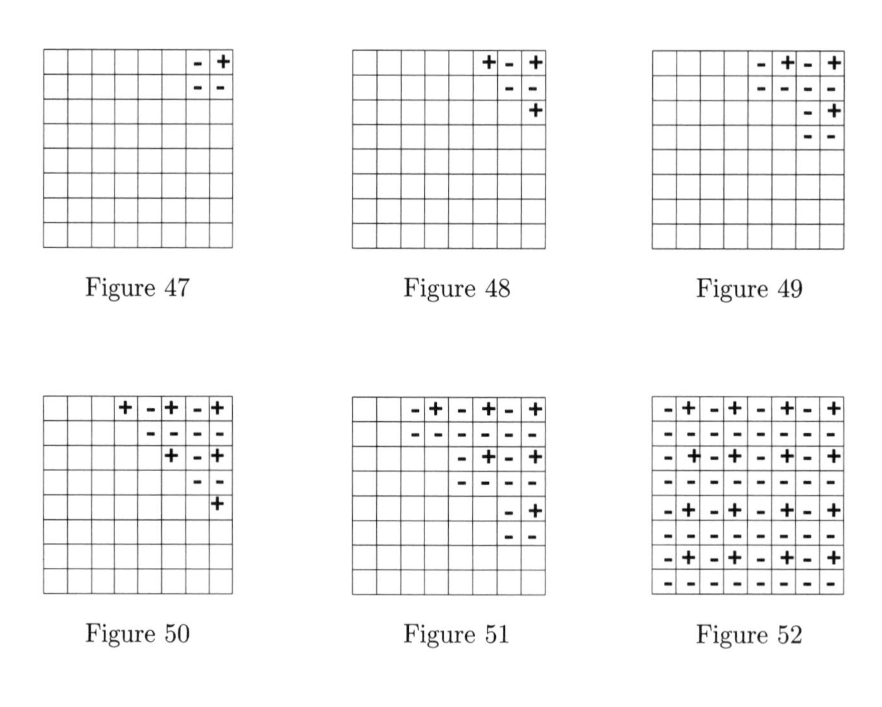

Figure 47 Figure 48 Figure 49

Figure 50 Figure 51 Figure 52

* * *

The method of finding winning positions just described is called *analysis from the endgame.* Applying it to the game with the castle (Problem 22) from the previous section, it is not hard to derive the set of winning positions for this game as well. Working as in Figures 53 and 54, we soon arrive at Figure 55.

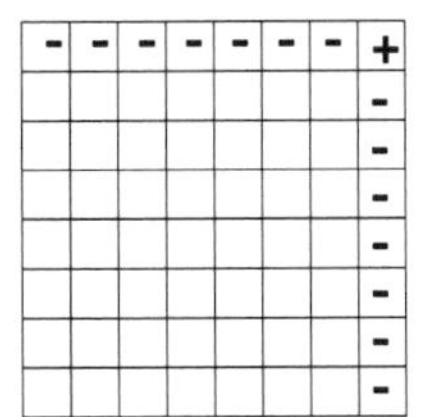

Figure 53

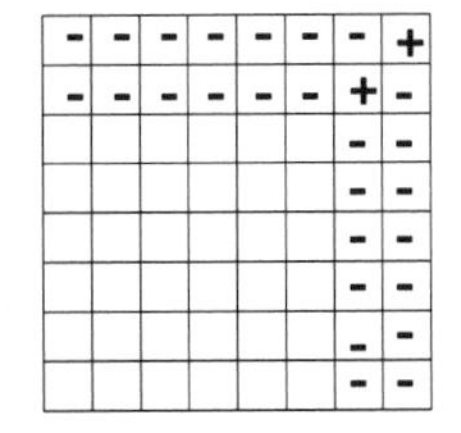

Figure 54

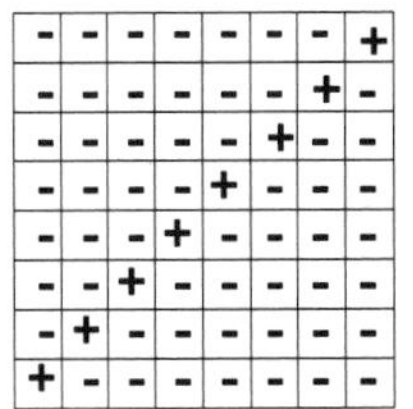

Figure 55

For teachers. Students often perform their own "analysis from the endgame" intuitively. That is, they can see to the end of the game from a few moves before, and begin to learn which of the last few possible moves are winning ones, then generalize this to the rest of the game. The best learning will occur if students make this discovery on their own (by playing the game), then are asked to articulate it.

Problem 30. A queen stands on square $c1$ of a chessboard. Players take turns moving the queen any number of squares to the right, upwards, or along a diagonal to the right and upwards. The player who can place the queen in square $h8$ wins.

Solution. Using analysis from the endgame, we obtain the configuration of pluses and minuses given in Figure 56. Thus, the first player wins; in fact, he has a choice of three initial moves. These are to squares $c5$, $e3$, or $d1$.

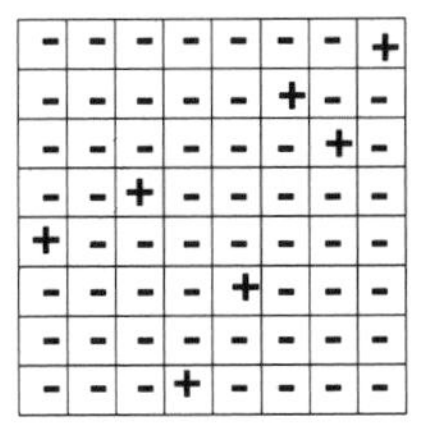

FIGURE 56

For teachers. This game can serve as a good introduction to analysis from the endgame. Student exercises can then be created, for example, by replacing the square checkerboard in Problems 22, 23, 30 with a rectangular board of any dimensions, or with a board of some other unusual form. For instance, one might solve Problem 22 on a checkerboard with the middle four squares removed (or, with some other squares removed). The arrangement of pluses and minuses on this sort of checkerboard is shown in Figure 57, in which the missing squares are shaded.

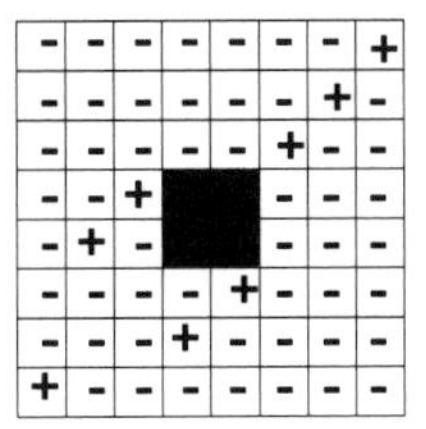

FIGURE 57

* * *

The following problem provides a chance for students to practice the technique of reformulating a game.

Problem 31. Of two piles of stones, one contains 7 stones, and the other 5. Players alternate taking any number of stones from one of the piles, or an equal number from each pile. The player who cannot move loses.

Solution. We can restate the situation in this problem as one which occurs on the usual chessboard. First we assign coordinates to each square, by numbering the rows from 0 to 7, starting at the top, and the columns from 0 to 7, starting at the right. Each position of the original game is characterized by an ordered pair of numbers: the number of stones in the first pile, followed by the number of stones in the second. To each such position we assign the square whose coordinates are these numbers. Now we note that a move in the original game corresponds to a queen's move on the chessboard, upwards, to the right, or on a diagonal upwards and to the right. This restatement of the problem makes the game identical to that of Problem 30. Notice that we can use the same technique to reformulate the games in Problems 10 and 20.

* * *

Problem 32. A knight is placed on square $a1$ of a chessboard. Players alternate moving the knight either two squares to the right and one square up or down, or two squares up and one square right or left (and usual knight moves but in restricted directions). The player who cannot move loses.

Problem 33. (a) There are two piles of 7 stones each. In each turn, a player may take a single stone from one of the piles, or a stone from each pile. The player who cannot move loses.

(b) In addition to the moves described above, players are allowed to take a stone from the first pile and place it on the second pile. Other rules remain the same.

Problem 34. There are two piles of 11 matches each. In one turn, a player must take two matches from one pile and one match from the other. The player who cannot move loses.

Problem 35. This game begins with the number 0. In one turn, a player can add to the current number any natural number from 1 through 9. The player who reaches the number 100 wins.

Problem 36. This game begins with the number 1. In one turn, a player can multiply the current number by any natural number from 2 through 9. The player who first names a number greater than 1000 wins.

Problem 37. This game begins with the number 2. In one turn, a player can add to the current number any natural number smaller than it. The player who reaches the number 1000 wins.

Problem 38. This game begins with the number 1000. In one turn, a player can subtract from the current number any natural number less than it which is a power of 2 (note that $1 = 2^0$). The player who reaches the number 0 wins.

CHAPTER 8

Problems for the First Year

As was emphasized in the preface, the first part of this book presents the basic topics for sessions of an "olympiad" mathematical circle (for students of age 11–13). However, these topics do not exhaust all the themes available for students of this age. In the present chapter we will try to fill this gap, at least partly.

For teachers. We would also like to say that we do not recommend preparing a session using only problems pertaining to a single topic. You can also use non-standard problems, which require something new and unusual, fresh ideas, or just the overcoming of technical difficulties. Since such problems are important for olympiads, contests, et cetera, we have gathered them together in this chapter.

§1. Logical problems

For teachers. When dealing with young students keep in mind that the most important goal is to teach them consistent and clear thinking; that is, how not to confuse cause and consequence; how to analyze cases carefully, without skipping any; how to build a chain of propositions and lemmas properly. The following problems in logic can help you in handling this.

1. Peter's mom said: "All champions are good at math." Peter says: "I am good at math. Therefore I am a champion!" Is his implication right or wrong?

2. There are four cards on the table with the symbols A, B, 4, and 5 written on their visible sides. What is the minimum number of cards we must turn over to find out whether the following statement is true: "If an even number is written on one side of a card then a vowel is written on the other side"?

3. A sum of fifteen cents was paid by two coins, and one of these coins was not a nickel. Find the values of the coins.

4. Assume that the following statements are true:

a) among people having TV sets there are some who are not mathematicians;

b) non-mathematicians who swim in swimming pools every day do not have TV sets.

Can we claim that not all people having TV sets swim every day?

5. During a trial in Wonderland the March Hare claimed that the cookies were stolen by the Mad Hatter. Then the Mad Hatter and the Dormouse gave testimonies which, for some reason, were not recorded. Later on in the trial it was found out that the cookies were stolen by only one of these three defendants, and, moreover, only the guilty one gave true testimony. Who stole the cookies?

* * *

6. In a box, there are pencils of at least two different colors, and of two different sizes. Prove that there are two pencils that differ both in color and in size.

7. There are three urns containing balls: the first one contains two white balls, the second—two black balls, and the third—a white ball and a black ball. The labels WW, BB, and WB were glued to the urns so that the contents of no urn corresponds to its label. Is it possible to choose one urn so that after drawing a ball from it one can always determine the contents of each urn?

8. Three people—A, B, and C—are sitting in a row in such a way that A sees B and C, B sees only C, and C sees nobody. They were shown 5 caps—3 red and 2 white. They were blindfolded, and three caps were put on their heads. Then the blindfolds were taken away and each of the people was asked if they could determine the color of their caps. After A, and then B, answered negatively, C replied affirmatively. How was that possible?

9. Three friends—sculptor White, violinist Black, and artist Redhead—met in a cafeteria. "It is remarkable that one of us has white hair, another one has black hair, and the third has red hair, though no one's name gives the color of their hair" said the black-haired person. "You are right," answered White. What color is the artist's hair?

* * *

The next eight problems take place on an island where all the inhabitants are either "knights" who always tell the truth or "knaves" who always lie (for many more such problems see [**16**]).

10. Person A said "I am a liar." Is he an inhabitant of our island?

11. What one question might be asked of an islander to find out where a road leads—to the city of knights or to the city of knaves?

12. What one question might be asked of an islander to find out whether she has a pet crocodile?

13. Assume that in the language of the island the words "yes" and "no" sound like "flip" and "flop", but we do not know which is which. What one question might be asked of an islander to find out whether he is a knight or a knave?

14. What one question might be asked of an islander so that the answer is always "flip"?

15. An islander A, in the presence of another islander B, said: "At least one of us is a knave." Is A a knight or a knave? What about B?

16. There are three people, A, B, and C. Among them is a knight, a knave, and a stranger (a normal person), who sometimes tells the truth and sometimes lies.

A said: "I am a normal person."

B said: "A and C sometimes tell the truth."

C said: "B is a normal person."

Who among them is a knight, who is a knave, and who is a normal person?

17. Several islanders met at a conference, and each of them told the others: "You are all knaves." How many knights might there be at that conference?

§2. Constructions and weighings

Mathematical and logical problems whose solution consists of a particular construction; that is, creating an example, are very common and useful. The students should understand that a construction may serve as a complete solution to problems of a certain type (such as those starting with the words "Is it possible to ... ?"). Such problems are usually quite attractive to younger students, and they can spend a lot of time trying to find a constructive solution to a tricky question or puzzle.

18. There are two egg timers: one for 7 minutes and one for 11 minutes. We must boil an egg for exactly 15 minutes. How can we do that using only these timers?

19. There are two buttons inside an elevator in a building with twenty floors. The elevator goes 13 floors up when the first button is pressed, and 8 floors down when the second one is pressed (a button will not function if there are not enough floors to go up or down). How can we get to the 8th floor from the 13th?

20. The number 458 is written on a blackboard. It is allowed either to double the number on the blackboard, or to erase its last digit. How can we obtain the number 14 using these operations?

21. Cards with the numbers 7, 8, 9, 4, 5, 6, 1, 2, and 3 are laid in a row in the indicated order. It is permitted to choose several consecutive cards and rearrange them in the reverse order. Is it possible to obtain the arrangement 1, 2, 3, 4, 5, 6, 7, 8, 9 after three such operations?

22. The numbers 1 through 16 are placed in the boxes of a 4×4 table as shown in Figure 58 (a). It is permitted to increase all the numbers in any row by 1 or decrease all the numbers in any column by 1. Is it possible to obtain the table shown in Figure 58 (b) using these operations?

1	2	3	4
5	6	7	8
9	10	11	12
13	14	15	16

(a)

1	5	9	13
2	6	10	14
3	7	11	15
4	8	12	16

(b)

FIGURE 58

23. Is it possible to write the numbers 1 through 100 in a row in such a way that the (positive) difference between any two neighboring numbers is not less than 50?

24. Divide a set of stones which weigh 1g, 2g, 3g, ... , 555g into three heaps of equal weight.

25. Fill the boxes of a 4×4 table with non-zero numbers so that the sum of the numbers in the corners of any 2×2, 3×3, or 4×4 square is zero.

26. Is it possible to label the edges of a cube using the numbers 1 through 12 in such a way that the sums of the numbers on any two faces of the cube are equal?

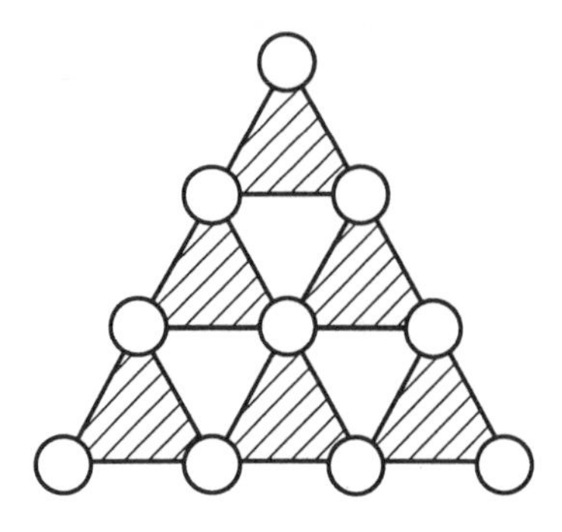

FIGURE 59

27*. Is it possible to place the numbers 0 through 9 in the circles in Figure 59 without repetitions so that all the sums of the numbers in the vertices of the shaded triangles are equal?

28. Prove that one can cross out several digits at the beginning and several at the end of the 400-digit number 84198419...8419 in such a way that the sum of the remaining digits is 1984.

29. Find a two-digit number, the sum of whose digits does not change when the number is multiplied by any one-digit number.

30. Do there exist two consecutive natural numbers such that the sums of their digits are both divisible by 7?

31. Do there exist several positive numbers, whose sum is 1, and the sum of whose squares is less than 0.01?

32. A castle consists of 64 identical square rooms, having a door in every wall and arranged in an 8×8 square. All the floors are colored white. Every morning a painter walks through the castle recoloring floors in all the rooms he visits from white to black and vice versa. Is it possible that some day the rooms will be colored as a standard chessboard is?

33. Can one place a few dimes on the surface of a table so that each coin touches exactly three other coins?

34. In a warehouse N containers marked 1 through N are arranged in two piles. A forklift can take several containers from the top of one pile and place them on the top of the other pile. Prove that all the containers can be arranged in one pile in increasing order of their numbers with $2N - 1$ such operations of the forklift.

* * *

There are many problems involving weighing which are closely related to construction problems. In solving these problems, we must not neglect even the simplest or most unlikely cases. Arguments like "We will consider the worst case" are usually very vague and unacceptable.

In all the problems of this set we consider "a weighing" as performed on a standard balance with two pans but without arrows or weights, unless otherwise specified.

35. There are 9 coins, one of which is counterfeit (it is lighter than the others). Find the counterfeit coin using two weighings.

36. There are 10 bags with coins. One of them contains only counterfeit coins, each of which is 1 gram lighter than a genuine coin. Using only one weighing on a balance with an arrow showing the difference between weights on the pans, find the "counterfeit" bag.

37. There are 101 coins, and only one of them differs from the other (genuine) ones by weight. We have to determine whether this counterfeit coin is heavier or lighter than a genuine coin. How can we do this using two weighings?

38. There are 6 coins; two of them are counterfeit and are lighter than the genuine coins. Using three weighings, determine both counterfeit coins.

39. There are 10 bags with coins, and some of these bags contain only counterfeit coins. A counterfeit coin is 1g lighter than a genuine coin. One of the bags is known to be filled with the genuine coins. Using one weighing on a balance with one pan and with an arrow showing the weight on the pan, determine which bags are "counterfeit" and which are not.

40. There are 5 coins, three of which are genuine. One is counterfeit and heavier than a genuine coin, and another one is counterfeit and lighter than a genuine coin. Using three weighings, find both counterfeit coins.

41. There are 68 coins of different weight. Using 100 weighings, find the heaviest and the lightest of the coins.

42. There are 64 stones of different weight. Using 68 weighings, find the heaviest and the second heaviest stones.

43. We have 6 weights: two green, two red, and two white. In each pair one of the weights is heavier. All the heavy weights have the same weight, and all the light weights have the same weight. Using two weighings, determine which weights are the heavy ones.

44. There are 6 coins, two of which are counterfeit: they are 0.1g heavier than the genuine coins. The pans of a balance are out of equilibrium only if the difference of weights is at least 0.2g. Find both counterfeit coins using four weighings.

45. a) There are 16 coins. One of them is counterfeit: it differs in weight from a genuine coin, though we do not know whether it is heavier or lighter. Find the counterfeit coin using four weighings.

b)*There are 12 coins. One of them is counterfeit: it differs by weight from a genuine coin, though we do not know whether it is heavier or lighter. Find the counterfeit coin using three weighings.

46*. Fourteen coins were presented in court as evidence. The judge knows that exactly 7 of these are counterfeit and weigh less than the genuine coins. A lawyer claims to know which coins are counterfeit and which are genuine, and she is required to prove it. How can she accomplish this using only three weighings?

§3. Problems in geometry

The problems in this section can be split naturally into two sets. The first set (Problems 47–57) continues the previous section: it is dedicated to geometric constructions. The second set contains more "standard" geometry problems.

47. Draw a broken line made up of 4 segments passing through all 9 points shown in Figure 60.

FIGURE 60

48. Cut a square into 5 rectangles in such a way that no two of them have a complete common side (but may have some parts of their sides in common).

49. Is it possible to draw a closed 8-segment broken line which intersects each segment of itself exactly once?

50. Is it possible to cut a square into several obtuse triangles?

51. Is it true that among any 10 segments there always are 3 which can form a triangle?

52. A king wants to build 6 fortresses and connect each pair of them by a road. Draw a scheme of fortresses and roads such that there are only 3 crossroads, each formed by 2 intersecting roads.

53. Is it possible to choose 6 points on the plane and connect them by disjoint segments (that is, by segments which do not have common inner points) so that each point is connected with exactly 4 other points?

54. Can we tile the plane with congruent pentagons?

55. Cut a 3×9 rectangle into 8 squares.

56. Prove that a square can be dissected into 1989 squares.

57. Cut an arbitrary triangle into 3 parts such that they can be rearranged to form a rectangle.

* * *

58. Points M and K are given on sides AB and BC of triangle ABC respectively. Segments AK and CM meet at point O. Prove that if $OM = OK$ and $\angle KAC = \angle MCA$, then triangle ABC is isosceles.

59. Altitude AK, angle bisector BH, and median CM of triangle ABC meet at one point O, and $AO = BO$. Prove that triangle ABC is equilateral.

60. In hexagon $ABCDEF$ triangles ABC, ABF, FED, CDB, FEA, and CDE are congruent. Prove that diagonals AD, BE, and CF are equal.

61. Altitude CH and median BK are drawn in acute triangle ABC. If $BK = CH$, and $\angle KBC = \angle HCB$, prove that triangle ABC is equilateral.

62. Diagonals AC and BD of quadrilateral $ABCD$ meet at point O. The perimeters of triangles ABC and ABD are equal, as are the perimeters of triangles ACD and BCD. Prove that $AO = BO$.

63. Prove that the star shown in Figure 61 cannot be drawn to satisfy the following inequalities: $BC > AB$, $DE > CD$, $FG > EF$, $HI > GH$, $KA > IK$.

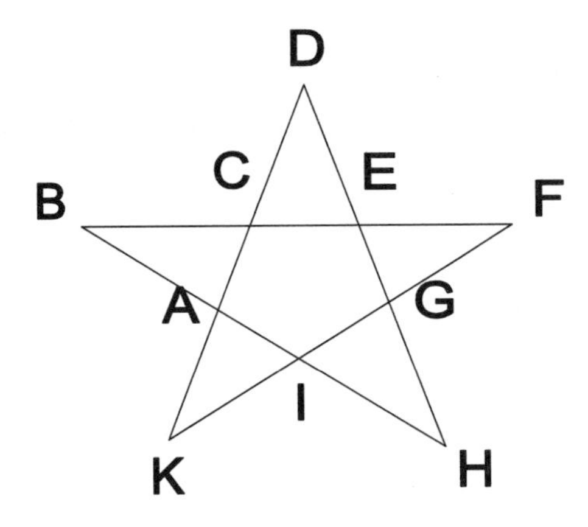

FIGURE 61

§4. Problems on integers

This topic has already been discussed in the chapter "Divisibility and Remainders". However, there are many nice problems dealing with integers, so many that we consider it necessary to gather some of them in this section. For instance, the set of Problems 70–84 is just an extension of the chapter on divisibility. Other problems bring in new themes.

64. If every boy in a class buys a muffin and every girl buys a sandwich, they will spend one cent less than if every boy buys a sandwich and every girl buys a muffin. We know that the number of boys in the class is greater than the number of girls. Find the difference.

65. 175 Humpties cost more than 126 Dumpties. Prove that you cannot buy three Humpties and one Dumpty for one dollar.

66. In a class every boy is friends with exactly three girls, and every girl is friends with exactly two boys. It is known that there are only 19 desks (each holding at most two students), and 31 of the students in the class study French. How many students are there?

67. Two teams played each other in a decathlon. In each event the winning team gets 4 points, the losing team gets 1 point, and both teams get 2 points in case of a draw. After all 10 events the two teams have 46 points together. How many draws were there?

68. Four friends bought a boat. The first friend paid half of the sum paid by the others; the second paid one third of the sum paid by the others; the third paid one quarter of what was paid by the others, and the fourth friend paid 130 dollars. What was the price of the boat, and how much did each of the friends pay?

69. The road connecting two mountain villages goes only uphill or downhill. A bus always travels 15 mph uphill and 30 mph downhill. Find the distance between the villages if it takes exactly 4 hours for the bus to complete a round trip.

* * *

70. Do there exist natural numbers a and b such that $ab(a-b) = 45045$?

71. Let us denote the sum of three consecutive natural numbers by a, and the sum of the next three consecutive natural numbers by b. Can the product ab be equal to 111111111?

72. Prove that the last non-zero digit of the number 1985! is even.

73. The natural numbers x and y satisfy the relation $34x = 43y$. Prove that the number $x + y$ is composite.

74. Do there exist non-zero integers a and b such that one of them is divisible by their sum while the other is divisible by their difference?

75. The prime numbers p and q, and natural number n satisfy the following equality:

$$\frac{1}{p} + \frac{1}{q} + \frac{1}{pq} = \frac{1}{n}.$$

Find these numbers.

76. Prove that a natural number written using one 1, two 2's, three 3's, ..., nine 9's cannot be a perfect square.

77. Each of the natural numbers a, b, c, and d is divisible by $ab - cd$. Prove that $ab - cd$ equals either 1 or -1.

78. In a certain country, banknotes of four types are in circulation: 1 dollar, 10 dollar, 100 dollar, and 1000 dollar bills. Is it possible to pay one million dollars using exactly half a million notes?

79. The number 1 is written on a blackboard. After each second the number on the blackboard is increased by the sum of its digits. Is it possible that at some moment the number 123456 will be written on the blackboard?

80. Prove that the number 3999991 is not prime.

81. a) Find a seven-digit number with all its digits different, which is divisible by each of those digits.

b) Does there exist an eight-digit number with the same property?

82. We calculate the sum of the digits of the number 19^{100}. Then we find the sum of the digits of the result, et cetera, until we have a single digit. Which digit is this?

83. Prove that the remainder when any prime number is divided by 30 is either 1 or a prime number.

84. Does there exist a natural number such that the product of its digits equals 1980?

* * *

85. A natural number ends in 2. If we move this digit 2 to the beginning of the number, then the number will be doubled. Find the smallest number with this property.

86. Given a six-digit number $\overline{abcdef}$ such that $\overline{abc} - \overline{def}$ is divisible by 7, prove that the number itself is also divisible by 7.

87. Find the smallest natural number which is 4 times smaller than the number written with the same digits but in the reverse order.

88. A three-digit number is given whose first and last digits differ by at least 2. We find the difference between this number and the reverse number (the number written with the same digits but in the reverse order). Then we add the result to its reverse number. Prove that this sum is equal to 1089.

* * *

89. Which number is greater: 2^{300} or 3^{200}?

90. Which number is greater: 31^{11} or 17^{14}?

91. Which number is greater: 50^{99} or $99!$?

92. Which number is greater: $888\ldots88 \times 333\ldots33$ or $444\ldots44 \times 666\ldots67$ (each of the numbers has 1989 digits)?

93. Which type of six-digit numbers are there more of: those that can be represented as the product of two three-digit numbers, or those that cannot?

* * *

94. Several identical paper triangles are given. The vertices of each one are marked with the numbers 1, 2, and 3. They are piled up to form a triangular prism. Is it possible that all the sums of the numbers along the edges of the prism are equal to 55?

95. Can one place 15 integers around a circle so that the sum of every 4 consecutive numbers is equal either to 1 or 3?

96*. Find a thousand natural numbers such that their sum equals their product.

97. The numbers 2^{1989} and 5^{1989} are written one after another. How many digits in all are there?

98. A bus ticket (whose number in Russia consists of 6 arbitrary digits) is called "lucky" if the sum of its first three digits equals the sum of the last three. Prove that the number of "lucky" tickets equals the number of tickets with the sum of their digits equal to 27.

§5. Miscellaneous

99. Fourteen students in a class study Spanish, and eight students study French. We know that three students study both languages. How many students are there in the class if every one of them is studying at least one language?

100. The plane is colored using two colors. Prove that there are two identically colored points exactly 1 meter apart.

101. A straight line is colored using two colors. Prove that we can find a segment of non-zero length with its endpoints and midpoint colored the same.

102. A 8×8 square is formed by 1×2 dominos. Prove that some pair of them forms a 2×2 square.

103. A 3×3 table is filled with numbers. It is allowed to increase each number in any 2×2 square by 1. Is it possible, using these operations, to obtain the table shown in Figure 62 from a table filled with zeros?

4	9	5
10	18	12
6	13	7

FIGURE 62

104. We call a bus overcrowded if there are more than 50% of the maximum allowable number of passengers inside. Children ride in several buses to a summer camp. Which is greater: the percentage of overcrowded buses or the percentage of children riding in the overcrowded buses?

105. Problem lists for the all-city olympiads in grades 6–11 are compiled so that each list contains eight questions, and there are exactly three questions in each grade which are not used in the other grades' olympiads. What is the maximum possible number of questions used by the problem committee?

106. All the students in a school are arranged in a rectangular array. After that, the tallest student in each row was chosen, and then among these John Smith happened to be the shortest. Then, in each column, the shortest student was chosen, and Mary Brown was the tallest of these. Who is taller: John or Mary?

107. Thirty chairs stand in a row. Every now and then a person comes and sits in one of the free chairs. After that, if any of the neighboring chairs is occupied, one of the person's neighbors stands up and leaves. What is the maximum number of chairs that can be occupied simultaneously, if originally

a) all the chairs are free?

b) ten chairs are taken?

108. Three pawns are placed on the vertices of a pentagon. It is allowed to move a pawn along any diagonal of the pentagon to any free vertex. Is it possible that after several such moves one of the pawns occupies its original position while the other two have changed their places?

109. None of the numbers a, b, c, d, e, or f equals zero. Prove that there are both positive and negative numbers among the numbers ab, cd, ef, $-ac$, $-be$, and $-df$.

110*. Professor Rubik splits his famous $3 \times 3 \times 3$ cube with an ax. What is the minimum possible number of blows he needs to split the cube into 27 small cubes if it is allowed to put some pieces on the top of others between blows?

111. The boxes of a sheet of graph paper are colored using eight colors. Prove that one can find a figure such as shown in Figure 63 which contains two boxes of the same color.

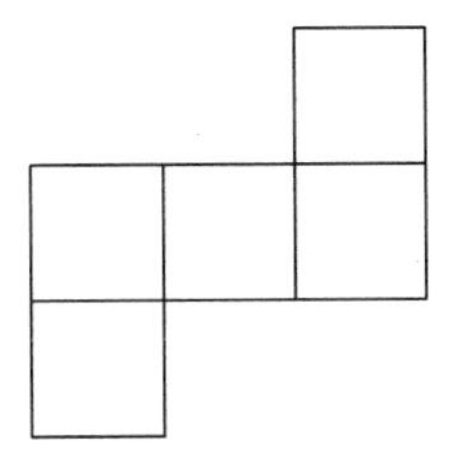

FIGURE 63

112. A six-digit number is given. How many seven-digit numbers are there which will produce that number if one digit is crossed out?

113. How many bus tickets do you have to buy in a row to be sure you have purchased a "lucky" ticket? (See Problem 98 for the definition of the "lucky" ticket. Bus tickets are numbered consecutively, and the ticket 999999 is followed by the ticket 000000).

114*. There was a volleyball tournament in which each team played every other team exactly once. We say that team A is better than team B if A defeated B, or if there is a team C such that A defeated C and C defeated B. Prove that the team which won the tournament is better than any other team.

115*. A 20×30 rectangle is cut from a sheet of graph paper. Is it possible to draw a straight line which intersects the interiors of 30 boxes of the rectangle?

116. The natural numbers 1 through 64 are written in squares of a chessboard, and each number is written exactly once. Prove that numbers in some pair of neighboring squares differ by at least 5.

CHAPTER 9

Induction

I. S. Rubanov

§1. Process and method of induction

(An Introduction for Teachers). Almost everyone has once had fun arranging dominoes in a row and starting a wave. Push the first domino and it topples the second, the second will topple the third and so forth until all the dominoes are toppled. Now let us change the set of dominoes into an infinite series of propositions: P_1, P_2, P_3, ... , numbered by positive integers. Assume that we can prove that:

(B): the first proposition of the series is true;

(S): the truth of every proposition in the series implies the truth of the next one.

Then, in fact, we have already proved all the propositions in the series. Indeed, we can "push the first domino", i.e., prove the first statement (B), and then statement (S) means that each domino, in falling, topples the next one. Whatever the "domino" (proposition) we choose, it will be eventually hit by this wave of "falling dominos" (proofs).

This is a description of the method of mathematical induction (MMI). Theorem (B) is called *base of induction*, and theorem (S) is *the inductive step.* Our reasoning with the wave of falling dominoes shows that step (S) is but a shortened form of the chain of theorems shown in the figure below:

$$P_1 \longrightarrow P_2 \longrightarrow P_3 \longrightarrow \dots \longrightarrow P_k \longrightarrow P_{k+1} \longrightarrow \dots$$

We will call theorems in this chain "steps", and the process of their successive proof—"the process of induction". This process can be visually represented as a wave of proofs, running from statement to statement along a chain of theorems.

Psychologically, the essence of induction is in its process. How can we teach this? We will try to show you in a dialog between teacher ("T") and student ("S"), which roughly resembles a session of a real mathematical circle. At the end of the dialog some methodological comments for the teacher are given (references to these comments are indicated in the text of the dialog).

* * *

Problem 1. T: One box was cut off from a 16×16 square of graph paper. Prove that the figure obtained can be dissected into trominos of a certain type—"corners" (see Figure 64.)

S: But this is easy—any "corner" has three boxes, and $16^2 - 1$ is divisible by 3.

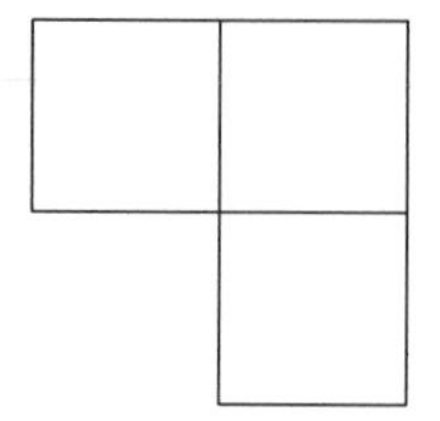

FIGURE 64

T: If it is so easy, could you cut a 1×6 band into "corners"? Six is also divisible by 3!

S: Well ... Actually, I should not have said that. I don't exactly know how to solve this problem.(1)

T: OK, you cannot solve *this* problem. Perhaps you can think of another problem which is similar yet easier?

S: Well, you can take another square, of smaller size, say, 4×4.

T: Or 2×2?(2)

S: But there is nothing to prove in this case—when you cut out any box what you get is just a "corner". What sense does that make?

T: Try now to solve the problem about the 4×4 square.

S: Uh-huh. A 4×4 square can be cut into four 2×2 squares. It is clear what to do with the one with the cut box. What about the other three?

T: Try to cut a "corner" from them, located in the center of the big square (see Figure 65).

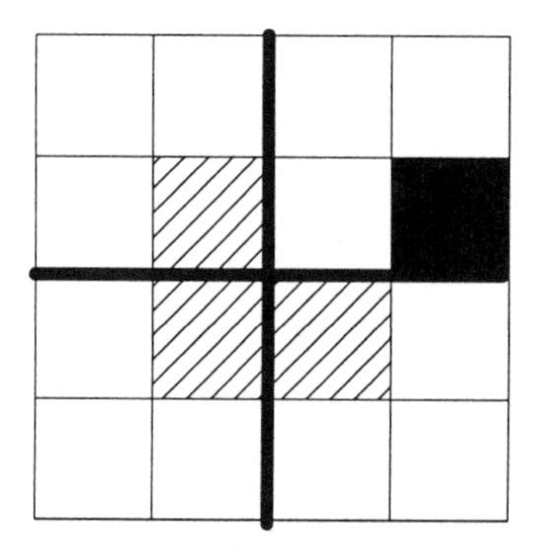

FIGURE 65

S: Got it! Each of them would lack one box and turn into a "corner". So we can solve the problem for a 4×4 square too. Now?

T: Try an 8×8 square. It can be dissected into four 4×4 squares. Make use of this.

S: Well, we can reason as we did before. One of those squares has the "missing" box in it. And we have already proved that this one can be cut into "corners". The three other squares will lack one box after we cut out one "corner" in the center of 8×8 square—so we will be able to dissect them, too.

T: Do you see now how to solve the original problem?

S: Sure. We cut the 16×16 square into four 8×8 squares. One of them contains the cut box. We have just proved that it can be dissected into "corners", right? Then we cut out a "corner" in the center of the big square and we get three more 8×8 squares, each without one box, and each can be cut into "corners". That's it!

T: Not yet. We solved this problem by building "bridges" from similar but simpler questions. Could we build such bridges once more, to other, more complicated questions?(3)

S: Of course. Let us prove that one can dissect a 32×32 square into "corners". We just divide it mentally into four 16×16 squares ...

T: There you are! But ... is it possible to go further?

S: Certainly. Having proved the proposition for a 32×32 square we can now derive, in the very same way, a method of dissection for a 64×64 square, then for a 128×128 square and so on ...

T: Thus, we have an infinite chain of propositions about squares of different sizes. Can we say that we have proved them all?

S: Yes, we have. First, we proved the first statement in the chain—about a 2×2 square. Then we derived the second proposition from it, then the third from the second, et cetera. It seems quite clear that

<u>going along this chain we will reach any of its statements; therefore, all of them are true.</u>

T: Right. It looks like a "wave of proofs" running along the chain of theorems: $2 \times 2 \longrightarrow 4 \times 4 \longrightarrow 8 \times 8 \longrightarrow \dots$. It is quite evident that the wave will not miss any statement in this chain.

* * *

<u>Methodological remark.</u> A few comments on the previous dialog.

Comment №1. When the student "proved" the statement of the problem using divisibility by 3, the teacher faced a typical classroom problem—how to explain the nature of the error, without giving away too much. The teacher overcomes this with a counterexample, prepared beforehand. It is always useful to be aware of such obstacles and know some ways to avoid them. This must be done easily, without major distraction from the flow of solution.

Comment №2. This retort is not accidental. The student can hardly think about the 2×2 case as something important—it's not a problem at all (we will come across this psychological moment several times). However, the teacher knows this case is easier to start with.

Comment №3. The following "step-by-step" scheme appears in this part of the dialog:

$$2 \times 2 \longrightarrow 4 \times 4 \longrightarrow 8 \times 8 \longrightarrow 16 \times 16 .$$

We have here the beginning of the induction process: the base 2×2 and the first three steps. It is essential that we have made enough induction steps for the student to notice an analogy. Now, after the hint, he is able to develop the whole process of induction.

In fact, there are other inductive solutions to this question but they would not yield any educational benefit, since the notion of induction in them is not as clear as in the solution given above. Thus, the teacher leads the student away from these, using directive hints. The teacher here has played his part precisely: sometimes he leads away from a deceptive analogy and helps to save the student's energy. Unobtrusiveness is very important: the more the student does on his/her own the better.

Let us sum up the results. The student (but more often this is the responsibility of the teacher) explained the scheme of MMI. The underlined sentence ("going along this chain we will reach any of its statements") is but an informal statement of the principle of mathematical induction which is the cornerstone of MMI. You can read about the formal side of it in any of the books [**76, 78, 79**]. We must say, though, that it would not be wise to talk about this at the very beginning of the discussion. It may be premature or even harmful since formalization of this intuitively clear statement may give rise to feelings of misunderstanding and uncertainty. On the contrary, one must use all means to make this scheme as evident and vivid as possible. Aside from the "wave" and dominoes (see Figure 66), other useful analogies include climbing a staircase, zipping a zipper, et cetera.

FIGURE 66

* * *

Now let us go on with our dialog:

T: So, we have proved an infinite series of statements about the possibility of dissecting squares into "corners". Now, we write them all down, without any "et cetera's".

S: But ... we will certainly run out of paper.

T: Yes, we would, if we wrote each statement separately. But all the statements look alike. Only the size of squares differ. This fact allows us to encode the whole chain in just one line:

(*) A $2^n \times 2^n$ square with one box cut out can be dissected into "corners".

Here we have the variable n. Each statement in our chain can be obtained by replacing n with a number. For instance, $n = 5$ gives us a proposition about the 32×32 square. And what is the tenth proposition in the chain?

S: We substitute $n = 10$ to get the statement about $2^{10} \times 2^{10}$, i.e., the 1024×1024 square.

T: Look at this: a variable is such a common thing, but it is really powerful—it allows us to fold an infinite chain into one short sentence. So, what is "a variable"?

S: Well ... it is just a letter ... an unknown ...

T: Remember: this "letter" denotes an empty space, a room, where you can put various numbers or objects. You can also call it a "placeholder". Those numbers or objects that are allowed to be put into the "room" are called its possible values. For example, the values of the variable n in (*) are the natural numbers (positive integers). Because of this, sentence (*) replaces the infinite chain of statements.

Now we must recall the proof of the infinite chain (*). Let us number all the statements: P_1 is the one about the 2×2 square, P_2 is about the 4×4 square, and so on.

First we proved proposition P_1. Then we dealt with the infinite chain of similar theorems: if P_1 is proved, then P_2 is true; if P_2 is proved, then P_3 is true, et cetera. Let us try to encode this chain also: "For any natural n ...

S: ... if P_n is true, then P_{n+1} is also true."

T: And now, please, decode this phrase: what do P_n and P_{n+1} denote?

S:

(**) "Whichever natural number n is, if it is already proved that the $2^n \times 2^n$ square without one box can be cut into "corners", then it is also true that the $2^{n+1} \times 2^{n+1}$ square without one box can be cut into "corners"."

T: Can you prove that?

S: I think so. We mentally divide the $2^{n+1} \times 2^{n+1}$ square into four $2^n \times 2^n$ squares. One of these lacks one box, and can be dissected into "corners" by assumption. Then we cut out one "corner" in the center of the big square so that it contains one box from each of the other three $2^n \times 2^n$ squares. After that, we can use the assumption again!

T: Absolutely. Note that as soon as you proved the general theorem (**), you proved all the theorems from the chain encoded by (**). For example, if $n = 1$, we get our old proof stating that the possibility of dissecting the 2×2 square implies the possibility of dissecting the 4×4 square. Therefore, just as (**) can be considered as encoding a whole chain of theorems, your reasoning can be considered as encoding a whole "wave of proofs" of those theorems. I believe you got it: it is useful and easier to prove a chain of similar theorems in this convoluted way. But first you must learn how to express a chain of theorems this way.

* * *

The method we applied in solving Problem 1 is what we call the METHOD OF MATHEMATICAL INDUCTION (MMI). What is its essence?

First, we regard statement (*) not as one whole fact but as an infinite series of similar propositions.

Second, we prove the first proposition in the series—this is called the "base of the induction."

Third, we derive the second proposition from the first, the third (in the same way) from the second, et cetera. That was the "inductive step"; (**)—is its shortened (convoluted) form. Since, step by step, we can reach any proposition from the base, they are all true.

* * *

"A method is an idea applied twice"
(G. Polya)

To learn MMI successfully it is usually necessary to replay the scenario above for several different questions. Consider now four more "key problems".

Problem 2. Prove that number $111\ldots 11$ (243 ones) is divisible by 243.

Hint. This question may be generalized to the proposition that a number written with 3^n ones is divisible by 3^n.

Base: 111 is divisible by 3. Students often start with the statement that 111,111,111 is divisible by 9—our base sounds too easy to them.

Here we have two obstacles

a) an attempt to generalize the divisibility tests for 3 and 9 and use an incorrect "test" for divisibility by 27;

b) reasoning of the sort: "if a number is divisible by 3 and 9, then it is divisible by $27 = 3 \times 9$."

The correct kind of inductive step is to divide the number written with 3^{n+1} ones by the number written with 3^n ones and check that the result is a multiple of 3.

Problem 3. Prove that for any natural number n, greater than 3, there exists a convex n-gon with exactly 3 acute angles.

Comment. This question is a good key problem if students already know the fact that a convex polygon cannot have more than 3 acute angles. The base $n = 4$ can be checked by direct construction.

Inductive step: let us "saw off" one of the non-acute angles. Then the number of angles in the polygon increases by 1 and all the acute angles are retained (see Figure 67).

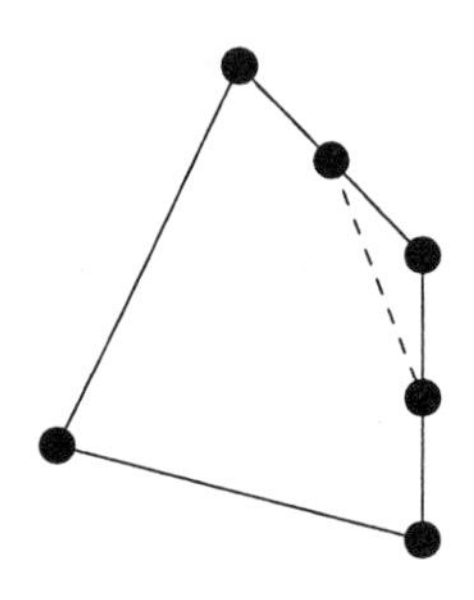

FIGURE 67

Another way to do this—to build a new angle on one of the sides—is a bit more difficult. There are also other solutions (using inscribed polygons and so on) but most are more difficult for students to make precise. Perhaps the teacher can even give a hint about "sawing off" an angle.

The statement of the question is obviously true for $n = 3$, but we will not gain anything by starting the induction from 3, because the method fails when you try to make the step from $n = 3$ to $n = 4$.

Our third question gives an example of construction by induction. You can read about it in more detail in [**79**].

Problem 4. ("Tower of Hanoi") Peter has a children's game. It has three spindles on a base, with n rings on one of them. The rings are arranged in order of their size (see Figure 68). It is permitted to move the highest (smallest) ring on any spindle onto another spindle, except that you cannot put a larger ring on top of a smaller one. Prove that

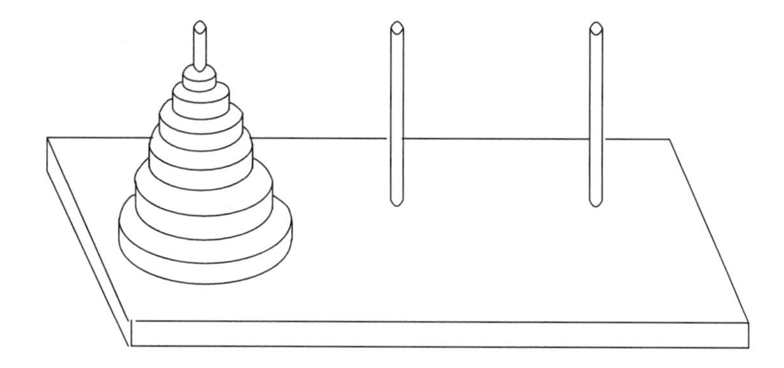

FIGURE 68

a) It is possible to move all the rings to one of the free spindles;
b) Peter can do so using $2^n - 1$ moves.
c)*It is not possible to do so using fewer moves.

Hint. a), b): The base (n=1) is easy.

Inductive step: We have $n = k + 1$ rings. By the inductive assumption we can move all but the largest ring to the third spindle using $2^k - 1$ moves. Then we move the remaining ring to the second spindle. After that we can move all the rings from the third spindle to the second using $2^k - 1$ moves. In all, we have made $(2^k - 1) + 1 + (2^k - 1) = 2^{k+1} - 1$ moves. It is useful to do the first few steps of the induction "manually", even using a physical model.

c) This question must be used with care—it is more difficult than the others given here. The main idea of the proof is that to move the widest ring to the second spindle, we must first move all the other rings to the third spindle.

Problem 5. The plane is divided into regions by several straight lines. Prove that one can color these regions using two colors so that any two adjacent regions have different colors (we call two regions *adjacent* if they share at least one line segment).

Hint. Here we encounter another obstacle: no explicit variable for induction is given in the statement. Thus, we should start the solution by revealing this hidden variable. To do this, we can rewrite the statement as follows: "There are n straight lines on a plane ... ." The base can be $n = 1$ or $n = 2$ (either will work). The inductive step: remove for a moment the $(k + 1)$st line, color the map obtained, then restore the removed line and reverse the colors of all the regions on one side of the line.

For teachers. The first few key problems can be discussed according to the scenario of the dialog above; that is, growing the chain from one particular proposition. Students should understand the essence of the process of induction and the connection between chains of theorems and propositions using integer variables.

If students are not well prepared one can skip the idea of constructing a chain of inductive steps. This can be introduced later, at a second stage, whose goal is to teach the students to work with the inductive step in its convoluted form. While doing so it would be wise to give questions in a general form (like in Problems 3 and 4). There we already have a chain of statements and the solution may start right from the "unfolding", as follows: "Here we have a convoluted chain of theorems. What is the first theorem? The fifth? The 1995th?" However, the chain of inductive steps should be developed and convoluted according to the old scheme, until students get accustomed to it and understand well the connection between a long chain and its convoluted form.

* * *

To sum up their experience with key problems, students should have a clear

General Plan for Solution by the Method of Mathematical Induction

1. Find, in the statement of the question, a series of similar propositions. If variables are hidden you should reveal them by reformulating the question. If there is no chain, try to grow that chain so that the question will be a part of it.
2. Prove the first proposition (base of the induction).
3. Prove that for any natural number n the truth of the nth proposition implies the truth of the $(n+1)$st proposition (inductive step).
4. If the base and the step are proved, then all the propositions in the series are proved simultaneously, since you can reach any of them from the base by moving "step-by-step".

The last item in this scheme is the same for all the problems, so it is often skipped. However, knowing it is vital. Also, the first item is not emphasized and is natural for those who are used to MMI; nevertheless we recommend that the students pay close attention to it for a while.

§2. MMI and guessing by analogy

We continue our dialog.

Problem 6. Into how many parts do n straight lines divide a plane if no two of them are parallel and no three meet at the same point? (Figure 69 shows an example where $n = 5$.)

S: Let us try to follow the scheme. Do we have a chain? It seems so: into how many parts does one line divide a plane? 2 lines? 3 lines ... ?

The base is evident: one line dissects a plane into 2 parts (half-planes).

T: Or zero lines—into one part.

S: By all means. Item three—the inductive step ... !?

T: I can understand your embarrassment: we run into a new difficulty. In the previous problems we dealt with chains of statements, not with chains of questions. But we will get a chain of statements if we give hypothetical, unproved answers to these questions.

S: How can we?

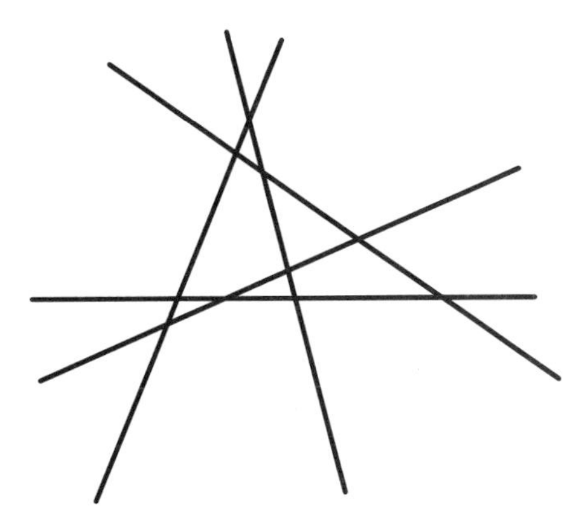

FIGURE 69

T: Try to guess a rule, a function giving the number of parts L_n in terms of the number of lines n. Physicists would do an experiment. We can experiment too, calculating the numbers L_n for small values of n. Go ahead!

S: OK. So, $L_0 = 1$, $L_1 = 2$, $L_2 = 4$, $L_3 = 7$, $L_4 = 11$. I must think a little ... .Ah, I got it! When you add the nth line the number of parts increases by n. Hence, $L_n = 1 + (1 + 2 + \ldots + n)$. I did it!

T: No, not yet. Don't forget that you have only guessed it, not proved it. You have checked your result only for $n = 0, 1, 2, 3, 4$. For all other values of n this is just a guess based on your conjecture that adding the nth line increases the number of parts by n. What if this is wrong? The only guarantee is a proof.

S: ... by the method of mathematical induction.

T: But we should enhance our plan from §1 by another item:

1a. If there is a chain of questions rather than a chain of statements in a mathematical problem, insert your hypothetical answers. You can guess the answers by experimenting with the first few questions in the chain. However, after you are sure the answers are correct, don't forget to prove them rigorously.

S: Now I know how to get over this. We have already proved the base, right? To prove the inductive step is easy: the nth line intersects the other lines at $n - 1$ points, which divide the line into n parts. Therefore the nth line divides n of the old parts of the plane into new parts.

* * *

The process of guessing by analogy, just demonstrated by our student, is a very powerful and, sometimes, very dangerous tool: it is tempting to mistake the regularity one finds as a proof. The two examples below can serve as good medicine for this disease.

Problem 7. Is it true that the number $n^2 + n + 41$ is prime for any natural number n?

Hint. The answer is no: For $n = 40$ we have $40^2 + 40 + 41 = 41^2$, and for $n = 41$ $41^2 + 41 + 41 = 41 \cdot 43$. But anyone trying to find an answer by "experimenting" with small values of n would come to the opposite conclusion, since this formula gives prime numbers for n from 1 through 39. This famous example was given by Leonard Euler.

Problem 8*. A set of n points is taken on a circle and each pair is connected by a segment. It happens that no three of these segments meet at the same point. Into how many parts do they divide the interior of the circle?

Hint. For $n = 1, 2, 3, 4, 5$ we obtain 1, 2, 4, 8, and 16 respectively. This result provokes a guess to the formula 2^{n-1}. However, in fact, the number of parts equals $\frac{n(n-1)(n-2)(n-3)}{24} + \frac{n(n-1)}{2} + 1$.

Other similar examples can be found in [**78**].

§3. Classical elementary problems

Among classical MMI problems in elementary mathematics three large groups can be distinguished: proofs of identities, proofs of inequalities and proofs of divisibility questions. Though their solutions by MMI seem to be quite simple, students usually encounter some obstacles of a psychological as well as of a methodical nature. We begin by discussing these.

T: Let us talk more about Problem 6. Do you like the way the formula $1 + (1 + 2 + 3 + \ldots + n)$ looks?

S: Not much. It is too bulky. It would be better to get rid of this ellipsis (the three dots).

T: No problem. You can prove by MMI that $1 + 2 + 3 + \ldots + n = n(n + 1)/2$.

S: But ... to use MMI you need a chain of statements ...

T: Take a close look: there is variable n in the formula. As we know, this is a good sign of a convoluted set of problems. Substitute, for instance, 1995 for n.

S: We get $1 + 2 + \ldots + 1995 = 1995 \cdot 1996/2$.

T: That is, a numerical equation. Our convoluted set of problems consists of all these equations (for $n = 1, 2, 3, \ldots, 1995$)! To prove the formula means to show that all these numerical equations are true. If we do this, we say that this equation is "true for all admissible values of the variable" and it is called an *identity*. If an identity contains an integer variable you can try to prove it by induction.

S: What if our equation is not true for some n?

T: Then it is not an identity and we will not be able to prove it—the proof of either the base or the inductive step will not go through. Actually, to distinguish between identities and other, arbitrary equations with variables, you must preface identities with phrases like "for any natural number n ... it is true that ... ", but this is not the usual practice. It is implied that the reader knows from the context whether an identity or a conditional equation is being discussed.

S: Well, let us apply MMI. Base: $n = 1$. So we must prove that ... $1 + 2 + \ldots + 1 = 1 \cdot 2/2 = 1$?!

T: No, no. We must prove that $1 = 1 \cdot 2/2$. You were puzzled by the formula $1+2+\ldots+n$. This is quite good and convenient, but for $n = 1$ its "tail" $2+\ldots+n$ makes no sense and, in fact, does not exist at all.

S: OK, so the base is clear. Let us move to the second equation in the series. We must show that $1 + 2 = 2 \cdot 3/2$. This is easy: $3 = 3$. Now, move to the third equation: $1 + 2 + 3 = 3 \cdot 4/2$. This is easy too: $6 = 6$. To the fourth ... it is just another simple calculation. So, what now? Must we check each equation directly? We haven't got any steps!

T: Try to rewrite the step in a general, convoluted form.

S(after a while): I cannot do that either.

For teachers. To a person who has mastered MMI well enough, the proof of identities may seem rather trivial. However, our dialog shows two sources of problems for students. First, students often do not accept an identity with an integer variable as a chain of statements. This is probably because simple numerical identities are not considered independent propositions. Also, what is interesting in a statement such as $1+2+3=3\cdot 4/2=6$?

Second, it is next to impossible to see how the general form of the inductive step looks. Indeed, when you check the equations $1+2=2\cdot 3/2$, $1+2+3=3\cdot 4/2$, and so on, there is no connection between two successive facts—you just check them.

That is why identities, despite their simplicity, cannot serve as key questions. To start learning and teaching mathematical induction from these will create trouble (this is not very important for really gifted students—they will manage to learn the method in any case). On the other hand, identities are very useful for practice, because their proofs are usually short and clear.

T: Well, I will help you. Imagine that we follow the steps of the induction, one after another and the wave of proofs have reached the kth statement. What is that statement?

S: We obtain $1+2+3+\ldots+k=k(k+1)/2$. (#)

T: Exactly. Now, tell me, please, what is the next statement, which the wave has not yet reached?

S: Certainly, $n=k+1$ and we get $1+2+\ldots+(k+1)=(k+1)(k+2)/2$.

T: Good. Let us write this as follows:

$$1+2+3+\ldots+k+(k+1)=\frac{1}{2}(k+1)(k+2). \qquad (\#\#)$$

Now, tell me what would be the next step of induction?

S: That's clear: to derive (##) from (#).

T: Assume that we learned how to derive (##) from (#) for any natural number k. Then we would have all the steps of induction proved at once. This means that the inductive step states that:

For any natural k the equation $1+2+\ldots+k=k(k+1)/2$ implies the equation $1+2+\ldots+(k+1)=(k+1)(k+2)/2$.

In other words: (#) is given, and we must prove (##) (if k is an arbitrary natural number). For convenience we denote the left sides of (#) and (##) as S_k and S_{k+1} respectively.

S: Proposition (##) shows that $S_{k+1}=S_k+(k+1)$ (that is why the teacher has written the next-to-last summand!). Now we have already learned that $S_k=k(k+1)/2$. Thus we have

$$S_{k+1}=\frac{1}{2}k(k+1)+(k+1)=\frac{1}{2}\big[k(k+1)+2(k+1)\big]=\frac{1}{2}(k+1)(k+2).$$

T: Remember the helpful idea that we used to prove the inductive step: the left side of equation (##) was expressed with the left side of (#) and the latter was substituted into the right side of (#).

For teachers. Another difficulty now arises in connection with identities. It may not be clear to a student how to make a step "in letters". The teacher in our dialog showed how to overcome that. It is important that he used another letter—different from that used in the statement of the identity—to denote the variable. The point is that the letter k plays the role not of a variable but of a constant (though arbitrary) number marking the place that the wave of our inductive proof reached at the given moment. It will become a variable later, in the general statement of the inductive step.

Quite often, the variables in the statement of the proposition and in the step are both denoted by the same letter. While stating the step theorem, phrases like "... now we substitute $n+1$ in place of n" are used. This is not advisable in the beginning of the study since it disorients most students conceptually (it is hard to see a chain in the statement of the inductive step) as well as technically (it is not that easy for a beginner to substitute $n+1$ for n).

Now we can say goodbye to the characters in our dialog and go on to deal with problems. Problems 9–16 are about identities with the natural number n as their variable.

Problem 9. Show that $1+3+\ldots+(2n-1)=n^2$.

Problem 10. Show that $1^2+2^2+\ldots+n^2=n(n+1)(2n+1)/6$.

Problem 11. Show that $1\cdot 2+2\cdot 3+\ldots+(n-1)\cdot n=(n-1)n(n+1)/3$.

Problem 12. Show that

$$\frac{1}{1\cdot 2}+\frac{1}{2\cdot 3}+\ldots+\frac{1}{(n-1)n}=\frac{n-1}{n}.$$

Problem 13. Show that

$$1+x+x^2+\ldots+x^n=(x^{n+1}-1)/(x-1).$$

Problem 14. Show that

$$\frac{1}{a(a+b)}+\frac{1}{(a+b)(a+2b)}+\ldots+\frac{1}{(a+(n-1)b)(a+nb)}=\frac{n}{a(a+nb)},$$

where a and b are any natural numbers.

Problem 15. Show that

$$\frac{m!}{0!}+\frac{(m+1)!}{1!}+\ldots+\frac{(m+n)!}{n!}=\frac{(m+n+1)!}{n!(m+1)},$$

where m, $n=0, 1, 2, \ldots$.

Problem 16. Show that

$$\left(1-\frac{1}{4}\right)\left(1-\frac{1}{9}\right)\ldots\left(1-\frac{1}{n^2}\right)=\frac{n+1}{2n}.$$

Comments. In Problems 9–15 the proof of the inductive step is exactly the same as in the dialog. However, in Problem 16, it may be proved more easily by representing the $(k+1)$st left side not as a sum but as the product of the kth left side and $(1-\frac{1}{k^2})$. This trick may also be useful in proving certain inequalities (see below).

In Problem 11 the base is not $n=1$ but $n=2$. Students should see that this doesn't influence the process of induction.

In Problem 15 induction is possible on either of the two variables. It is instructive to carry out and compare both proofs. Remember to start from zero!

Problems 11 and 12 are special cases of Problems 15 (for $m = 2$) and 14 (for $a = b = 1$) respectively. Given other values of m, a, and b we obtain any number of exercises like 11 and 12. It would be wise to let good students try to find the statement of the general problem which generates these exercises.

Most of the identities 9–16 have good non-inductive proofs which are not too difficult. Problem 9 has a neat geometric proof (see Figure 70). Identity 11 can be obtained from identities 9 and 10. Identity 13 can be proved by division of $x^{n+1} - 1$ by $x - 1$, and identity 16 by direct calculation. To prove identity 12 it suffices to note that its left side equals

$$\left(1 - \frac{1}{2}\right) + \left(\frac{1}{2} - \frac{1}{3}\right) + \ldots + \left(\frac{1}{n-1} - \frac{1}{n}\right),$$

and that this sum "telescopes".

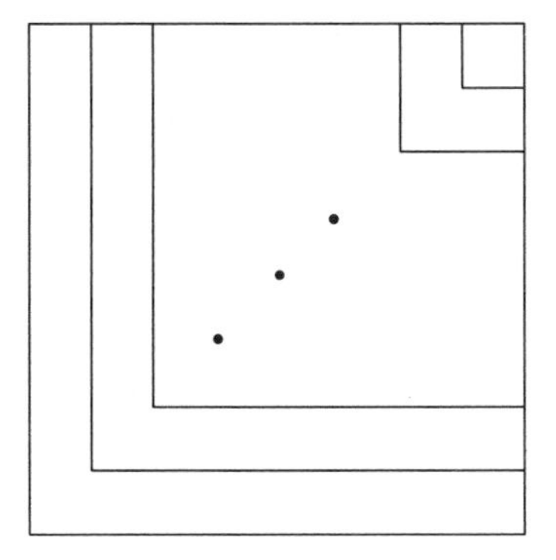

FIGURE 70

This device works for other identities too.

Discussion of these alternative proofs can be very helpful to students who have already mastered MMI.

Divisibility questions constitute the next natural step in our study. The techniques of forming statements and inductive steps are similar to those for identities: we usually find the increment of the expression under consideration and prove that it is divisible by a given number. Problems 17–19 have simple alternative solutions (using modular arithmetic). The rather difficult Problem 22 may serve as the source of a number of exercises like 18–19.

Prove that for any natural number n

Problem 17. $n^3 + (n+1)^3 + (n+2)^3$ is divisible by 9.

Problem 18. $3^{2n+2} + 8n - 9$ is divisible by 16.

Problem 19. $4^n + 15n - 1$ is divisible by 9.

Problem 20. $11^{n+2} + 12^{2n+1}$ is divisible by 133.

Problem 21. $2^{3^n} + 1$ is divisible by 3^{n+1}.

Problem 22*. $ab^n + cn + d$ is divisible by the positive integer m given that $a + d$, $(b - 1)c$, and $ab - a + c$ are divisible by m.

Our trio of standard MMI themes is completed by questions involving inequalities. Here the proofs of the inductive steps are usually more varied (see [**78**]). Prove the following inequalities:

Problem 23. $2^n > n$, where n is an arbitrary natural number.

Problem 24. Find all natural numbers n such that

a) $2^n > 2n + 1$; b) $2^n > n^2$.

Problem 25.

$$\frac{1}{n+1} + \frac{1}{n+2} + \dots + \frac{1}{2n} > \frac{13}{24}, \qquad n = 2, 3, \dots .$$

Problem 26. $2^n > 1 + n\sqrt{2^{n-1}}$, $n = 2, 3, \dots$.

Problem 27. Prove that the absolute value of the sum of several numbers does not exceed the sum of the absolute values of these numbers.

Problem 28. $(1 + x)^n > 1 + nx$, where $x > -1$, $x \neq 0$, and $n = 2, 3, \dots$.

Problem 29.

$$\frac{1 \cdot 3 \cdot 5 \dots (2n-1)}{2 \cdot 4 \cdot 6 \dots 2n} \leq \frac{1}{\sqrt{2n+1}},$$

where n is any natural number.

Hints. 23, 24: To prove the inductive step you may show that for any n, the increment of the left side of the inequality is greater than the increment of the right side.

24b: Use 24a to prove the step.

25: Prove that the left side of the inequality is monotonically increasing.

27: Induction can proceed on the number of summands.

28, 29: See the hint to Problem 16.

§4. Other models of MMI

So far we have been dealing with the basic version of MMI. When this is well learned we can try other, more complicated forms of induction. Some of these can be considered corollaries of the basic form, but it is more natural from a methodological point of view to discuss them separately, keeping in mind the image of "a wave of proofs".

First, consider the method "Induction from all $n \leq k$ to $n = k + 1$", sometimes called "strong induction".

In the usual method of MMI, the inductive step consists of deriving proposition P_{k+1} from the previous proposition P_k. Sometimes, however, to show the truth of P_{k+1} we must use more than one (or even all) of the previous statements P_1 through P_k. This is certainly valid, since the wave has reached P_k and, therefore, all the propositions in the chain preceding it are also already proved. Thus the statement of the inductive step is:

(S'): For any natural k the truth of P_1, P_2, ... , and P_k implies the truth of P_{k+1}.

* * *

Consider an example.

Problem 30. Prove that every natural number can be represented as a sum of several distinct powers of 2.

Solution. First, let us prove the base. If the number given equals 1 or 2, then the existence of the required representation is simple.

Now denote the given number by n and find the largest power of 2 not exceeding n. Let it be 2^m; that is, $2^m \leq n < 2^{m+1}$. The difference $d = n - 2^m$ is less than n and also less than 2^m, since $2^{m+1} = 2^m + 2^m$. By the induction hypothesis, d can be represented as a sum of several different powers of 2, and it is clear that 2^m is too big to be included. Thus, adding 2^m we get the required expression for n. The induction is complete.

Problem 31. Prove that any polygon (not necessarily convex) can be dissected into triangles by disjoint diagonals (they are allowed to meet only at vertices of the polygon).

Hint. Use an induction on the number of sides. The inductive step is based on a lemma stating that each polygon (except a triangle) has at least one diagonal which lies completely within the polygon. Such a diagonal dissects the polygon into two polygons with fewer sides.

* * *

Another scheme of MMI is demonstrated by

Problem 32. It is known that $x + 1/x$ is an integer. Prove that $x^n + 1/x^n$ is also an integer (for any natural n).

Solution. We have $(x^k + 1/x^k)(x + 1/x) = x^{k-1} + 1/x^{k-1} + x^{k+1} + 1/x^{k+1}$ and hence $x^{k+1} + 1/x^{k+1} = (x^k + 1/x^k)(x + 1/x) - (x^{k-1} + 1/x^{k-1})$. So we see that the $(k+1)$st sum is an integer if the two preceding sums are integers. Thus the process of induction will go as usual if we check that the first two sums, $x + 1/x$ and $x^2 + 1/x^2$, are integers. This is left to the reader.

Comment. A peculiarity of this version of MMI is that the inductive step is based on two preceding propositions, not one. Thus, the base in this case consists of the first two propositions in the series (it is natural to use the word *base* for that starting segment of the chain in which the statements must be checked directly).

Problem 33. The sequence $a_1, a_2, \ldots, a_n, \ldots$ of numbers is such that $a_1 = 3$, $a_2 = 5$, and $a_{n+1} = 3a_n - 2a_{n-1}$ for $n > 2$. Prove that $a_n = 2^n + 1$ for all natural numbers n.

Hint. See the more general Problem 43.

Remark. In Problem 33 and the next three problems we will encounter not only proof by induction but also definitions by induction: all elements of the given sequences, except for the first few, are defined by induction, using the preceding elements. Sequences defined in this way are called *recursive*; see [**75**] and [**77**] for more details. See also [**79**], Chapter 2, about definitions, constructions, and calculations using induction.

Problem 34. The sequence (a_n) is such that: $a_1 = 1$, $a_2 = 2$, $a_{n+1} = a_n - a_{n-1}$ if $n > 2$. Prove that $a_{n+6} = a_n$ for all natural numbers n.

Problem 35. The sequence of Fibonacci numbers is defined by: $F_1 = F_2 = 1$ and $F_{n+1} = F_n + F_{n-1}$ if $n \geq 2$. Prove that any natural number can be represented as the sum of several different Fibonacci numbers.

Problem 36*. Prove that the nth Fibonacci number is divisible by 3 if and only if n is divisible by 4.

Hint. It is not easy to prove this fact alone by induction. Prove a more general statement about the repetition of remainders of Fibonacci numbers modulo 3 (with period 8). If you want to know more about Fibonacci numbers, see [**75**].

Problem 37. A bank has an unlimited supply of 3-peso and 5-peso notes. Prove that it can pay any number of pesos greater than 7.

Hint. Try induction on the number of pesos the bank must pay. The base consists of three facts: $8 = 5 + 3$, $9 = 3 + 3 + 3$, $10 = 5 + 5$. Inductive step: if the bank can pay k, $k+1$, and $k+2$ pesos, then it can pay $k+3$, $k+4$, and $k+5$ pesos. This induction with a compound base may be split into three standard inductions using the following schemes:

$$8 \to 11 \to 14 \to \dots, \quad 9 \to 12 \to 15 \to \dots, \text{ and } 10 \to 13 \to 16 \to \dots.$$

Note that a similar splitting is impossible in Problems 32–36.

There also exists a non-inductive solution to this problem based on the equations $3n + 1 = 5 + 5 + 3(n - 3)$ and $3n + 2 = 5 + 3(n - 1)$, but it is not easier than the solution above.

The following three questions are very close to Problem 37.

Problem 38. It is allowed to tear a piece of paper into 4 or 6 smaller pieces. Prove that following this rule you can tear a sheet of paper into any number of pieces greater than 8.

Problem 39. Prove that a square can be dissected into n squares for $n \geq 6$.

Problem 40. Prove that an equilateral triangle can be dissected into n equilateral triangles for $n \geq 6$.

Comments. 38: If we tear a piece of paper into 4 or 6 smaller pieces, then the number of pieces increases by 3 or 5 respectively. Now we use the method of solution from Problem 37.

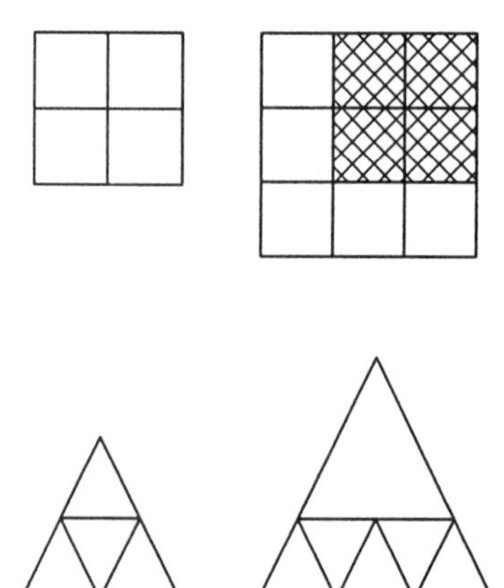

FIGURE 71

39, 40: A square (equilateral triangle) can be dissected into 4 or 6 squares (equilateral triangles) as shown in Figure 71. Thus Problem 39 can be reduced to Problem 38. There exist other non-inductive solutions based on the possibility of cutting a square (equilateral triangle) into any even number (greater than 2) of squares (or equilateral triangles) greater than 2—see Figure 72.

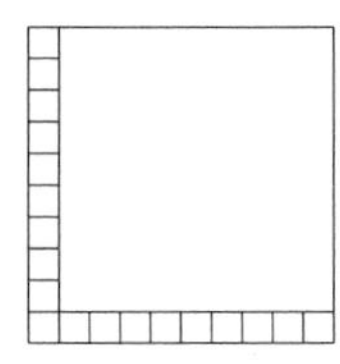

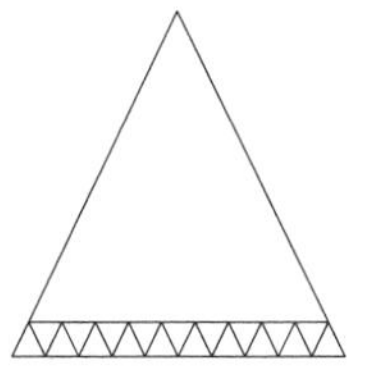

FIGURE 72

Other schemes of induction are even more exotic. An example is the method of "ramifying induction" which enables us to give a proof of a remarkable inequality for the arithmetic and geometric means.

Problem 41.* Prove that for any non-negative numbers $x_1, x_2, \dots, x_n$

$$\frac{x_1 + x_2 + \ldots + x_n}{n} \geq \sqrt[n]{x_1 x_2 \ldots x_n}\,.$$

Sketch of proof. The base: $n = 2$ is rather simple. Then you must use steps from $n = 2^k$ to 2^{k+1} in order to prove the inequality for all n equal to power of 2. And finally, you prove that if the inequality is true for any n numbers, then it is true for any $n - 1$ numbers. The wave of proofs spreads in accordance with the scheme in Figure 73.

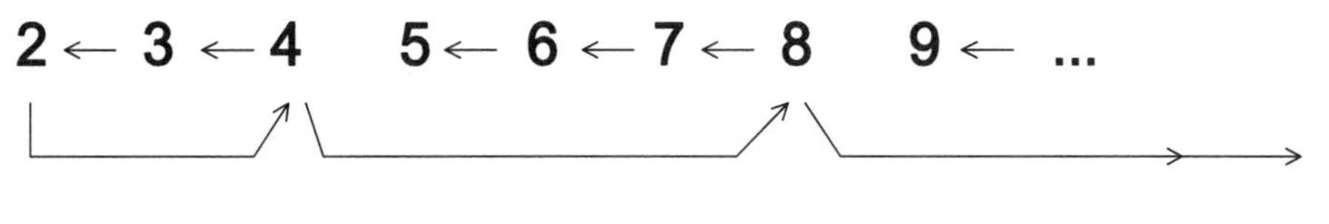

FIGURE 73

See details in [**78**] (example 24), and also in the chapter "Inequalities".

Schemes involving "backwards induction" (over negative integers) and "double (or, 2-dimensional) induction" (for theorems involving two natural parameters) are illustrated in Problems 43 and 44.

§5. Problems with no comments

Problem 42. Two relatively prime natural numbers m, n, and the number 0 are given. A calculator can execute only one operation: to calculate the arithmetic mean of two given natural numbers if they are both even or both odd. Prove that using this calculator you can obtain all the natural numbers 1 through n, if you can enter into the calculator only the three numbers initially given or results of previous calculations.

Problem 43. For the sequence $a_1, a_2, \ldots$ from Problem 33 we can define elements $a_0, a_{-1}, a_{-2}, \ldots$ so that the equation $a_{n+1} = 3a_n - 2a_{n-1}$ will hold true for any integer n (positive or negative). Prove that the equality $a_n = 2^n + 1$ will still be true for all integers n.

Problem 44. Prove that $2^{m+n-2} \geq mn$ if m and n are positive integers.

Problem 45.* Several squares are given. Prove that it is possible to cut them into pieces and arrange them to form a single large square.

Problem 46.* Prove that among any 2^{n+1} natural numbers there are 2^n numbers whose sum is divisible by 2^n.

Problem 47. What is the greatest number of parts into which n circles can dissect a plane? What about n triangles?

Remark. Compare Problem 6. Examples of the required dissections can also be done by induction.

Problem 48. Several circles are drawn on a plane. A chord then is drawn in each of them. Prove that this "map" can be colored using three colors so that the colors of any two adjacent regions are different.

* * *

Problem 49.* Prove by "reductio ad absurdum" that the principle of mathematical induction stated in the very beginning of the present chapter is equivalent to the following "well order principle": in any non-empty set of natural numbers there exists a least element. Try to rewrite the solution of one of the previous problems (say, Problem 46) using this principle and compare it to the proof by induction.

For more about the well order principle and its applications, see [**19**], pp.88–96.

* * *

Conclusion. The method of mathematical induction is a very helpful and useful idea. You will find its applications in various places in this book, as well as in other mathematical contexts. However, we would like to warn you against an "addiction" to it. You should not think that any question with statements and/or proofs using the words "et cetera" or "similarly" is a problem to be solved by MMI. Proofs by induction for many of those questions (you will see some of them in the chapters "Graphs–2" and "Inequalities") look rather artificial compared to other proofs involving such simple methods as direct calculation or recursive reasoning. It is not advisable to use such unnatural examples when teaching the nature of MMI, although they can be used well after the method is completely mastered.

CHAPTER 10

Divisibility–2: Congruence and Diophantine Equations

§1. Congruence

In the chapter "Divisibility and remainders" we discussed the concept of remainders. We noticed that in solving many problems on divisibility we dealt mostly not with the numbers themselves but with their remainders when divided by some fixed number.

Thus it is natural to give the following definition: integers a and b are called *congruent modulo* m if they have equal remainders when divided by m. This is written: $a \equiv b \pmod{m}$.

For example, $9 \equiv 29 \pmod{10}$, $1 \equiv 3 \pmod{2}$, $16 \equiv 9 \pmod{7}$, $3 \equiv 0 \pmod{3}$, $2n+1 \equiv 1 \pmod{n}$. Note that A is divisible by m if and only if $A \equiv 0 \pmod{m}$.

Problem 1. Prove that $a \equiv b \pmod{m}$ if and only if $a - b$ is divisible by m.

Solution. If $a \equiv b \pmod{m}$, then let r be the common remainder when a or b is divided by m.

$$a = mk_1 + r, \quad b = mk_2 + r .$$

Thus $a - b = m(k_1 - k_2)$ is divisible by m.

Conversely, if $a-b$ is divisible by m, then we divide a and b by m with remainder. We have $a = mk_1 + r_1$, $b = mk_2 + r_2$. Hence $a - b = m(k_1 - k_2) + r_1 - r_2$ is, by assumption, divisible by m. Therefore $r_1 - r_2$ is divisible by m. Since $|r_1 - r_2| < m$, we have $r_1 = r_2$.

This problem allows us to give another definition of congruence: integers a and b are congruent modulo m if $a - b$ is divisible by m.

From now on we will use either definition.

<u>**For teachers.**</u> Before giving definitions of congruence it is important to check whether your students remember how to work with remainders (for example, by giving them a few problems similar to problems from §2 in the chapter "Divisibility and remainders").

It is remarkable that our new definition leads us to easier proofs of the basic properties of remainders.

Problem 2. If $a \equiv b \pmod{m}$ and $c \equiv d \pmod{m}$, prove that $a+c \equiv b+d \pmod{m}$.

Solution. Since $a-b$ is divisible by m and $c-d$ is divisible by m, $(a-b)+(c-d) = (a+c) - (b+d)$ is divisible by m.

Problem 3. If $a \equiv b \pmod{m}$ and $c \equiv d \pmod{m}$, prove that $a-c \equiv b-d \pmod{m}$.

Problem 4. If $a \equiv b \pmod{m}$ and $c \equiv d \pmod{m}$, prove that $ac \equiv bd \pmod{m}$.

Solution. We have $ac-bd = ac-bc+bc-bd = (a-b)c+b(c-d)$ which is divisible by m.

Problem 5. If $a \equiv b \pmod{m}$ and n is any natural number, then $a^n \equiv b^n \pmod{m}$.

Methodological remark. Statements and proofs of properties of remainders look more attractive and simple when written using the language of congruences. For instance, without the new notation, the statement of Problem 2 would have read: the sum of two numbers and the sum of their remainders modulo m have equal remainders modulo m.

Basically, Problems 2–5 state that congruences modulo a given number may be added, subtracted, multiplied, and exponentiated, like equations. We delay the question of division of congruences until §4.

Before going further we show how the solution of a problem can be explained in the language of congruences.

Problem 6. Prove that n^2+1 is not divisible by 3 for any integer n.

Solution. It is clear that each integer n is congruent modulo 3 either to 0, or to 1, or to 2.

If $n \equiv 0 \pmod{3}$, then $n^2 \equiv 0 \pmod{3}$—(multiplication of congruences) and $n^2+1 \equiv 1 \pmod{3}$—(addition of congruences).

If $n \equiv 1 \pmod{3}$, then $n^2+1 \equiv 2 \pmod{3}$.

If $n \equiv 2 \pmod{3}$, then $n^2+1 \equiv 2 \pmod{3}$.

Thus we never get $n^2+1 \equiv 0 \pmod{3}$.

Let us consider another problem, which shows us that using negative integers in congruences can be quite helpful. As a matter of fact, in the arithmetic of remainders we can deal with negative integers in the same way as with positive ones.

Problem 7. Reduce 6^{100} modulo 7.

Solution. Since $6-(-1) = 6+1 = 7$, we can say that $6 \equiv -1 \pmod{7}$. Raising this congruence to the power of 100 we have $6^{100} \equiv (-1)^{100} \pmod{7}$; that is, $6^{100} \equiv 1 \pmod{7}$.

Here are several more problems using the same idea.

Problem 8. Prove that $30^{99}+61^{100}$ is divisible by 31.

Problem 9. Prove that

a) $43^{101}+23^{101}$ is divisible by 66.

b) a^n+b^n is divisible by $a+b$, if n is odd.

Problem 10. Prove that $1^n+2^n+\ldots+(n-1)^n$ is divisible by n for odd n.

Problem 11. Prove that there exist infinitely many natural numbers that cannot be represented as a sum of three cubes.

Problem 12. Prove that 10^{3n+1} cannot be represented as a sum of the cubes of two integers.

Problems 11 and 12 are much more difficult than the preceding problems, since to solve them one must know which number m to divide by.

Let us analyze the solution of Problem 12. The cube of a natural number is always congruent to either 0, or 1, or -1 modulo 7 (check this!). Therefore the sum of two cubes is always congruent to one of the following numbers: -2, -1, 0, 1, 2. Since $10 \equiv 3 \pmod 7$ we have $10^3 \equiv -1 \pmod 7$ and 10^{3n+1} is congruent to either 3 or -3 modulo 7. Thus it cannot have the required representation.

* * *

Let us fix a natural number n. Then the infinite set $\mathbb{Z}$ of all integers quite naturally falls into n classes: two numbers are in one class if they have equal remainders modulo n (i.e., are congruent modulo n). For instance, if $n = 2$, we have two classes: even and odd numbers. When solving problems on divisibility it is often sufficient to check the truth of statements not for every integer but only for one (arbitrary!) representative from each class. Remember that in problems of §2 of the chapter "Divisibility and remainders" we usually used as representatives all the positive remainders modulo some number, and in Problems 11 and 12 of this chapter it was more helpful to choose other representatives (that is, some of the chosen numbers were negative).

The next two problems illustrate the same idea.

Problem 13. Prove that among any 51 integers there are 2 with squares having equal remainders modulo 100.

Problem 14. Call a natural number n "convenient", if $n^2 + 1$ is divisible by 1000001. Prove that among the numbers 1, 2, ..., 1000000 there are evenly many "convenient" numbers.

A concluding series of problems:

Problem 15. a) Can the perfect square of a natural number end with 2 (that is, can its units digit equal 2)?

b) Is it possible to write the square of a natural number using only the digits 2, 3, 7, 8 (perhaps with repetitions)?

Problem 16. Find a number, which, when added to $(n^2 - 1)^{1000} \cdot (n^2 + 1)^{1001}$ makes the result divisible by n.

Problem 17. Find the remainder when the number $10^{10} + 10^{100} + 10^{1000} + \dots + 10^{10000000000}$ is divided by 7.

Problem 18. How many natural numbers n not greater than 10000 are there such that $2^n - n^2$ is divisible by 7?

Problem 19. Denote by k the product of the first several prime numbers (but more than one prime number). Prove that the number a) $k - 1$; b) $k + 1$ cannot be a perfect square.

Problem 20. Does there exist a natural number n such that $n^2 + n + 1$ is divisible by 1955?

Problem 21. Prove that $11^{n+2} + 12^{2n+1}$ is divisible by 133 for any natural n.

Solution.

$$\begin{aligned} 11^{n+2} + 12^{2n+1} &= 121 \cdot 11^n + 12 \cdot 12^{2n} \\ &= 133 \cdot 11^n - 12 \cdot 11^n + 12 \cdot 12^{2n} \\ &\equiv 12(12^{2n} - 11^n) \\ &= 12(144^n - 11^n) \equiv 0 \pmod{133} . \end{aligned}$$

This solution shows that not only beautiful ideas but also simple "manual" calculations may give us a neat proof.

Problem 22*. Let n be a natural number such that $n+1$ is divisible by 24. Prove that the sum of all the divisors of n is also divisible by 24.

Problem 23*. The sequence of natural numbers $a_1, a_2, a_3, \ldots$ satisfies the condition $a_{n+2} = a_{n+1}a_n + 1$ for all n.

a) If $a_1 = a_2 = 1$, prove that no member of the sequence is divisible by 4.

b) Prove that $a_n - 22$ is composite for all $n > 10$, no matter what a_1 and a_2 are.

This is the end of our list of problems about congruence for now. We should emphasize that almost all the following problems of this chapter, in fact, continue this theme.

§2. Decimal representation and divisibility tests

The tests for divisibility by 10, 2, 5, 4 are familiar to most people. Working with congruences, we can state and prove far stronger propositions.

Problem 24. Prove that any natural number is congruent to its last digit modulo a) 10; b) 2; c) 5.

Solution. Subtract the last digit from the given number. We obtain a number ending in zero. This number is divisible by 10 and, therefore, by 2 and 5.

For teachers. Before starting the topic of divisibility tests it is necessary that the students understand the identity

$$\boxed{\overline{a_1a_2\ldots a_n} = a_1 10^{n-1} + a_2 10^{n-2} + \ldots + a_{n-1}10^1 + a_n}$$

where the "overlined" row of digits denotes the natural number written with these digits in the indicated order. For example, $\overline{ab} = 10a + b$, where a is the tens digit and b is the units digit of number $\overline{ab}$.

Problem 25. Prove that $\overline{a_1a_2a_3\ldots a_{n-1}a_n} = \overline{a_{n-1}a_n} \pmod 4$.

Problem 26. State and prove analogous tests of divisibility for 2^n and 5^n.

Now, a few problems whose solutions use the indicated divisibility tests.

Problem 27. The last digit of the square of a natural number is 6. Prove that its next-to-last digit is odd.

Solution. Since the last digit of its square is 6, the given natural number was even. The square of an even number is divisible by 4. Hence, the number formed by its two last digits must be divisible by 4. It is easy to write all two-digit numbers which end with 6 and are multiples of 4: 16, 36, 56, 76, 96. All their tens digits are odd.

Problem 28. The next-to-last digit of the square of a natural number is odd. Prove that its last digit is 6.

Problem 29. Prove that a power of 2 cannot end with four equal digits.

Problem 30. Find at least one 100-digit number without zeros in its decimal representation, which is divisible by the sum of its digits.

Solution. Let us find, by trial and error, a number with the sum of its digits equal to 125. Divisibility by 125 is determined by the last three digits of a number. Thus, the number 111...11599125 will do (the number begins with 94 ones).

Let's discuss tests of divisibility by 3 and 9, which can also be stated in a more general form.

Problem 31. Prove that any natural number is congruent to the sum of its digits modulo a) 3; b) 9.

Solution. Consider the number

$$\overline{a_1a_2\dots a_n} = a_1 10^{n-1} + a_2 10^{n-2} + \dots + a_{n-1}10^1 + a_n\,.$$

Clearly, $10 \equiv 1 \pmod 9$. Thus $10^k \equiv 1 \pmod 9$ for any natural k. So we have

$$a_1 10^{n-1} + a_2 10^{n-2} + \dots + a_{n-1}10^1 + a_n \equiv a_1 + a_2 + \dots + a_n \pmod 9\,.$$

The reasoning for divisibility by 3 is completely analogous.

The next set of problems is connected with these tests.

Problem 32. Is it possible to write a perfect square using only the digits a) 2, 3, 6; b) 1, 2, 3 exactly 10 times each?

Problem 33. The sum of the digits was calculated for the number 2^{100}, then the sum of the digits was calculated for the resulting number and so on, until a single digit is left. Which digit is this?

Problem 34. Prove that if you reverse the order of the digits in any natural number and subtract the result from the initial number, then the difference is divisible by 9.

Problem 35. Write one digit to the left and one to the right of the number 15 so that the number obtained is divisible by 15.

Problem 36. How many four-digit numbers with the two middle digits 97 are divisible by 45?

Problem 37. Find the least natural number divisible by 36 which has all 10 digits in its decimal representation.

Problem 38. Prove that the product of the last digit of the number 2^n and of the sum of all its digits but the last is divisible by 3.

Problem 39. Can the sum of the digits of a perfect square be equal to 1970?

Problem 40. A three-digit number was decreased by the sum of its digits. Then the same operation was carried out with the resulting number, et cetera, 100 times in all. Prove that the final number is zero.

Problem 41*. Let A be the sum of the digits of 4444^{4444}, and B the sum of the digits of A. Find the sum of the digits of B.

Ideas very close to those used in the proof of the test of divisibility by 9 allow us to prove another remarkable divisibility test.

Problem 42. Prove that

$$\overline{a_1a_2\dots a_n} \equiv a_n - a_{n-1} + \dots + (-1)^{n-1}a_1 \pmod{11}\,.$$

Hint. $10 \equiv -1 \pmod{11}$, so even powers of 10 are congruent to 1, and odd powers to -1, modulo 11.

Another set of problems is devoted to this test.

Problem 43. Prove that the number 111...11 ($2n$ ones) is composite.

Problem 44. Prove that the number $\overline{a_1 a_2 \dots a_n a_n \dots a_2 a_1}$ is composite.

Problem 45. Let a, b, c, d be distinct digits. Prove that $\overline{cdcdcdcd}$ is not divisible by $\overline{aabb}$.

Problem 46. A is a six-digit number with digits 1, 2, 3, 4, 5, and 6 each used one time. Prove that A is not divisible by 11.

Problem 47. Prove that the difference between a number with oddly many digits and the number written with the same digits in reverse order is divisible by 99.

* * *

Divisibility tests are only one way to link divisibility properties of a number with its decimal representation. This is demonstrated in the following set of problems.

Problem 48. Is it possible to form two numbers using only the digits 2, 3, 4, 9 such that one of them is 19 times greater than the other?

Problem 49. The sum of the two digits a and b is divisible by 7. Prove that the number $\overline{aba}$ is also divisible by 7.

Problem 50. The sum of the digits of a three-digit number equals 7. Prove that this number is divisible by 7 if and only if its two last digits are equal.

Problem 51. a) The six-digit number $\overline{abcdef}$ satisfies the property that $\overline{def} - \overline{abc}$ is divisible by 7. Prove that the number itself is divisible by 7.

b) State and prove a test for divisibility by 7.

c) State and prove a test for divisibility by 13.

Problem 52. a) The six-digit number $\overline{abcdef}$ is such that $\overline{abc} + \overline{def}$ is divisible by 37. Prove that the number itself is divisible by 37.

b) State and prove a test for divisibility by 37.

Problem 53. Is there a three-digit number $\overline{abc}$ (where $a \neq c$) such that $\overline{abc} - \overline{cba}$ is a perfect square?

Problem 54. Find the smallest number written only with ones which is divisible by 333...33 (one hundred 3's in the representation).

Problem 55. Is it possible for the sum of the first several natural numbers to be divisible by 1989?

Problem 56. Find all natural numbers which become 9 times greater if you insert a zero between their units digit and their tens digit.

Solution. Write our number as $10a + b$, where b is the units digit, and a is some natural number. We get the equation $100a+b = 9(10a+b)$ and, therefore, $10a = 8b$; that is, $5a = 4b$. Hence b is divisible by 5. Investigating the two cases $b = 0$, $b = 5$, we find that the only answer is 45.

<u>**Methodological remark.**</u> We have just seen that sometimes it can be very useful to write an equation for the digits of the required number.

Problem 57. Zero was inserted between the digits of a two-digit number divisible by 3, and the result was increased by twice its hundreds digit. The number obtained happens to be 9 times greater than the initial one. Find the original number.

Problem 58. Find a four-digit number which is a perfect square, whose first two digits are equal and whose last two digits are equal.

Problem 59. Find all three-digit numbers, any power of which ends with three digits forming the original number.

Problem 60. The two digits 4 and 3 are written to the right of a natural number, then the operation is repeated many times (for instance, 51 generates 5143, then 514343, et cetera). Prove that eventually we will have a composite number.

Problem 61*. Prove that all the numbers in the series

$$10001, 100010001, 1000100010001, \ldots$$

are composite.

§3. Equations in integers and other problems

A well-known problem asserts that the sum of N pesos can always be paid with 3- and 5-peso bills if $N > 7$ (see the chapter "Induction", Problem 37). Translated into the language of equations, this means that the equation

$$3x + 5y = N \qquad (*)$$

always has a solution in non-negative integers x and y for natural numbers $N > 7$. In this section we will find integer solutions for similar equations and others, though we will usually look for integer solutions without any additional restrictions.

* * *

Problem 62. Solve the equation $3x + 5y = 7$ in integers.

Let us analyze the solution. This will give us an opportunity to solve other problems analogously.

First, we find one particular solution (this idea can often help in solving mathematical problems). Note that $3 \cdot 2 + 5 \cdot (-1) = 1$. Multiplying this equation by 7, we have $3 \cdot 14 + 5 \cdot (-7) = 7$, and $x_0 = 14$, $y_0 = -7$ is one solution (one among many). Thus,

$$3x + 5y = 7; \quad 3x_0 + 5y_0 = 7\ .$$

Subtracting one equation from the other and denoting $x - x_0$ and $y - y_0$ by a and b respectively we have

$$3a + 5b = 0\ .$$

So we see that b must be divisible by 3 and a by 5. Let $a = 5k$. Then $b = -3k$, where k can be any integer. Then, we have the set of solutions:

$$\begin{matrix} x - x_0 = 5k \\ y - y_0 = -3k \end{matrix} \quad \text{i.e.,} \quad \begin{matrix} x = 14 + 5k \\ y = -7 - 3k \end{matrix}\ ,$$

where k is an arbitrary integer. Certainly, no other solutions exist, since our transformations always give us equivalent equations.

For teachers. We hope your students also realize this fact clearly. Without a full understanding of this part of the solution—that the pairs (x, y) exhaust the set of solutions of equation $(*)$ for $N = 7$—it is next to impossible to go ahead.

Problem 63. Find all integer roots of the equation $3x - 12y = 7$.

This barrier overridden, we analyze briefly the "very difficult" Problem 64.

Problem 64. Solve the equation $1990x - 173y = 11$.

The coefficients in the equation are large enough to make it difficult to find a particular solution. However, it is not hard to see that the numbers 1990 and 173 are relatively prime, and this helps.

Lemma. The greatest common divisor (gcd) of these numbers can be represented as $1990m - 173n$ for some integers m and n.

You can prove this lemma using the fact that all the numbers we obtain while calculating the gcd via Euclid's algorithm (see the chapter "Divisibility and remainders") can be represented in this form. This is not easy, but we leave it to the reader.

More specifically Euclid's algorithm gives us $m = 2$, $n = 23$. Therefore, we get $(2, 23)$ as one particular solution of the equation $1990m - 173n = 1$. Hence $x_0 = 2 \cdot 11 = 22$, $y_0 = 23 \cdot 11 = 253$ is a solution to the equation

$$1990x - 173y = 11 .$$

As in Problem 62, we have that

$$x = x_0 + 173k = 22 + 173k,$$
$$y = y_0 + 1990k = 253 + 1990k, \qquad \text{where } k \text{ is any integer.}$$

Problem 65. Find all integer roots of the equation $21x + 48y = 6$.

Remark. Now, generally speaking, we can solve any equation of the form $Ax + By = C$ (this is the so-called *general linear Diophantine equation with two variables*) in integers x and y.

Theorem. If the coefficients A and B in the linear Diophantine equation are relatively prime, then there are integers x_0 and y_0 such that $Ax_0 + By_0 = C$ and all the roots of the equation can be given by the following formulas:

$$x = x_0 + Bk, \ y = y_0 - Ak .$$

Exercise. Try to state the general result and prove it rigorously. (Do not forget about the case when A and B are not relatively prime and C is not divisible by $\gcd(A, B)$.)

Problem 66. Solve the equation $2x + 3y + 5z = 11$ in integers. (By the way, does it have any solutions in natural numbers?)

Problem 67*. A pawn stands on one of the boxes of a band of graph paper of unit width which is infinite in both directions. It can move m boxes to the right or n boxes to the left. Which m and n satisfy the property that the pawn can move onto the box just to the right of the starting box (in several moves)? What is the minimum number of moves needed to do this?

* * *

An equation with more than one variable, for which integer solutions are required, is called a *Diophantine equation* after the famous Greek mathematician Diophantos of Alexandria, who investigated such equations in very early times. Let us examine some more complicated Diophantine equations.

Solve, for integer values of the variables:

Problem 68. $(2x + y)(5x + 3y) = 7$.

Problem 69. $xy = x + y + 3$.

Problem 70. $x^2 = 14 + y^2$.

Problem 71. $x^2 + y^2 = x + y + 2$.

Solutions to all these problems are connected with a very common idea—case-by-case analysis. Certainly it is impossible to write down all pairs of integers and check for each of them whether it satisfies the equation or not. However, some simple transformations can reduce this analysis to just a short job.

Here is a solution to Problem 69. Since $xy - x - y = 3$, we have $(x-1)(y-1) = 4$. It remains to analyze all the representations of 4 as a product of two factors. **Answer:** $(x, y) = (5, 2)$, $(2, 5)$, $(0, -3)$, $(-3, 0)$, $(3, 3)$, $(-1, -1)$.

In Problems 70 and 71 we can also transform the given equation. Thus, we have:

IDEA ONE: Use an appropriate transformation of the equation, then a case-by-case analysis.

However, even in the very next problem this idea will not work.

Problem 72. $x^2 + y^2 = 4z - 1$.

It is clear that one cannot transform the equation to a more tractable type; it is also impossible to analyze all the appropriate triples of integers. This new representative of our "Diophantine zoo" is remarkable in that it has no integer solutions.

Indeed, which remainders can perfect squares give when divided by 4? (The choice of modulo 4 was determined by the form of the given equation.) The only possible remainders are 0 and 1. Since the sum of two such remainders cannot give remainder -1, we have no solutions for this equation.

Thus we have:

IDEA TWO: Consider remainders modulo some natural number.

Problem 73. $x^2 - 7y = 10$.

Problem 74. $x^3 + 21y^2 + 5 = 0$.

Problem 75. $15x^2 - 7y^2 = 9$.

Problem 76. $x^2 + y^2 + z^2 = 8t - 1$.

Problem 77. $3^m + 7 = 2^n$.

Solution to Problem 74. Because x^3 can be congruent modulo 7 only to 0, 1 and -1, $x^3 + 21y^2 + 5$ must be congruent to 5, 6, or 4, and, therefore, cannot be zero.

Methodological remark. You probably have already noticed that "idea two" allows us only to prove the absence of roots. Indeed, if an equation has solutions modulo 7 or modulo 3, then it does not mean that the equation has at least one integer solution. For example, the equation $2x^2 - y^3 = 6$ has roots modulo 7 ($x \equiv 0$

and $y \equiv 1$) and modulo 3 ($x \equiv 0$ and $y \equiv 0$), but it has no integer roots (this can be easily shown using remainders when divided by 8).

Let us struggle now with Problem 77. We can see at once that a solution exists: $m = 2$, $n = 4$. Does it make sense to consider remainders? Do not jump to conclusions! The left side is congruent to 1 modulo 3. Since $2 \equiv -1 \pmod 3$, this implies that n is even; that is, $n = 2k$. So we have $3^m + 7 = 4^k$. Now residues modulo 4 can help us. We have $4^k - 7 = 1 \pmod 4$. Hence, $3^m \equiv 1 \pmod 4$, which implies that m is also even; that is, $m = 2p$. The equation becomes $3^{2p} + 7 = 2^{2k}$. What now? We use "idea one":

$$7 = 2^{2k} - 3^{2p} = (2^k - 3^p)(2^k + 3^p) .$$

Therefore, $2^k + 3^p = 7$, $2^k - 3^p = 1$, and we obtain the unique solution $k = 2$, $p = 1$; that is, $m = 2$, $n = 4$.

We used both ideas, or methods, tried earlier. Such a combination of ideas is a very common phenomenon in mathematics.

Here is another Diophantine equation whose solution uses a similar combination:

Problem 78. $3 \cdot 2^m + 1 = n^2$.

Solution. Since $n^2 \equiv 1 \pmod 3$, it is clear that n is not divisible by 3. So $n = 3k+1$ or $n = 3k+2$. We investigate each case.

a) If $n = 3k + 2$, then $3 \cdot 2^m + 1 = 9k^2 + 12k + 4$. Simplifying, we get

$$2^m = 3k^2 + 4k + 1 = (3k + 1)(k + 1) .$$

The only factors of a power of 2 are other powers of 2. Therefore, $k + 1$ and $3k + 1$ are powers of 2. The values $k = 0$ and $k = 1$ do fit and we have the solutions $n = 2$, $m = 1$ and $n = 5$, $m = 3$ respectively. However, if $k \geq 2$, then $4(k + 1) > 3k + 1 > 2(k + 1)$. This inequality shows that $k + 1$ and $3k + 1$ cannot be powers of 2 simultaneously.

b) $n = 3k + 1$. Proceeding analogously we find one more root: $n = 7$, $m = 4$.

Aside from the ideas of case-by-case analysis and factorization, the idea of estimation was used:

IDEA THREE: When solving Diophantine equations, inequalities and estimates may be of use.

Problem 79. $1/a + 1/b + 1/c = 1$.

Problem 80. $x^2 - y^2 = 1988$.

Problem 81. Prove that the equation $1/x - 1/y = 1/n$ has exactly one solution in natural numbers if and only if n is prime.

Problem 82. $x^3 + 3 = 4y(y + 1)$.

We cannot discuss, in this small section, the many other interesting and complicated methods of solving Diophantine equations. Instead, we refer you to [**55, 58, 25**].

Below are two extra problems which are much more difficult than those above. You might like to see the hints before you try to solve them on your own.

Problem 83*. $x^2 + y^2 = z^2$.

Problem 84*. $x^2 - 2y^2 = 1$.

§4. Fermat's "little" theorem

This section is devoted to a remarkable and non-trivial fact in number theory that was stated and proved by the famous XVII century French mathematician Pierre de Fermat. But, before we begin, we discuss (as was promised in §1) the question of division of congruences.

Problem 85. Let $ka \equiv kb \pmod{m}$ where k and m are relatively prime. Then $a \equiv b \pmod{m}$.

Solution. Since $ka \equiv kb \pmod{m}$, $ka - kb = k(a - b)$ is divisible by m. Since k and m are relatively prime, $a - b$ is divisible by m; that is, $a \equiv b \pmod{m}$.

It is easy to find examples to show that it is necessary for k and m to be relatively prime. Indeed, $5 \cdot 3 \equiv 5 \cdot 7 \pmod{10}$ but 3 and 7 are not congruent modulo 10.

In any case, the following is true:

Problem 86. If $ka \equiv kb \pmod{kn}$, then $a \equiv b \pmod{n}$.

Now we are ready to state and prove Fermat's "little" theorem.

Theorem. Let p be prime number and A be a number not divisible by p. Then $A^{p-1} \equiv 1 \pmod{p}$.

Proof. Consider the $p - 1$ numbers: $A, 2A, 3A, \ldots, (p-1)A$. We can show that no two of these numbers have the same remainder when divided by p. Indeed, if $kA \equiv nA \pmod{p}$, then $k = n \pmod{p}$ (see Problem 85). This is impossible if k and n are unequal and both are less than p. Therefore, among the remainders of these $p - 1$ numbers when divided by p each of the numbers 1 through $p - 1$ is represented exactly once. Multiplying them together, we have

$$A \cdot 2A \cdot 3A \ldots (p-1)A \equiv 1 \cdot 2 \cdot 3 \ldots (p-1) \pmod{p};$$

that is, $(p-1)! \cdot A^{p-1} \equiv (p-1)! \pmod{p}$. Now p is prime, which implies that $(p-1)!$ and p are relatively prime. Using the result of Problem 85 again we obtain $A^{p-1} \equiv 1 \pmod{p}$. □

Corollary. Let p be a prime number. Then for any integer A we have $A^p \equiv A \pmod{p}$.

Fermat's "little" theorem is not just an unexpected and "neat" fact. It also provides a very powerful tool for solving many problems in arithmetic. Some of these are given below.

Problem 87. Find the remainder when 2^{100} is divided by 101.

Problem 88. Find the remainder when 3^{102} is divided by 101.

Solution. Since 101 is prime, we get $3^{100} \equiv 1 \pmod{101}$. Thus $3^{102} = 9 \cdot 3^{100} \equiv 9 \pmod{101}$.

For teachers. Such computational exercises using Fermat's theorem can become quite routine for students.

Problem 89. Prove that $300^{3000} - 1$ is divisible by 1001.

Problem 90. Find the remainder when 8^{900} is divided by 29.

Problem 91. Prove that $7^{120} - 1$ is divisible by 143.

Solution. Let us prove that $7^{120} - 1$ is divisible by 11 and 13. Indeed, $(7^{12})^{10} \equiv 1 \pmod{11}$ and $(7^{10})^{12} \equiv 1 \pmod{13}$.

Problem 92. Prove that the number $30^{239} + 239^{30}$ is not prime.

Problem 93. Let p be a prime number, and suppose a and b are arbitrary integers. Prove that $(a+b)^p = a^p + b^p \pmod{p}$.

Try to contrive two proofs: one utilizing Fermat's "little" theorem and the other using the binomial theorem (see the chapter "Combinatorics–2").

Problem 94. The sum of the numbers a, b, and c is divisible by 30. Prove that $a^5 + b^5 + c^5$ is also divisible by 30.

Problem 95. Let p and q be different primes. Prove that

a) $p^q + q^p = p + q \pmod{pq}$.

b) $\left[\dfrac{p^q + q^p}{pq}\right]$ is even if $p,\ q \neq 2$, where $[x]$ denotes the greatest integer function.

Problem 96. Let p be prime, and suppose p does not divide some number a. Prove that there exists a natural number b such that $ab \equiv 1 \pmod{p}$.

Problem 97. (Wilson's theorem). Let p be prime. Prove that $(p-1)! \equiv -1 \pmod{p}$.

Problem 98. Let n be a natural number not divisible by 17. Prove that either $n^8 + 1$ or $n^8 - 1$ is divisible by 17.

Problem 99. a) Let p be a prime not equal to 3. Prove that the number $111\ldots11$ (p ones) is not divisible by p.

b) Let $p > 5$ be a prime. Prove that the number $111\ldots11$ ($p-1$ ones) is divisible by p.

Problem 100. Prove that for each prime p the difference

$$111\ldots11222\ldots22333\ldots33\ldots888\ldots88999\ldots99 - 123456789$$

(in the first number each non-zero digit is written exactly p times) is divisible by p.

CHAPTER 11

Combinatorics–2

This chapter is a direct continuation of the chapter "Combinatorics–1". The present material draws on results explained in that chapter.

* * *

For teachers. Students should solve again a few problems involving the combinatorial ideas discussed earlier and go on only if these problems do not produce any difficulties. If they do, we recommend going back to the chapter "Combinatorics–1".

The contents of this chapter are connected with one very important combinatorial object which we begin to study right now.

§1. Combinations

Let us begin with a simple problem.

Problem 1. Two students must be chosen out of a group of thirty for a mathematical contest. In how many ways can this be done?

Solution. You can choose the first participant of the contest in 30 ways. No matter who the first was, the second can be chosen in 29 ways. But now each pair is counted exactly twice. Thus the answer is $30 \cdot 29/2 = 435$ ways. Note that we have merely repeated the solution to Problem 22 from the chapter "Combinatorics–1".

Assume now that we must choose a team not of two people but of k, and that the group consists of n students, not of 30. The number of ways this can be done is called *the number of combinations of k elements taken from n elements* and is denoted $\binom{n}{k}$ (to be read as "n choose k").

For instance, $\binom{2}{1} = 2$, $\binom{3}{2} = 3$, $\binom{n}{1} = n$, $\binom{n}{n} = 1$. Note that $\binom{n}{0}$ also can be interpreted naturally: there is only one way to choose nobody (zero people) out of n. That is, $\binom{n}{0} = 1$ for all n.

It is remarkable that some properties of these numbers can be proved by simple combinatorial reasoning, not using a formula for the calculation of $\binom{n}{k}$.

Property 1. $\binom{n}{n-k} = \binom{n}{k}$.

Proof. Note that a choice of k contest participants is equivalent to the choice of $n-k$ students who will not take part in the contest. Thus, the number of ways to choose k people out of n equals the number of ways to choose $n-k$ people out of n; that is, $\binom{n}{n-k} = \binom{n}{k}$.

Property 2. $\binom{n+1}{k} = \binom{n}{k} + \binom{n}{k-1}$.

Proof. Assume that there are $n+1$ students in the group. Consider one of them and denote him or her by A. Let us divide the set of all possible teams into two subsets: those teams containing A, and the others which do not contain A. The cardinality of the first set is $\binom{n}{k-1}$—since we must complement the team with $k-1$ more students chosen from the n students remaining. The number of teams in the second set equals $\binom{n}{k}$. Now from the remaining n students we must choose the entire team. Thus $\binom{n+1}{k} = \binom{n}{k} + \binom{n}{k-1}$.

Methodological remark. This reasoning allowed us to prove a rather significant fact without any calculations. This phenomenon is quite characteristic for combinatorics. Often just a few minutes of thinking and understanding the combinatorial sense of a question may let us avoid cumbersome calculations. This is why we consider it necessary to discuss proofs like those above in detail.

Let us now find a formula to calculate $\binom{n}{k}$.

Problem 2. How many ways are there to choose a team of three students out of a group of 30?

Solution. The first student can be chosen in 30 ways, the second in 29 ways, and the third in 28 ways. Thus we have $30 \cdot 29 \cdot 28$ ways. However, each team was counted several times: the same trio of students can be chosen in different ways. For instance, choosing student A first, then student B, and, finally, C is the same as choosing C first, then A, and then B. Since the number of permutations of 3 elements is $3!$, each team was counted exactly $3! = 6$ times. Therefore, $\binom{30}{3}$ equals $(30 \cdot 29 \cdot 28)/3!$.

In just the same way we can deduce a formula for calculating $\binom{n}{k}$, for arbitrary n and k:

$$\binom{n}{k} = n(n-1)(n-2)\dots(n-k+1)/k!.$$

For teachers. The numbers $\binom{n}{k}$ are central in this chapter. Thus, it is important to make sure that all students understand what we are counting here, and how the numbers of combinations can be calculated. Before revealing the general formula, proofs of a few problems similar to Problem 2 might be discussed.

Let us deal with a few more problems.

Problem 3. In how many ways can one choose 4 colors out of 7 given colors?
Answer. $\binom{7}{4} = 35$.

Problem 4. One student has 6 math books, and another has 8 books. How many ways are there to exchange 3 books belonging to the first student with 3 books belonging to the second?

Solution. The first student can choose his 3 books in $\binom{6}{3}$ ways, and the second in $\binom{8}{3}$ ways. Thus, the number of possible exchanges is $\binom{6}{3} \cdot \binom{8}{3} = 1120$.

Problem 5. There are 2 girls and 7 boys in a chess club. A team of four persons must be chosen for a tournament, and there must be at least 1 girl on the team. In how many ways can this be done?

Solution. There must be either 1 or 2 girls on the team. In the latter case 2 boys can be added to the team in $\binom{7}{2}$ ways. If there is only 1 girl on the team (there are two ways to choose her), then the team can be completed by adding 3 boys in $\binom{7}{3}$ different ways. Therefore, in all there are $\binom{7}{2} + 2 \cdot \binom{7}{3} = 91$ possible teams.

Problem 6. How many ways are there to divide 10 boys into two basketball teams of 5 boys each?

Solution. The first team can be chosen in $\binom{10}{5}$ ways. This choice completely determines the second team. However, this calculation counts each pair of complementing teams—say, A and B—two times: the first time, when A is chosen as the first team, and the second time, when B is chosen as the first team. Thus, the answer is $\binom{10}{5}/2$.

Methodological remark. After learning these formulas it is not obligatory to express the answers as decimal numbers. It is not bad if the expressions $\binom{n}{k}$ are present in answers.

Notice that the formula for $\binom{n}{k}$ makes the first property of symmetry—$\binom{n}{n-k} = \binom{n}{k}$—rather unclear. However, we can make the formula look more symmetric by multiplying its numerator and denominator by $(n-k)!$:

$$\begin{aligned}\binom{n}{k} &= \frac{n(n-1)\dots(n-k+1)(n-k)!}{k!(n-k)!} \\ &= \frac{n(n-1)\dots(n-k+1)(n-k)\dots 3\cdot 2\cdot 1}{k!(n-k)!} \\ &= \frac{n!}{k!(n-k)!}.\end{aligned}$$

Now the first property is quite evident.

Exercise. Prove the second property of $\binom{n}{k}$, using the formula above.

For teachers. We recommend spending at least one session of a math circle introducing definitions, properties, and the formula for the numbers $\binom{n}{k}$. It would also be helpful to solve several easy problems at this session, then to give problems connected with this theme during the next sessions.

Problem 7. Ten points are marked on a plane so that no three of them are on the same straight line. How many triangles are there with vertices at these points?

Problem 8. A special squad consists of 3 officers, 6 sergeants, and 60 privates. In how many ways can a group consisting of 1 officer, 2 sergeants, and 20 privates be chosen for an assignment?

Problem 9. Ten points are marked on a straight line, and 11 points are marked on another line, parallel to the first one. How many

a) triangles;

b) quadrilaterals

are there with vertices at these points?

Problem 10. A set of 15 different words is given. In how many ways is it possible to choose a subset of no more than 5 words?

Problem 11. There are 4 married couples in a club. How many ways are there to choose a committee of 3 members so that no two spouses are members of the committee?

Problem 12. There are 31 students in a class, including Pete and John. How many ways are there to choose a soccer team (11 players) so that Pete and John are not on the team together?

Problem 13. How many ways are there to rearrange the letters in the word "ASUNDER" so that vowels will be in alphabetical order, as well as consonants? Example: DANERUS (A–E–U, D–N–R–S).

Problem 14. We must choose a 5-member team from 12 girls and 10 boys. How many ways are there to make the choice so that there are no more than 3 boys on the team?

Problem 15. How many ways are there to put 12 white and 12 black checkers on the black squares of a chessboard?

Problem 16. a) How many ways are there to divide 15 people into three teams of 5 people each?

b) How many ways are there to choose two teams of 5 people each from 15 people?

Problem 17. In how many ways can you choose 10 cards from a deck of 52 cards so that

a) there is exactly one ace among the chosen cards?

b) there is at least one ace among the chosen cards?

Problem 18. How many six-digit numbers have 3 even and 3 odd digits?

Problem 19. How many ten-digit numbers have the sum of their digits equal to

a) 2;

b) 3;

c) 4?

Problem 20. A person has 6 friends. Each evening, for 5 days, he or she invites 3 of them so that the same group is never invited twice. How many ways are there to do this?

Problem 21. To participate in a sports lottery in Russia one must choose 6 out of the 45 numbers printed on a lottery card (all the printed cards are identical).

a) How many ways are there to fill in the lottery card?

b) After the end of the lottery, its organizers decided to count the number of ways to fill in the lottery card so that exactly 3 of the 6 chosen numbers are among the 6 winning numbers. Help them to find the answer.

§2. Pascal's triangle

This section is remarkable for its combination of almost all the ideas explained earlier in this chapter, leading us to some very beautiful combinatorial facts.

To start, let us assume we know all the numbers $\binom{n}{k}$ for some fixed number n. Then the second property,

$$\binom{n+1}{k} = \binom{n}{k} + \binom{n}{k-1},$$

allows us to calculate easily the numbers $\binom{n+1}{k}$ for all k. This idea gives us the following construction.

Since $\binom{0}{0} = 1$, we write 1 in the center of the first line on a sheet of paper. In the next line we write the numbers $\binom{1}{0} = 1$ and $\binom{1}{1} = 1$ in such a way that $1 = \binom{0}{0}$ is over the gap between these numbers (see Figure 74).

FIGURE 74

The numbers $\binom{2}{0}$ and $\binom{2}{2}$ are also 1. We write them in the next line (see Figure 75) with $\binom{2}{1}$, which, by the second property, equals $\binom{1}{0} + \binom{1}{1}$ written between them (see Figure 76).

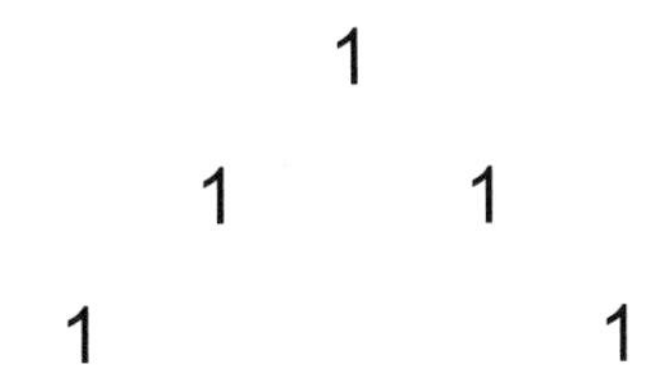

FIGURE 75

Thus, number $\binom{2}{1}$ equals the sum of the two numbers in the previous row, standing to the left and to the right of it.

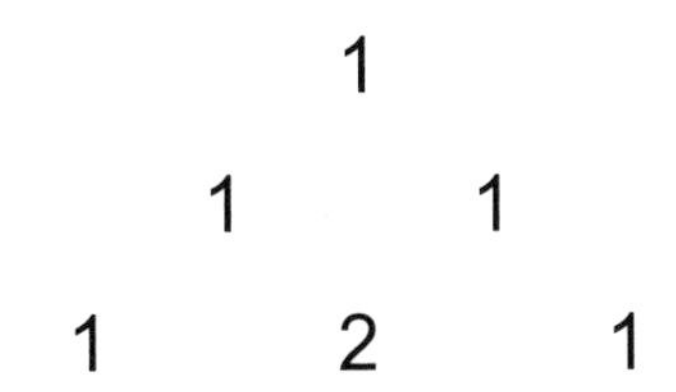

FIGURE 76

Using the same rule, we will fill all the following lines: first, we write the numbers $\binom{n}{0}$ and $\binom{n}{n}$ on the sides (they are always equal to 1), and then we write the sum of any two adjacent numbers of the previous row in a position between them in the next row.

Finally, we have the numerical triangle shown in Figure 77. It is called *Pascal's triangle.*

By construction, the number $\binom{n}{k}$ occupies the $(k+1)$st place in the $(n+1)$st row of this triangle. Therefore, it is more convenient to number the rows, and the places in the rows, starting from zero. Then we will have the number $\binom{n}{k}$ occupying the kth place in the nth row.

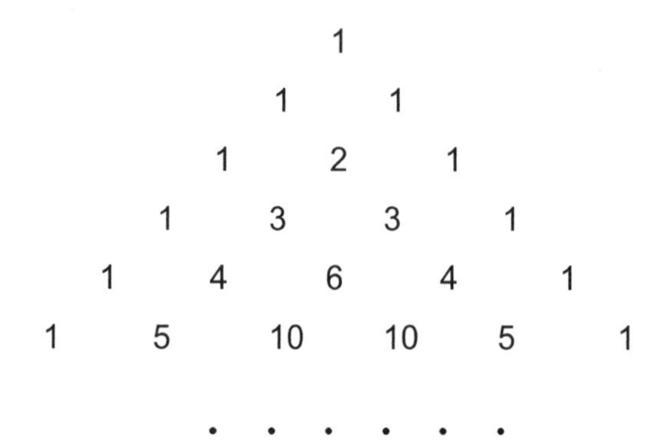

FIGURE 77

For teachers. Before going further, students should know the connection between the numbers $\binom{n}{k}$ and Pascal's triangle. The best exercise for this is direct calculation of numbers of combinations using the triangle-based procedure described above.

Now let us begin the investigation of properties of Pascal's triangle. Evaluate the sum of the numbers in its first few rows: 1, 2, 4, 8, 16. You may come to the very natural conjecture that the sum of the nth row is 2^n. We prove this by induction on n (see the chapter "Induction"). The base is already proved. To prove the inductive step, notice that each number of a row is taken as a summand in forming two adjacent numbers in the next row. Thus the sum of the numbers in the next row is exactly twice the sum of the numbers in the given row. This completes the inductive step.

It may also be proved that in every row of Pascal's triangle (except for the zeroth) the sum of the numbers in the even places equals the sum of numbers in the odd places.

Rewriting the proposition about the sum of the numbers in a row of Pascal's triangle, we get the following remarkable combinatorial identity:

$$\binom{n}{0} + \binom{n}{1} + \ldots + \binom{n}{n-1} + \binom{n}{n} = 2^n .$$

Here is a combinatorial proof. From the combinatorial point of view the identity states that the number of teams which can be chosen from n students equals 2^n if the cardinality of the team is arbitrary (using set theoretic language: the number of subsets of an n-element set equals 2^n).

We number all the students from 1 through n in an arbitrary order. Then for each possible team we construct a sequence of 0's and 1's in the following way: the first element of the sequence is 1 if the first student is on the team, and 0 otherwise. In the same way we define the second element of the sequence, the third, and so on. It is evident that different teams correspond to different sequences and vice versa. Thus the number of all teams is equal to the number of all sequences of 0's and 1's with n elements. Each element of such a sequence can be equal to either 0 or 1—that is, it can be chosen in two ways. Thus, the number of all such sequences equals $2 \times 2 \times \ldots \times 2 = 2^n$.

<u>**Methodological remark.**</u> The most important place in this reasoning was the construction of a correspondence between teams and sequences of 0's and 1's. Such a reformulation is often very helpful in the solution of many other combinatorial problems. As examples we submit here two more problems.

Problem 22. A person has 10 friends. Over several days he invites some of them to a dinner party so that the company never repeats (he may, for example, invite nobody on one of the days). For how many days can he follow this rule?

Problem 23. There are 7 steps in a flight of stairs (not counting the top and bottom of the flight). When going down, you can jump over some steps, perhaps even over all 7. How many ways are there to go down the stairs?

Before investigating the next property of Pascal's triangle, we analyze one problem which is remarkable in that the numbers $\binom{n}{k}$ arise unexpectedly in the course of its solution.

Problem 24. The map of a town is depicted in Figure 78. All its streets are one-way, so that you can drive only "east" or "north". How many different ways are there to reach point B starting from A?

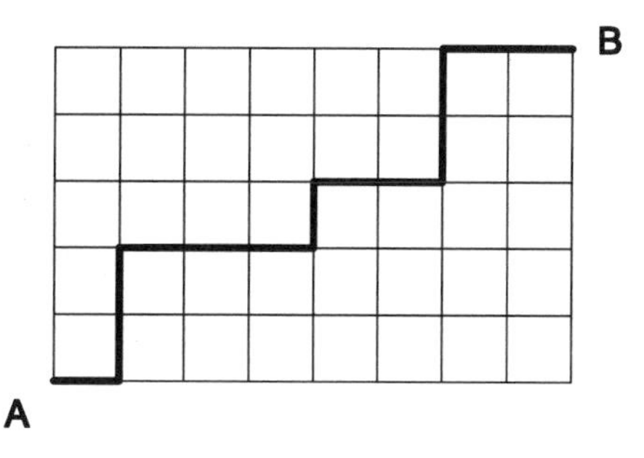

FIGURE 78

Solution. Let us call any segment of the grid connecting two neighboring nodes a "street". It is clear that each route from A to B consists of exactly 13 streets, 8 of which are horizontal and 5 are vertical. Given any route, we will construct a sequence of the letters N and E in the following way: when we drive "north", we add the letter N to the sequence, and when we drive "east", we add the letter E to the sequence. For instance, the route in Figure 78 corresponds to the sequence ENNEEENEENNEE. Each sequence constructed in this way contains 13 letters—8 letters E and 5 letters N. It remains to calculate the number of such sequences. Any sequence is uniquely determined by the list of the 5 places occupied by the letters N (or, alternatively, of the 8 places occupied by the letters E). Five places out of 13 can be chosen in $\binom{13}{5}$ ways. Thus the number of sequences, and, therefore, the number of routes, equals $\binom{13}{5}$.

The same reasoning for a $m \times n$ rectangle gives us the result $\binom{m+n}{m}$ or, equivalently, $\binom{m+n}{n}$.

Returning to Pascal's triangle, let us change its numbers to points (nodes) as shown in Figure 79. Let us write 1 next to the uppermost point S of the triangle,

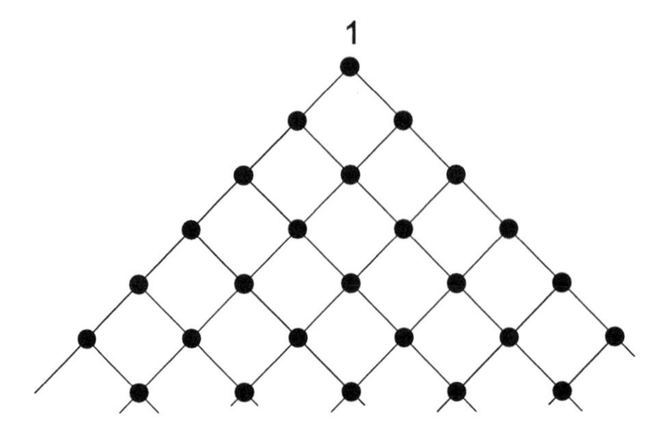

FIGURE 79

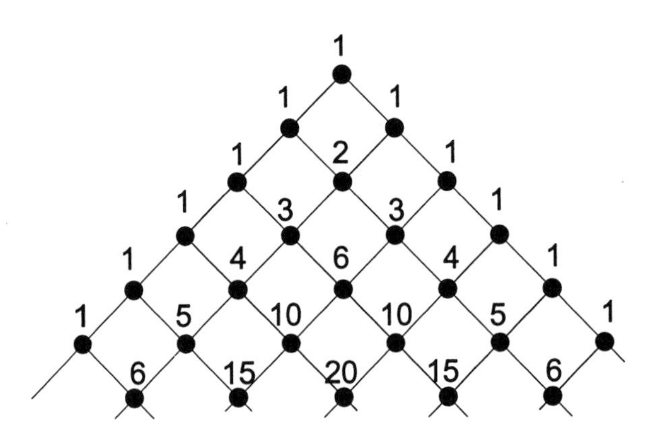

FIGURE 80

and, then, next to any other node—the number of ways one can reach this node from S, moving only downward. You can see (Figure 80) that we have Pascal's triangle again.

One proof of this fact is similar to the solution of the previous problem. However, here, as in the proof of the fact that the sum of the numbers in each row is a power of two, we can proceed by induction. Indeed, you can reach the kth node of the nth row only via either the $(k-1)$st node of the $(n-1)$st row or the kth node of the $(n-1)$st row (Figure 81). Thus, to find the required number of ways we must simply add the number of ways going to those two nodes of the previous row. Hence, using the assumption, the number of ways of going from S to the kth node of the nth row is $\binom{n-1}{k-1} + \binom{n-1}{k} = \binom{n}{k}$.

Methodological remark. Both properties of Pascal's triangle described above were proved in two ways—using "geometric" ideas and via direct combinatorial reasoning. It is useful to utilize both these approaches while solving various problems, especially combinatorial identities.

Some other properties of Pascal's triangle are given below as problems to be solved. They can be stated in terms of the triangle itself, or as combinatorial identities.

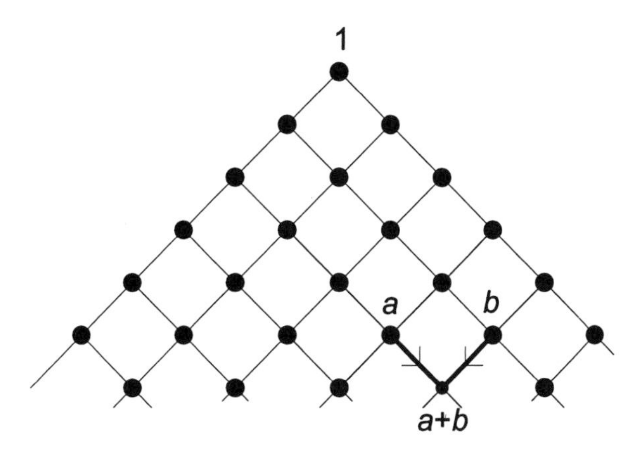

FIGURE 81

Problem 25. Prove that one can choose evenly many objects from a collection of n objects in 2^{n-1} ways.

Problem 26. Prove that

$$\binom{n}{0} - \binom{n}{1} + \ldots + (-1)^n \binom{n}{n} = 0\,.$$

For convenience, we introduce now the following definitions. We will call rays parallel to sides of Pascal's triangle its *diagonals*. More precisely, rays parallel to the right side are called *right diagonals* (one of them is selected in Figure 82), and those parallel to the left side are *left diagonals* (see Figure 83).

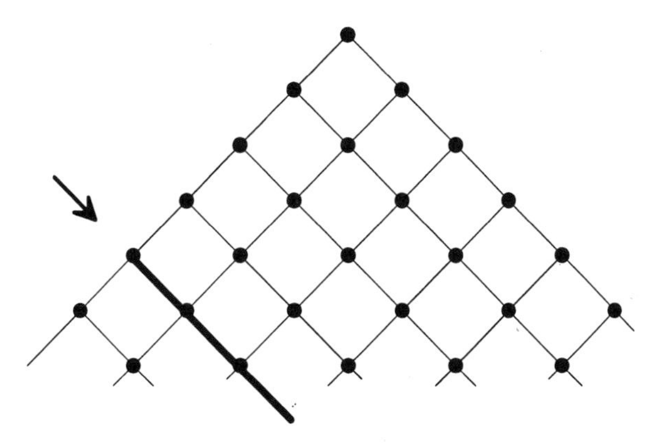

FIGURE 82

Problem 27. Prove that each number a in Pascal's triangle is equal to the sum of the numbers in the previous right diagonal, starting from its leftmost number through the number which is located in the same left diagonal as a (see Figure 84).

Problem 28. Prove that each number a in Pascal's triangle is equal to the sum of the numbers in the previous left diagonal, starting from the rightmost number through the number which is located in the same right diagonal as a (see Figure 85).

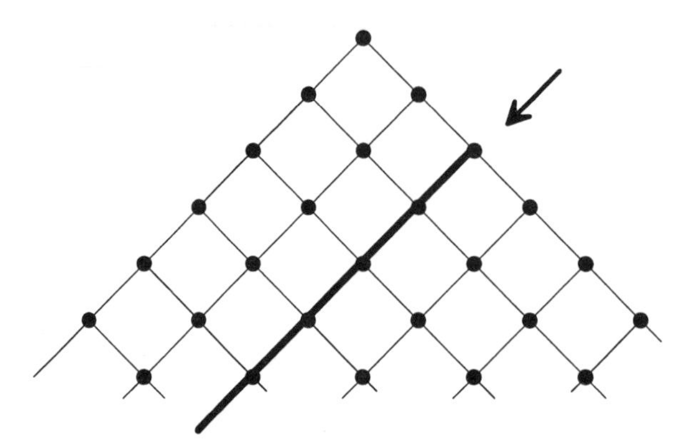

FIGURE 83

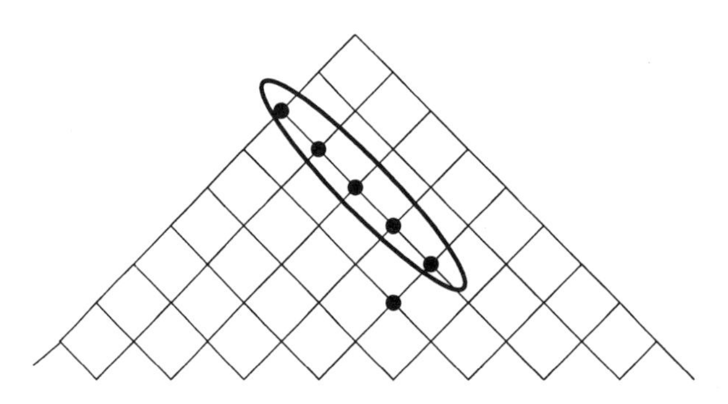

FIGURE 84

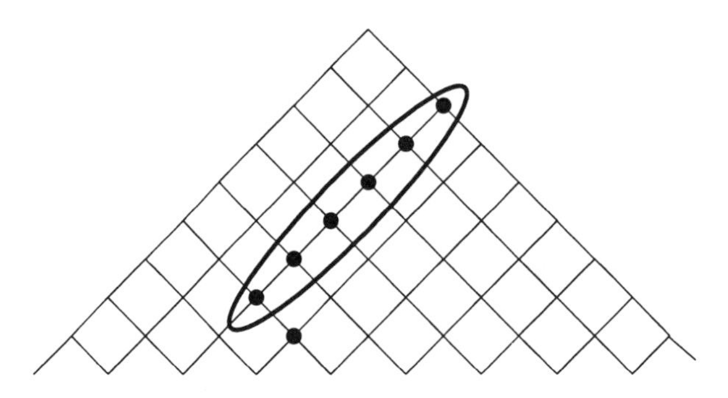

FIGURE 85

Problem 29. Prove that each number a in Pascal's triangle decreased by 1 is equal to the sum of the numbers within a parallelogram bounded by the sides of the triangle and diagonals going through a (the numbers on the diagonals are not included; see Figure 86).

Problem 30*. Prove that

$$\binom{n}{0}^2 + \binom{n}{1}^2 + \ldots + \binom{n}{n}^2 = \binom{2n}{n}.$$

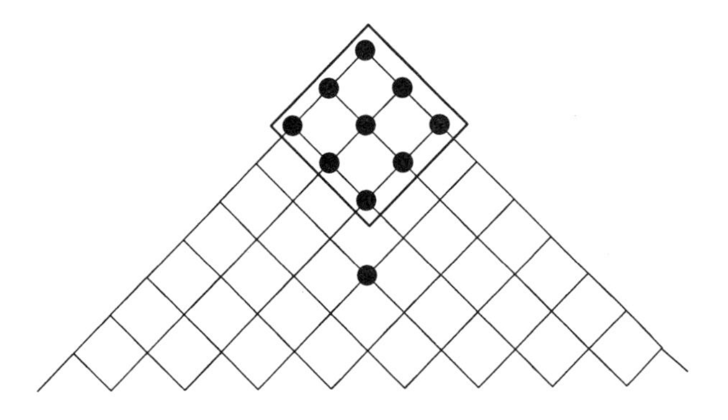

FIGURE 86

§3. Balls and walls

Let us begin with a discussion of two interesting problems. Each of these can be solved by direct, though technically complex, calculation (try this way, too, but later). On the other hand, simply restating the question allows us to reach the answer (which in each case is a number of combinations), rather easily.

Problem 31. Six boxes are numbered 1 through 6. How many ways are there to put 20 identical balls into these boxes so that none of them is empty?

Solution. Let us arrange the balls in a row. To determine the distribution of the balls in the boxes we must partition this row into six groups of balls using five walls: the first group for the first box, the second group for the second box, et cetera (see Figure 87). Thus, the number of ways to distribute our balls in the boxes equals the number of ways to put five walls into gaps between the balls in the row. Any wall can be in any of 19 gaps (there are $19 = 20 - 1$ gaps between 20 balls), and no two of them can be in the same gap (this would mean that one of the groups is empty). Therefore, the number of all possible partitions is $\binom{19}{5}$.

FIGURE 87

Exercise. How many ways are there to distribute n identical balls in m numbered boxes so that none of the boxes is empty?

Problem 32. Six boxes are numbered 1 through 6. How many ways are there to distribute 20 identical balls between the boxes (this time some of the boxes can be empty)?

Solution. Consider a row of 25 objects: 20 identical balls and 5 identical walls, which are arranged in an arbitrary order. Any such row corresponds without ambiguity to some partition of balls: balls located to the left of the first wall, go to the first box; balls between the first and the second wall go to the second box, et cetera (perhaps, some pair of walls are adjacent in the row, resulting in an empty box). Therefore, the number of partitions is equal to the number of all possible rows of 20 balls and 5 walls; that is, to $\binom{25}{5}$ (the row is completely determined by the 5 places occupied by the walls).

We should note that another solution to Problem 31 can be obtained as follows: put one ball in each box (to prevent empty boxes), then use the result of Problem 32 (with 14 balls instead of 20).

The ideas found during the process of solving the two previous problems show us how to solve the next, rather complicated, problem very neatly.

Problem 33. How many ways are there to represent the natural number n as a sum of

a) k natural numbers?

b) k non-negative integers?

Representations that differ in the order of the summands are different.

Hint. Represent n as the sum of n ones: $n = 1 + 1 + \ldots + 1$. Call these n ones "balls", and call the k summands from the statement "boxes". The answers are: a) $\binom{n-1}{k-1}$; b) $\binom{n+k-1}{n}$.

Methodological remark. The solutions just explained show us again how important a good reformulation of the statement of a problem can be. The reason for such a thorough discussion of the question of the distribution of balls in boxes is that similar reformulations (devising "walls", et cetera) are very useful in solving many quite different problems, not only in combinatorics, but in other fields of mathematics and, generally, science.

Problem 34. In how many ways can 12 pennies be put into five different purses so that none of them is empty?

Problem 35. A bookbinder must bind 12 identical books using red, green, or blue covers. In how many ways can he do this?

Problem 36. How many ways are there to cut a necklace (in the form of an unbroken circle) with 30 pearls into 8 parts (it is permitted to cut only between pearls)?

Problem 37. Thirty people vote for 5 candidates. How many possible distributions of their votes are there, if each of them votes for one candidate, and we consider only the numbers of votes given to each of the candidates?

Problem 38. There are 10 types of postcards in a post office. How many ways are there to buy

a) 12 postcards?

b) 8 postcards?

Problem 39. A train with m passengers must make n stops.

a) How many ways are there for passengers to get off the train at the stops?

b) Answer the same question if we take into account only the *number* of passengers who get off at each stop.

Problem 40. In a purse, there are 20 pennies, 20 nickels, and 20 dimes. How many ways are there to choose 20 coins out of these 60?

Problem 41. How many ways are there to put seven white and two black billiard balls in nine pockets? Some of the pockets may remain empty and the pockets are considered distinguishable.

Problem 42. In how many ways can three people divide among themselves six identical apples, one orange, one plum, and one tangerine (without cutting any fruit)?

Problem 43. How many ways are there to put four black, four white, and four blue balls into six different boxes?

Problem 44. A community with n members chooses its representative by voting.

a) In how many ways can "open" voting result, if everybody votes for one person (perhaps for himself/herself)? Open voting means that we take into account not only the numbers of votes, but also who votes for whom.

b) Answer the same question, if voting is not open (and the result consists only of the numbers of votes given to each of the members of the community).

Problem 45. How many ways are there to arrange five red, five blue, and five green balls in a row so that no two blue balls lie next to each other?

Problem 46*. How many ways are there to represent the number 1000000 as the product of three factors, if we consider products that differ in the order of factors as different?

Problem 47*. There are 12 books on a shelf. How many ways are there to choose five of them so that no two of the chosen books stand next to each other?

§4*. Newton's binomial theorem

For teachers. This section is marked with an asterisk because it is rather difficult for students of the middle grades. However, we have decided not to omit it, since its contents are closely related to the numbers $\binom{n}{k}$ and Pascal's triangle. It is also beyond any doubt that the binomial theorem is an indispensable element of mathematical education, although, its study can be postponed.

Remark. This topic is related to, but distinct from, combinatorics. However, it is relevant to give the statement and proof of the main theorem here, since it is used in combinatorics quite often and the proof itself is based on combinatorial ideas.

Everyone knows the identity:

$$(a+b)^2 = a^2 + 2ab + b^2 .$$

We will try to derive a formula for raising the binomial $(a+b)$ to an arbitrary positive integer power. Let us write a few successive powers of the binomial:

$$\begin{array}{lccccccc}
(a+b)^0 = & & & & 1 & & & \\
(a+b)^1 = & & & a & + & b & & \\
(a+b)^2 = & & a^2 & + & 2ab & + & b^2 & \\
(a+b)^3 = & a^3 & + & 3a^2b & + & 3ab^2 & + & b^3
\end{array}$$

It is obvious that the coefficients in the right parts of these identities form the corresponding rows of Pascal's triangle. We may suggest that the following identity holds true:

$$(a+b)^n = \binom{n}{0}a^n + \binom{n}{1}a^{n-1}b^1 + \binom{n}{2}a^{n-2}b^2 + \ldots + \binom{n}{n-1}a^1b^{n-1} + \binom{n}{n}b^n .$$

This is indeed true and this expansion is called *Newton's binomial theorem.* To prove it, we expand the product

$$(a+b)^n = (a+b)(a+b)(a+b)\dots(a+b)(a+b)$$

without grouping similar terms together and without changing the order of factors in each monomial. For example,

$$\begin{aligned}&(a+b)(\underline{a}+\underline{b})(\bar{a}+\bar{b})\\ &= a\underline{a}\bar{a} + a\underline{a}\bar{b} + a\underline{b}\bar{a} + a\underline{b}\bar{b} + b\underline{a}\bar{a} + b\underline{a}\bar{b} + b\underline{b}\bar{a} + b\underline{b}\bar{b}\,.\end{aligned}$$

Let us find the coefficient of the term $a^{n-k}b^k$ after reducing similar terms. It is clear that this coefficient equals the number of monomials which include b exactly k times (and include a exactly $n-k$ times). This number is equal to $\binom{n}{k}$, since this is the number of ways to choose k places for the letters b.

Exercise. Prove the binomial theorem by induction.

§5. Additional problems

For teachers. This section is auxiliary and optional: it does not contain substantial new material but is just a list of problems. By including this section in the chapter we pursue two goals. First, we enlarge the supply of problems for classes. Second, after learning basic combinatorial ideas, it is useful to keep reviewing them from time to time. These reviews can be organized as separate sessions or mathematical olympiads, "math battles", et cetera. The problems below can be used for such a review session or contest.

Problem 48. How many necklaces can be made using 5 identical red beans and 2 identical blue beans? (See Figure 88.)

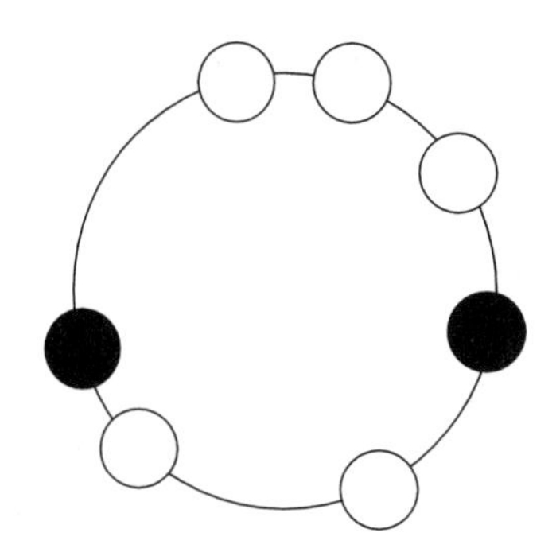

FIGURE 88

Problem 49. a) There are 30 members in a sports club, and the club's coach must choose 4 men for a 1000 meter run. How many ways are there to do that?

b) How many ways are there to choose a team of 4 members for the 100m + 200m + 300m + 400m relay?

Problem 50. How many six-letter "words" contain at least one letter A (if any sequence of letters is counted as a word)?

Problem 51. How many ways are there to draw a closed broken path made up of line segments with vertices coinciding with the vertices of a regular hexagon (the segments of the path are allowed to intersect each other)?

Problem 52. How many different four-digit numbers divisible by 4 can be written using the digits 1, 2, 3, and 4

a) if each digit can be used only once?

b) if each digit can be used any number of times?

Problem 53. A father has 2 apples and 3 pears. Each weekday (Monday through Friday) he gives one of the fruits to his daughter. In how many ways can this be done?

Problem 54. A theater group consists of 20 actors. How many ways are there to choose two groups of 6 actors each for the two performances of a play, so that none of the actors takes part in both performances?

Problem 55. Find the sum of all three-digit numbers that can be written using the digits 1, 2, 3, 4 (repetitions allowed).

Problem 56. How many ways are there to choose 6 cards from a complete deck of 52 cards in such a way that all four suits will be present?

Problem 57. How many ways are there to put three one dollar bills and ten quarters into 4 different boxes?

Problem 58. Find the number of integers from 0 through 999999 that have no two equal neighboring digits in their decimal representation.

Problem 59. How many ways are there to divide a deck of 36 cards, including 4 aces, into halves so that each half contains exactly 2 aces?

Problem 60. A rook stands on the leftmost box of a 1×30 strip of squares and can shift any number of boxes to the right in one move.

a) How many ways are there for the rook to reach the rightmost box?

b) How many ways are there to reach the rightmost box in exactly 7 moves?

Problem 61. Each side of a boat must be occupied by exactly 4 rowers. How many ways are there to choose a rowing team for the boat if we have 31 candidates, ten of whom want to be on the left side of the boat, twelve on the right side, and the other nine can sit on either side?

Problem 62*. Within a table of m rows and n columns a box is marked at the intersection of the pth row and the qth column. How many of the rectangles formed by the boxes of the table contain the marked box?

Problem 63*. A $10 \times 10 \times 10$ cube is formed of small unit cubes. A grasshopper sits in the center O of one of the corner cubes. At a given moment it can jump to the center of any of the cubes which has a common face with the cube where it sits, as long as the jump increases the distance between point O and the current position of the grasshopper. How many ways are there for the grasshopper to reach the unit cube at the opposite corner?

CHAPTER 12

Invariants

§1. Introduction

Teacher: "Let us do an experiment. As you see, there are 11 numbers on the blackboard—six zeros and five ones. You have to perform the following operation 10 times: cross out any two numbers. If they were equal, write another zero on the blackboard. If they were not equal, write a one. Do it in your notebooks in any order you wish. Done? Now I will tell you which number you have. Your result must be ... one!"

This short performance brings up a natural question: how did the teacher know which number the students would have at the end of the process described? Indeed, the operations could be performed in a number of different ways, but one thing is always the same: after each operation the sum of the numbers on the blackboard (or in the notebook) is always odd. This is quite easy to check, since this sum can increase or decrease only by 0 or 2. The original sum was odd, so, after 10 operations the only number remaining must be odd as well. Explaining this, one probably cannot help saying the magic word "invariant".

So, what is an "invariant"? Naturally, it is something that *is* invariant, that doesn't change—like the parity of the sum of the numbers in the last example.

Another example of an invariant:

Problem 1. There are only two letters in the alphabet of the Ao-Ao language: A and O. Moreover, the language satisfies the following conditions: if you delete two neighboring letters AO from any word, then you will get a word with the same meaning. Similarly, the meaning of a word will not change if you insert the combinations OA or AAOO any place in a word. Can we be sure that words AOO and OAA have the same meaning?

Solution. Note that for any permitted deletion or insertion of some combination of letters, the number of A's in the combination equals the number of O's. This means that the difference between the number of A's and the number of O's is invariant. Look at the example

$$\text{O} \to \text{OOA} \to \text{OAAOOOA} \to \text{OAOOA}.$$

In all these words the number of O's exceeds the number of A's by 1. Let us go back to the solution. The difference for the word AOO is (-1), and for the word OAA it is 1. Therefore, we cannot obtain the word OAA from the word AOO by using the permitted operations, and we cannot claim that these words are synonyms.

This solution illustrates the main idea of an invariant. We are given some objects, and are permitted to perform some operations on these objects. Then we are asked: is it possible to obtain one object from another using these operations?

To answer the question we construct a quantity that doesn't change under the given operations; in other words, it is invariant. If the values of this quantity are not equal for the two objects in question, then the answer is negative—we cannot obtain one object from another.

Let us investigate another problem:

Problem 2. A circle is divided into 6 sectors (see Figure 89), and a pawn stands in each of them. It is allowed to shift any two pawns to sectors bordering those they stand on at the moment. Is it possible to gather all pawns in one sector using such operations?

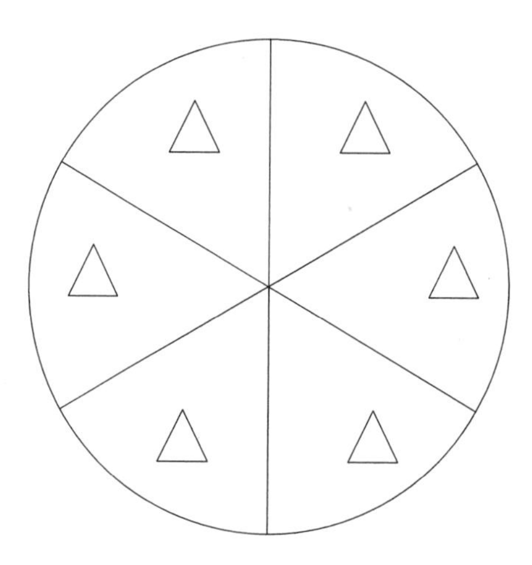

FIGURE 89

Solution. We number the sectors clockwise with the numbers 1 through 6 (see Figure 90) and for any arrangement of pawns inside the circle we consider the sum S of the numbers of the sectors occupied by pawns (counting multiplicities).

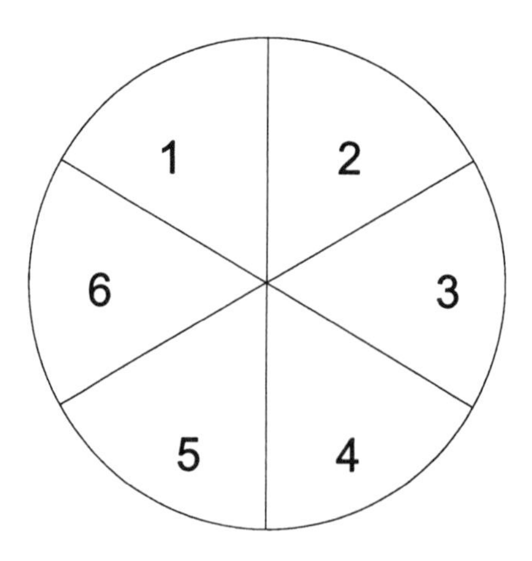

FIGURE 90

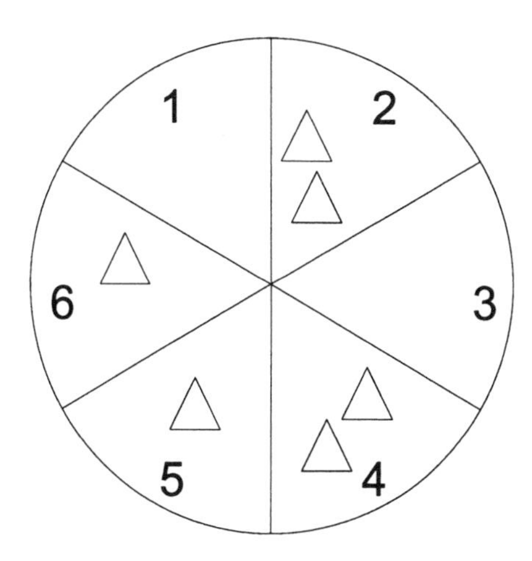

FIGURE 91

Example. For the arrangement in Figure 91 we have $S = 2+2+4+4+5+6 = 23$.

When you shift a pawn to a neighboring sector, the corresponding summand in sum S changes its parity (from odd to even, or from even to odd). Therefore, if we shift two pawns simultaneously, then the parity of S doesn't change at all—it is invariant. But for the arrangement in Figure 89 the value of S equals 21. If all the pawns are in one sector numbered A, then $S = 6A$. This is an even number, and 21 is odd. Thus, you cannot transform the initial arrangement into an arrangement with all the pawns in one sector.

Sometimes an invariant can be applied not to prove that some object cannot be obtained from a given one, but to learn which objects can be obtained from the given one. This is illustrated by the following problem.

Problem 3. The numbers 1, 2, 3, ... , 19, 20 are written on a blackboard. It is allowed to erase any two numbers a and b and write the new number $a + b - 1$. What number will be on the blackboard after 19 such operations?

Solution. For any collection of n numbers on the blackboard we consider the following quantity X: the sum of all the numbers decreased by n. Assume that we have transformed the collection as described in the statement. How would the quantity X change? If the sum of all the numbers except a and b equals S, then before the transformation $X = S + a + b - n$, and after the transformation $X = S+(a+b-1)-(n-1) = S+a+b-n$. So the value of X is the same: it is invariant. Initially (for the collection in the statement) we have $X = (1+2+\ldots+19+20)-20 = 190$. Therefore, after 19 operations, when there will be only one number on the blackboard, X will be equal to 190. This means that the last number, which is $X + 1$, is 191.

For teachers. If you hear the solution to this problem from one of your students, it will probably sound like this: at each step the sum of all the numbers decreases by 1. There are 19 steps, and originally, the sum is 210. Therefore, in the end the sum equals $210 - 19 = 191$.

This is a correct solution; however, you should explain that this problem is an "invariant" problem. The point is that in this case the invariant is so simple that it can be interpreted quite trivially. The next problem, though it is similar to Problem 3, does not allow for such a "simplification".

Problem 4. The numbers $1, 2, \ldots, 20$ are written on a blackboard. It is permitted to erase any two numbers a and b and write the new number $ab + a + b$. Which number can be on the blackboard after 19 such operations?

Hint. Consider as an invariant the quantity obtained by increasing each number by 1 and multiplying the results.

Here are a few more remarkable problems using the method of invariants.

Problem 5. There are six sparrows sitting on six trees, one sparrow on each tree. The trees stand in a row, with 10 meters between any two neighboring trees. If a sparrow flies from one tree to another, then at the same time some other sparrow flies from some tree to another the same distance away, but in the opposite direction. Is it possible for all the sparrows to gather on one tree? What if there are seven trees and seven sparrows?

Problem 6. In an 8×8 table one of the boxes is colored black and all the others are white. Prove that one cannot make all the boxes white by recoloring the rows and columns. "Recoloring" is the operation of changing the color of all the boxes in a row or in a column.

Problem 7. Solve the same problem for a 3×3 table (see Figure 92) if initially there is only one black box in a corner of the table.

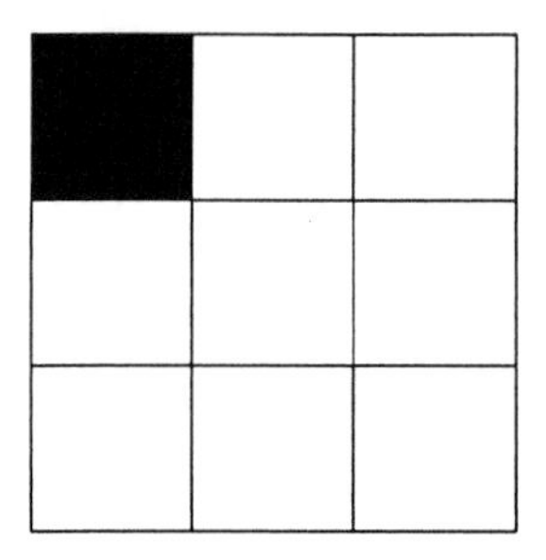

FIGURE 92

Problem 8. Solve the same problem for an 8×8 table if initially all four corner boxes are black and all the others are white.

Notice that Problem 6, unlike Problems 7 and 8, can be solved using only the idea of parity (of the number of black boxes).

Problem 9. The numbers $1, 2, 3, \ldots, 1989$ are written on a blackboard. It is permitted to erase any two of them and replace them with their difference. Can this operation be used to obtain a situation where all the numbers on the blackboard are zeros?

Problem 10. There are 13 gray, 15 brown, and 17 red chameleons on Chromatic Island. When two chameleons of different colors meet they both change their color to the third one (for instance, gray and brown both become red). Is it possible that after some time all the chameleons on the island are the same color?

Let us analyze the solution to Problem 10. How can we express the "numerical" meaning of the transformation? One way is to say that two chameleons of different colors "vanish" and two chameleons of the third color "appear". If we want to use a numerical invariant, we can easily think of a quantity depending only on the numbers (a, b, c), where a, b, and c are the numbers of gray, brown, and red chameleons respectively. The operation described means that the triple (a, b, c) turns into the triple $(a-1, b-1, c+2)$ or the triple $(a-1, b+2, c-1)$ or the triple $(a+2, b-1, c-1)$—depending on the initial color of the two chameleons that meet. It is clear that the differences between corresponding numbers in the old and new triples either do not change or change by 3, which means that the remainders of these differences when divided by 3 are invariant. Originally, $a - b = 13 - 15 = -2$, and if all the chameleons are red, we get $a - b = 0 - 0 = 0$. The numbers 0 and -2 give different remainders when divided by 3, which proves that all the chameleons cannot be red. The cases when all the chameleons are gray or brown are proved in just the same way.

For teachers. If the theme "Parity" has already been investigated and you analyzed solutions where parity played the role of an invariant, remind your students of this.

The theme "Invariants" is of rather an abstract character and even its very principle often remains vague and complicated for students. Thus one must pay special attention to the analysis of the logic of applying invariants in solving problems. It is especially important to analyze the simplest problems of the topic so that each student solves at least one problem independently. Try to illustrate the solutions by various examples, making the explanation as graphic and evident as possible. As always, introduce the new word "invariants" and the entire philosophy of invariant only after students have solved or at least investigated a few of the simplest problems using invariants.

Clearly, the main difficulty in solving problems using invariants is to invent the invariant quantity itself. This is a real art, which can be mastered only through the experience of solving similar problems. You should not restrain your fantasy. However, do not forget about the following simple rules:

a) the quantity we come up with must in fact be invariant;

b) this invariant must give different values for two objects given in the statement of a problem;

c) we must begin by determining the class of objects for which the quantity will be defined.

Here is another important example.

Problem 11. The numbers $+1$ and -1 are positioned at the vertices of a regular 12-gon so that all but one of the vertices are occupied by $+1$. It is permitted to change the sign of the numbers in any k successive vertices of the 12-gon. Is it possible to "shift" the only -1 to the adjacent vertex, if

a) $k = 3$;

b) $k = 4$;

c) $k = 6$?

Sketch of **solution**. The answer is negative in all three cases. The proof for all of them follows the same general scheme: we select some subset of vertices which satisfies the condition that there are evenly many selected vertices among any k successive ones (see Figure 93).

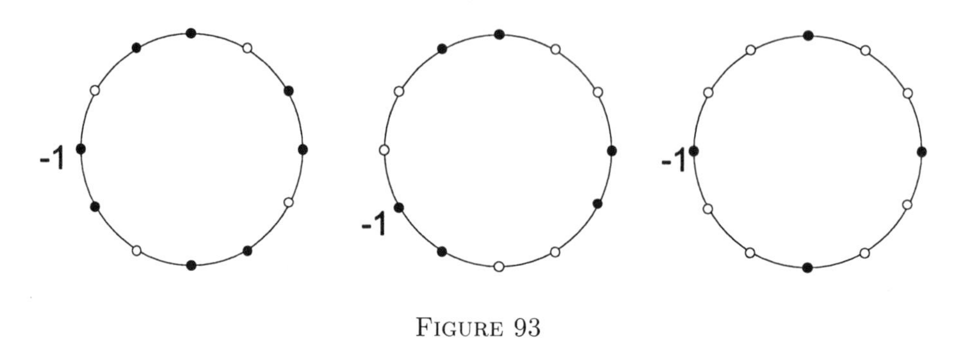

FIGURE 93

Check that this condition is true for the subsets shown in the figure.

For our invariant we take the product of the numbers on the selected vertices. Initially, it equals -1, but if the -1 has been "shifted" to the left adjacent vertex which is not among those selected, it is 1. Finally, the property of invariance for the quantity introduced follows from the property of the subset of selected vertices indicated above.

For teachers. This solution gives us a common idea in the method of invariants—to select some part of each object in which the changes caused by the permitted transformations can easily be described.

Comment. This idea also helps us to solve Problems 7 and 9.

By the way, you can ask your students one "tricky" question: We have proved that the -1 cannot be shifted to the left adjacent vertex. But can it be shifted to the right adjacent vertex?

§2. Colorings

Many problems involving invariants can be solved using one particular type of invariant: a so-called "coloring". The following is a standard example:

Problem 12. A special chess piece called a "camel" moves along a 10×10 board like a (1, 3)-knight. That is, it moves to any adjacent square and then moves three squares in any perpendicular direction (the usual chess knight's move, for example, can be described as of type (1, 2)). Is it possible for a "camel" to go from some square to an adjacent square?

Solution. The answer is no. Consider the standard chess coloring of the board in black and white. It is easy to check that a "camel" always moves from a square of one color to a square of the same color; in other words, the color of the square where the "camel" stands is invariant. Therefore, the answer is negative, since any two adjacent squares are always colored differently.

Here are some other problems using "coloring" methods in their solutions.

Problem 13. a) Prove that an 8×8 chessboard cannot be covered without overlapping by fifteen 1×4 polyminos and the single polymino shown in Figure 94.

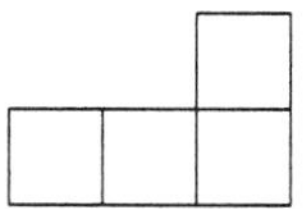

FIGURE 94

b) Prove that a 10×10 board cannot be covered without overlapping by the polyminos shown in Figure 95.

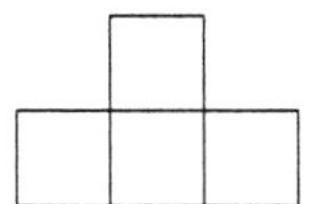

FIGURE 95

c) Prove that a 102×102 board cannot be covered without overlapping by 1×4 polyminos.

Hint to 13 b). Use the standard chess coloring of the board.

Problem 14. A rectangular board was covered without overlapping by 1×4 and 2×2 polyminos. Then the polyminos were removed from the board, but one 2×2 was lost. Instead, another 1×4 polymino was provided. Prove that now the board cannot be covered by the polyminos without overlapping.

Problem 15. Is it possible for a chess knight to pass through all the squares of a $4 \times N$ board having visited each square exactly once, and return to the initial square?

Let us analyze the solution to Problem 15. We color the squares of the $4 \times N$ board using four colors as shown in Figure 96. Assume that there exists such a "knight's tour". The coloring shown satisfies the condition that if a knight stands on a square of color 1 (2, respectively) then at the next move it will be on a square of color 3 (4, respectively).

1	2	1	2	1	2
3	4	3	4	3	4
4	3	4	3	4	3
2	1	2	1	2	1

FIGURE 96

Since the number of 1- and 2-colored squares equals the number of 3- and 4-colored squares, these pairs of colors alternate during the trip. Thus, each time the knight is on a square of color 3, it will go to a square of color 1 or 2 on the next move, and it is clear that it can go only to a square of color 1. Thus, colors 1 and 3 must alternate, which is impossible since in this case the knight would never visit the squares of color 2 or 4. This contradiction completes the proof.

For teachers. 1. A little fantasy will produce new "coloring" problems. We can investigate, for instance, some variations of polyminos and boards in Problem 13. Remember that a "coloring" method is usually used for proving a negative answer.

2. A little bit more about the "coloring" method itself: there are mathematical problems which can be solved using coloring, though they have nothing to do with the idea of invariant (see [**3**] and [**42**]). Moreover, some variations of this method can be considered as independent topics in a separate small session of a mathematical circle.

§3. Remainders as invariants

Below are seven more problems using the idea of invariants. They are remarkable in that the invariant in their solutions is a remainder modulo some natural number. This is a very common situation (see Problems 3, 7–9 which concern remainders modulo 2 (that is, parity), or Problem 11—modulo 3).

Problem 16. Prince Ivan has two magic swords. One of these can cut off 21 heads of an evil Dragon. Another sword can cut off 4 heads, but after that the Dragon grows 1985 new heads. Can Prince Ivan cut off all the heads of the Dragon, if originally there were 100 of them? (**Remark**. If, for instance, the Dragon had three heads, then it is impossible to cut them off with either of the swords.)

Problem 17. In the countries Dillia and Dallia the units of currency are the diller and the daller respectively. In Dillia the exchange rate is 10 dallers for 1 diller, and in Dallia the exchange rate is 10 dillers for 1 daller. A businessman has 1 diller and can travel in both countries, exchanging money free of charge. Prove that unless he spends some of his money, he will never have equal amounts of dillers and dallers.

Problem 18. Dr. Gizmo has invented a coin changing machine which can be used in any country in the world. No matter what the system of coinage, the machine takes any coin, and, if possible, returns exactly five others with the same total value. Prove that no matter how the coinage system works in a given country, you can never start with a single coin and end up with 26 coins.

Problem 19. There are three printing machines. The first accepts a card with any two numbers a and b on it and returns a card with the numbers $a+1$ and $b+1$. The second accepts only cards with two even numbers a and b and returns a card with the numbers $a/2$ and $b/2$ on it. The third accepts two cards with the numbers a, b and b, c respectively, and returns a card with the numbers a, c. All these machines also return the cards given to them. Is it possible to obtain a card with numbers 1, 1988, if we originally have only a card with the numbers 5, 19?

Problem 20. The number 8^n is written on a blackboard. The sum of its digits is calculated, then the sum of the digits of the result is calculated and so on, until we get a single digit. What is this digit if $n = 1989$?

Problem 21. There are Martian amoebae of three types (A, B, and C) in a test tube. Two amoebae of any two different types can merge into one amoeba of the third type. After several such merges only one amoeba remains in the test tube. What is its type, if initially there were 20 amoebae of type A, 21 amoebae of type B, and 22 amoebae of type C?

Problem 22. A pawn moves across an $n \times n$ chessboard so that in one move it can shift one square to the right, one square upward, or along a diagonal down and left (see Figure 97). Can the pawn go through all the squares on the board, visiting each exactly once, and finish its trip on the square to the right of the initial one?

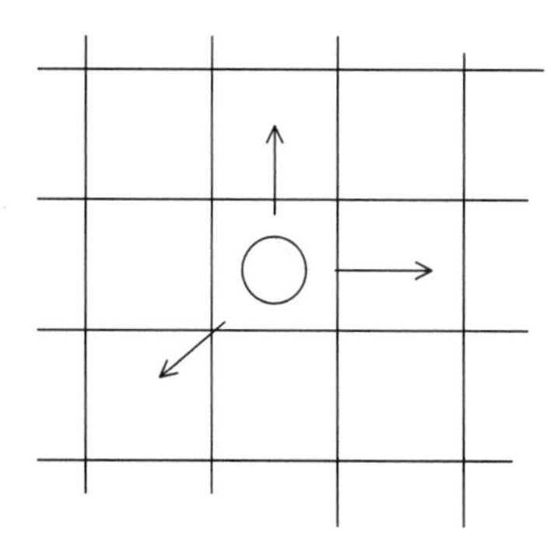

FIGURE 97

Let us try to simulate the process of solving Problem 19.

For teachers. It is very effective to present the solution as a short story telling how you came to it, how you thought of the invariant, and so on.

What do we have on the surface?—a few permitted operations are given and we are asked whether it is possible to obtain one given object from another. This picture definitely pushes us to find an invariant. Let us start the search.

The first operation maps (a, b) to $(a+1, b+1)$. What is invariant under this operation? Certainly, the difference between the numbers on the cards, since $(a+1)-(b+1) = a-b$. But the second operation changes the difference: $a/2-b/2 = (a-b)/2$—the difference is divided by two. The third operation adds the differences: $a-c = (a-b)+(b-c)$.

These observations make us think that it is not the difference between the numbers on the card which is invariant. However, it is very likely that the difference has something to do with this (so far unknown) invariant. So what can it be? Let us look more closely and try to obtain some cards from the given one.

$(5, 19) \to (6, 20)$
$(6, 20) \to (3, 10)$
$(3, 10) \to (20, 27)$
$(6, 20), (20, 27) \to (6, 27)$

Enough. Now we can observe the results of our work. We have the following cards: (5, 19), (6, 20), (3, 10), (20, 27), (6, 27). The differences for the pairs of numbers on the cards are: 14, 14, 7, 7, 21. Finally, we know what we must prove! The most plausible conjecture is that our difference $a-b$ is always divisible by 7. This fact can be proved quite easily. We need only consider the behavior of the difference under the operations permitted. It either does not change at all, or is multiplied by $\frac{1}{2}$, or two differences add up to give another one. But the difference for the card we want to obtain—(1, 1988)—equals $1 - 1988 = -1987$ and is not divisible by 7. This completes the solution, and the answer is negative.

* * *

The problems from this set are more difficult than most of Problems 1–23, but they can serve as good exercises for homework and further investigation.

Problem 23. The boxes of an $m \times n$ table are filled with numbers so that the sum of the numbers in each row and in each column is equal to 1. Prove that $m = n$.

Remark. Strange as it may seem, this is an "invariant" problem.

Problem 24. There are 7 glasses on a table—all standing upside down. It is allowed to turn over any 4 of them in one move. Is it possible to reach a situation when all the glasses stand right side up?

Problem 25. Seven zeros and one 1 are positioned on the vertices of a cube. It is allowed to add one to the numbers at the endpoints of any edge of the cube. Is it possible to make all the numbers equal? Or make all the numbers divisible by 3?

Problem 26. A circle is divided into six sectors and the six numbers 1, 0, 1, 0, 0, 0 are written clockwise, one in each sector. It is permitted to add one to the numbers in any two adjacent sectors. Is it possible to make all the numbers equal?

Problem 27. In the situation of Problem 20, find out which cards can be obtained from the card (5, 19) and which cards cannot.

Problem 28. There is a heap of 1001 stones on a table. You are allowed to perform the following operation: you choose one of the heaps containing more than 1 stone, throw away one stone from that heap and divide it into two smaller (not necessarily equal) heaps. Is it possible to reach a situation in which all the heaps on the table contain exactly 3 stones?

Problem 29. The numbers 1, 2, 3, ..., n are written in a row. It is permitted to transpose any two neighboring numbers. If 1989 such operations are performed, is it possible that the final arrangement of numbers coincides with the original?

Problem 30. A trio of numbers is given. It is permitted to perform the following operation on the trio: to change two of them—say, a and b—to $(a+b)/\sqrt{2}$ and $(a-b)/\sqrt{2}$. Is it possible to obtain the trio $(1, \sqrt{2}, 1+\sqrt{2})$ from the trio $(2, \sqrt{2}, 1/\sqrt{2})$, using such operations?

<u>**For teachers.**</u> 1. "Invariant" problems are very popular; for example, at least two or three problems in each St. Petersburg All-City Mathematical Olympiad can be solved using the idea of an invariant.

2. The idea of an invariant is widespread and permeates different fields of science. If your students are familiar with the basics of physics, then you can analyze as examples some corollaries of the law of conservation of energy, or the theorem of the conservation of momentum, et cetera.

3. Students must understand that if some invariant (or even several invariants) give the same values for two given objects, then it does not mean that the objects can be obtained from each other by the described operations. This is a standard

mistake which arises after the first acquaintance with invariants. Give your students some simple examples refuting this error.

4. Once more we recall the simplest and most standard invariants:

1) remainder modulo some natural number—Problems 3, 7–11, 17–22;

2) selecting a part of an object—Problems 7, 9, 12;

3) coloring—Problems 13–16;

4) some algebraic expression involving given variables—Problems 4, 26, 30.

CHAPTER 13

Graphs–2

This chapter continues the investigation of graphs begun in the first part of the present book. Why should we return to this topic? First, graphs are interesting and fruitful objects of study. Second, and most important, elementary reasoning about graphs allows us to get closer to more serious mathematics (this refers mostly to §§1 and 3 of the present chapter).

§1. Isomorphism

As we mentioned before, the same graph can be depicted in a number of ways. For instance, the same acquaintance scheme or system of airline routes may be pictured as figures which do not even resemble each other. Consider the following example: during a tournament involving five teams A, B, C, D, and E, team A played teams B, D, and E. In addition, team C played teams B and D, while D also played E. It is clear that both drawings in Figure 98 represent this situation.

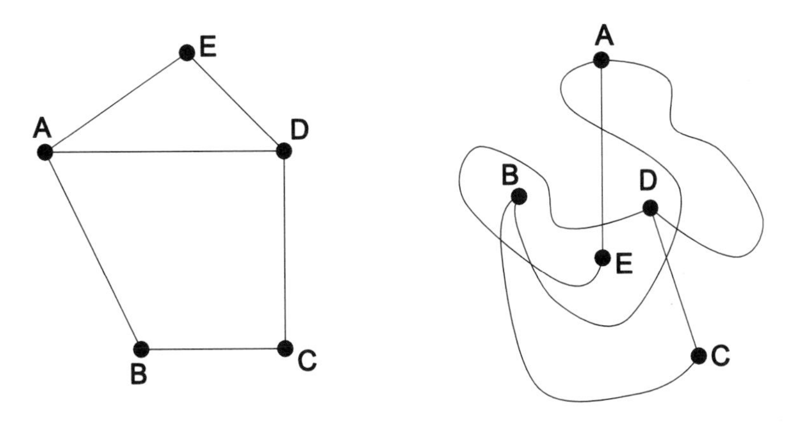

FIGURE 98

Now we give an exact definition.

Definition. Two graphs are called *isomorphic* if they have equally many vertices (say n) and the vertices of each graph can be numbered 1 through n in such a way that vertices of the first graph are connected by an edge if and only if the two vertices having the same numbers in the second graph are connected by an edge.

Now we can pay our debt and prove that the graphs shown in Figure 99 (which are copies of those in the chapter "Graphs–1") are not isomorphic.

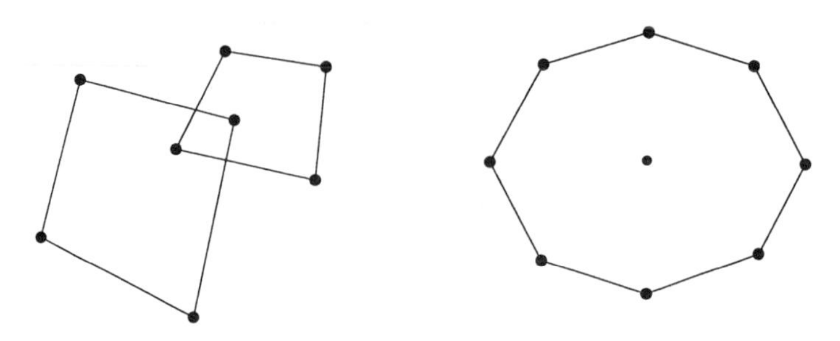

FIGURE 99

The point is that these graphs have different numbers of connected components: the first graph has three and the second has two.

We now show that isomorphic graphs must have equal numbers of connected components. It suffices to see that if two vertices of the first graph belong to one connected component, then they are connected by some path, which implies that the two corresponding vertices of the second graph are connected by some path as well and, therefore, also belong to one connected component.

Problem 1. Are the graphs in the pairs shown in Figure 100 isomorphic?

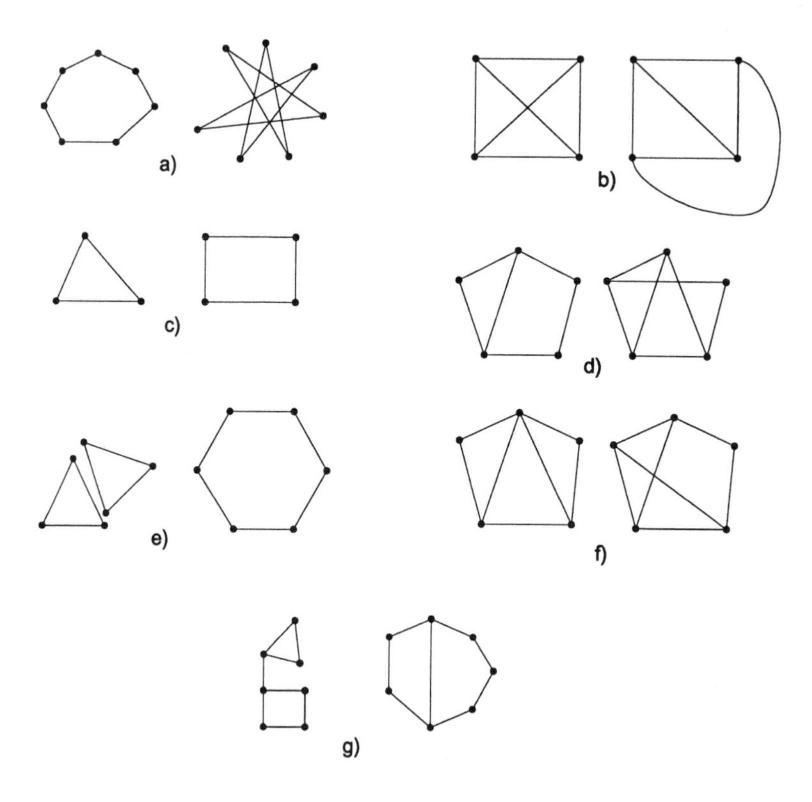

FIGURE 100

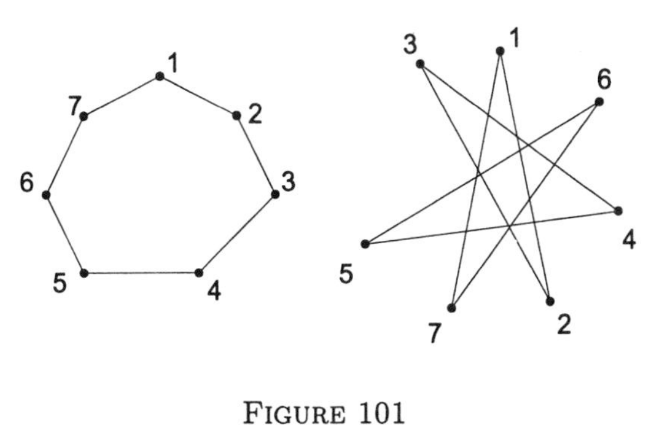

FIGURE 101

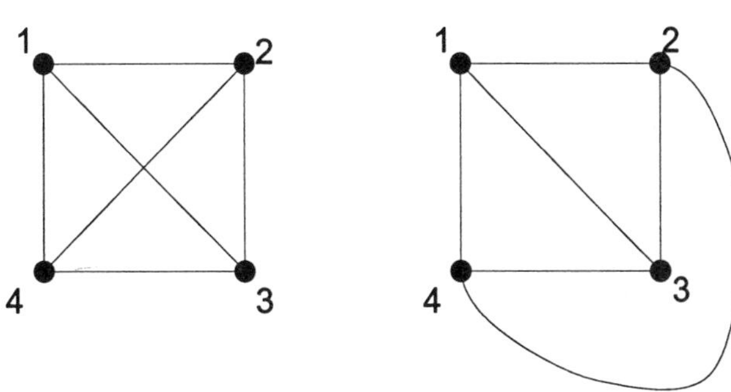

FIGURE 102

Solution. Figures 101 and 102 show that the graphs in the pairs labelled a) and b) are isomorphic. The graphs in the other pairs are not isomorphic.

Hints. c) The numbers of vertices are not equal. d) The numbers of edges are not equal. e) The numbers of connected components are not equal. f) The first graph has a vertex with four outgoing edges, but there is no such vertex in the second graph. g) There is an edge in the first graph such that after deleting this edge the graph will fall into two connected components. However, the second graph does not contain such an edge. This can be proved in another way: we consider closed paths which do not pass twice through any vertex. The first graph has two such closed paths: of length* 3 and of length 4. The second graph has three such paths: their lengths are 4, 5, and 7.

For teachers. Intuitively, students understand quite well when graphs are "identical". Thus it is interesting to listen to their own independent attempts to give a precise definition of isomorphism. Sometimes these "definitions" may involve something like "graphs are isomorphic (identical) if they have equal numbers of vertices and edges". Finding solutions to the different parts of Problem 1 can also lead to a very interesting discussion about various criteria for non-isomorphism.

One important concept was incidentally mentioned in the solution to Problem 1. Now we will give it a more accurate definition.

*The length of a path is the number of edges it consists of.

Definition. A *cycle* is any closed path in a graph which does not pass through the same vertex of the graph twice.

In discussing item g), we noticed that the graph shown in Figure 103 (a) had two cycles: 1–2–3–4–1 and 5–6–7–5, while the graph shown in Figure 103 (b) had three cycles: 1–5–6–7–1, 1–2–3–4–5–1, and 1–2–3–4–5–6–7–1.

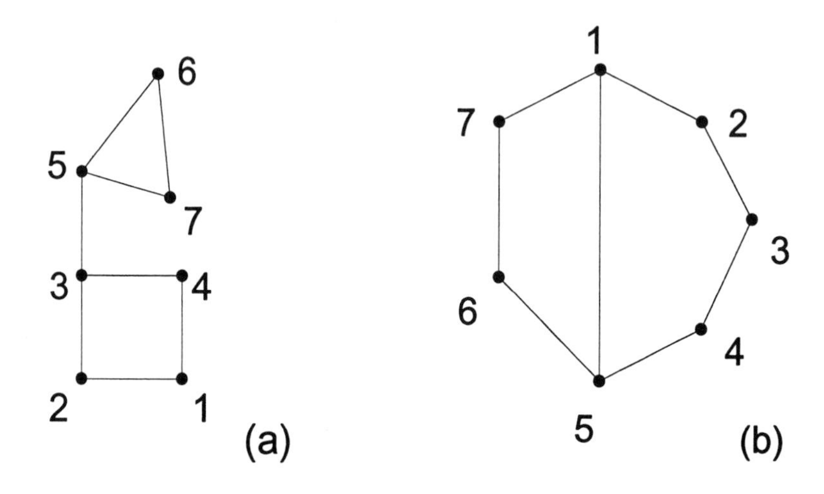

FIGURE 103

Here are two more problems connected with these concepts and definitions.

Problem 2. Prove that there does not exist a graph with 5 vertices with degrees equal to 4, 4, 4, 4, and 2.

Problem 3. Prove that there exists a graph with $2n$ vertices with degrees 1, 1, 2, 2, ... , n, and n.

Problem 4. Is it true that two graphs must be isomorphic, if

a) they both have 10 vertices and the degree of each equals 9?

b) they both have 8 vertices and the degree of each equals 3?

c) they are both connected, without cycles, and have 6 edges?

Problem 5. In a connected graph the degrees of four of the vertices equal 3 and the degrees of all other vertices equal 4. Prove that we cannot delete one edge in such a way that the graph splits into two isomorphic connected components.

§2. Trees

In this section we will discuss a certain type of graph which looks quite simple but plays an important part in the theory of graphs.

Definition. A *tree* is a connected graph without cycles.

For example, the graphs in Figure 104 are trees, while the graphs in Figures 105 and 106 are not.

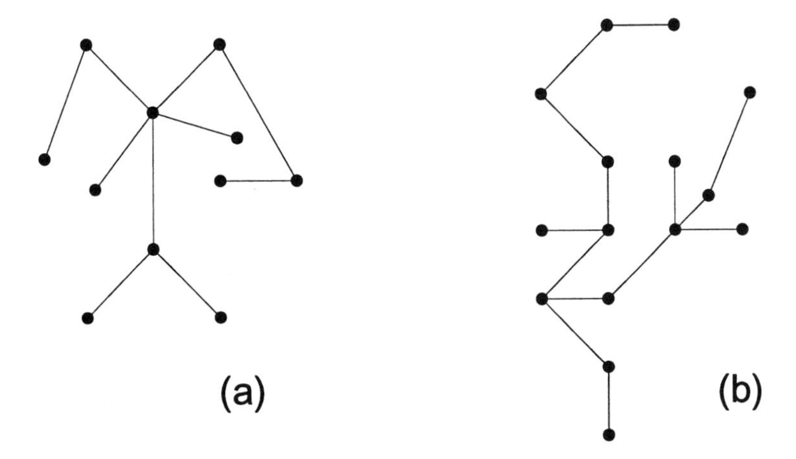

FIGURE 104

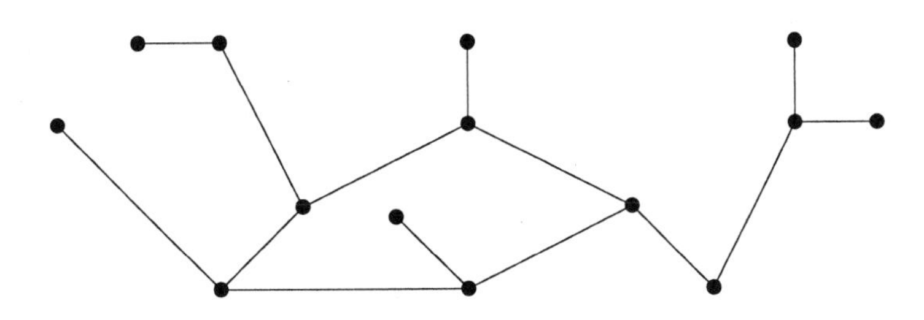

FIGURE 105

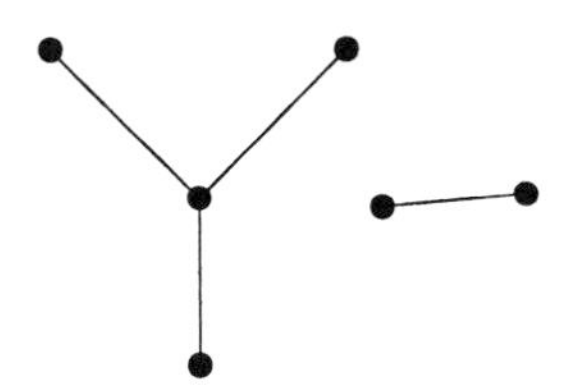

FIGURE 106

This type of graph was given its name because some of them really do resemble trees (see Figure 104 (b)).

When studying the properties of trees, the concept of a *simple path* turns out to be quite useful. We have already defined the concept of a path in the chapter "Graphs–1". A path is called "simple" if it does not include any of its edges more than once.

Problem 6. Prove that a graph in which any two vertices are connected by one and only one simple path is a tree.

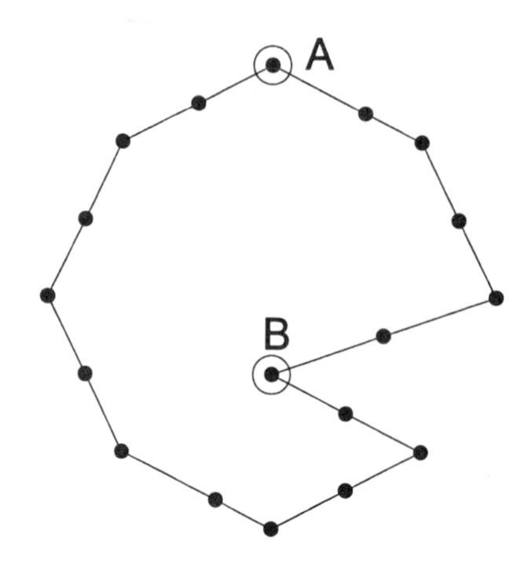

FIGURE 107

Solution. It is obvious that such a graph is connected. Let us assume that it has a cycle. Then any two vertices of this cycle are connected by at least two simple paths (see Figure 107). This contradiction proves that our assumption was wrong.

Now we will prove the converse proposition.

Problem 7. Prove that in any tree every two vertices are connected by one and only one simple path.

Solution. Assuming otherwise, suppose two vertices X and Y are connected by two different simple paths. It seems, at first, that by going from X to Y by the first path and then returning by the second path we obtain a cycle. Unfortunately, this is not completely true. The problem is that our paths may have common vertices (other than their common ends; see Figure 108), and by definition, the vertices of a cycle must not repeat. To extract a real cycle from this "improper" one, we must do the following:

1) going from X, choose the first vertex where our paths diverge (this is point A in Figure 108),

2) beyond this chosen vertex we must find, on path number 1, the first point that also belongs to path number 2 (vertex B in Figure 108).

Now the parts of our cycles between vertices A and B form a cycle.

<u>**For teachers.**</u> The first two sentences of this solution contain its basic idea. Further technicalities may be a bit too complicated for some students.

The results of Problems 6 and 7 give another possible definition of a tree, equivalent to the first.

Definition. A tree is a graph in which any two different vertices are connected by one and only one simple path.

In solving the following problems we will use either definition.

Problem 8. Prove that in any tree having at least one edge there exists a vertex which is an endpoint of exactly one edge (such a vertex is called a *pendant* vertex).

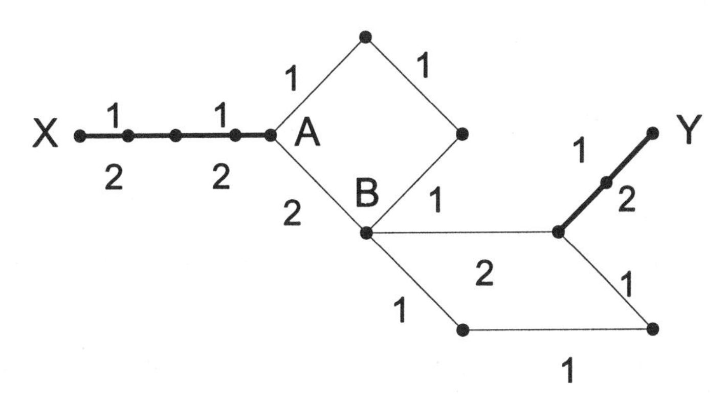

FIGURE 108

Solution. Consider an arbitrary vertex of the tree and move along any edge going out of it to another vertex. If this new vertex has degree 1, then we stay there; otherwise, we move along any other edge to another vertex and so on. It is clear that we cannot come to a vertex we have visited before—this would mean the existence of a cycle. On the other hand, since our graph has a finite number of vertices, our trip must end somewhere. But the vertex it will end in must be a pendant vertex!

The statement of Problem 8 is called The Pendant Vertex Lemma. This lemma will be used later in other solutions.

For teachers. Starting with this section, we will formulate our problems either in the language of graph theory, or more informally. Our experience shows that we should not make excessive use of either of these two forms. First, students must understand the formal language. Second, they must learn to see the real meaning of the problem behind its informal statement. Thus, it is better to use both languages freely, without being obsessed with one of them.

Problem 9. All the vertices of a graph have degree 3. Prove that the graph has a cycle.

Problem 10. Prove that if an edge (excluding its ends) is deleted from a tree, then the resulting graph is not connected.

Problem 11. There are 101 towns in Forestland. Some of them are connected by roads, and each pair of towns is connected by one and only one simple path. How many roads are there?

Solution. Translating the problem into more formal language, we can say that the graph of the roads of Forestland is a tree. This tree must have a pendant vertex. Let us delete it, together with its edge. The resulting graph is also a tree and so it has a pendant vertex, which we also delete, together with its only edge. Performing this operation 100 times we finally obtain a tree with one vertex and, of course, with no edges. Since we were deleting one edge per operation, we conclude that there were 100 edges.

In just the same way you can prove another, more general fact;

Theorem. In any tree, the number of vertices exceeds the number of edges by 1.

The converse theorem is true as well.

Problem 12. Prove that a connected graph in which the number of vertices exceeds the number of edges by 1 is a tree.

* * *

Problem 13. A volleyball net has the form of a rectangular lattice with dimensions 50×600. What is the maximum number of unit strings you can cut before the net falls apart into more than one piece?

Solution. We consider this volleyball net as a graph, its nodes as vertices, and the strings as edges. Our objective is to erase as many edges as possible while keeping the graph connected. We delete the edges one by one as long as we can. Notice that if the graph has a cycle, then we can delete any of the edges in this cycle. But a connected graph without cycles is a tree—thus, when we have obtained a tree, we cannot delete any more of the graph's edges!

Let us calculate the number of edges in our graph at this final moment. The number of vertices is the same as originally—that is, it equals $51 \cdot 601 = 30651$. On the other hand, a tree with this many vertices must have $30651 - 1 = 30650$ edges. At the very beginning we had $601 \cdot 50 + 600 \cdot 51 = 60650$ edges. Thus we can delete no more than 30000 edges—and it is easy to see that we actually can do this.

Methodological remark. We call your attention to the key idea of the solution—finding the "maximal" tree within our graph. Certainly, this "maximal" tree is not unique (see Figure 109). This method (selecting the "maximal" tree) will also help in solving the following three problems.

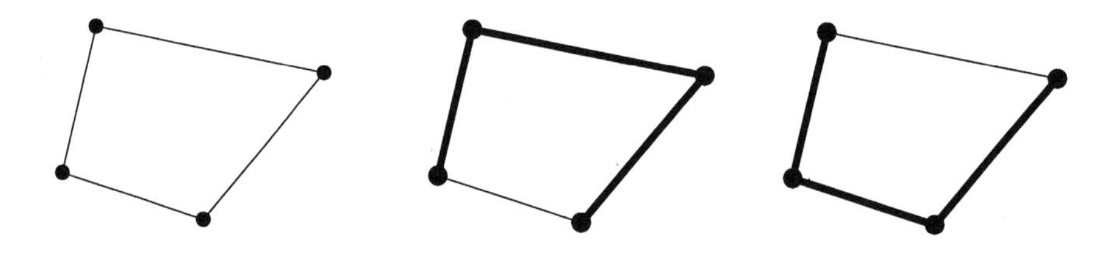

FIGURE 109

Problem 14. There are 30 towns in a country. Each of them is connected to every other by a single road. What is the maximum number of roads that can be closed in such a way that one can still reach each town from any other?

Problem 15. Prove that in any connected graph it is possible to delete a vertex, with all the edges leaving it, so that the graph remains connected.

Problem 16*. There are 100 towns in a country and some of them are connected by airlines. It is known that one can reach every town from any other (perhaps with several intermediate stops). Prove that you can fly around the country and visit all the towns making no more than

a) 198 flights;

b) 196 flights.

For teachers. A separate session can be devoted to the concept of a tree and problems connected with it. Problems using the idea of a maximal tree can be posed one by one after a thorough discussion of the idea.

§3. Euler's theorem

In this section we will prove a classical theorem named after the great XVIII century mathematician Leonard Euler. In connection with it we also discuss properties of an important type of graph, whose definition is given on the next line.

Definition. A graph that can be drawn in such a way that its edges do not intersect each other (except at their endpoints) is called *planar*.

For instance, the graph shown in Figure 110 is planar (a graph isomorphic to it is depicted in Figure 111), while the graph shown in Figure 112 is not (this fact will be proved a bit later).

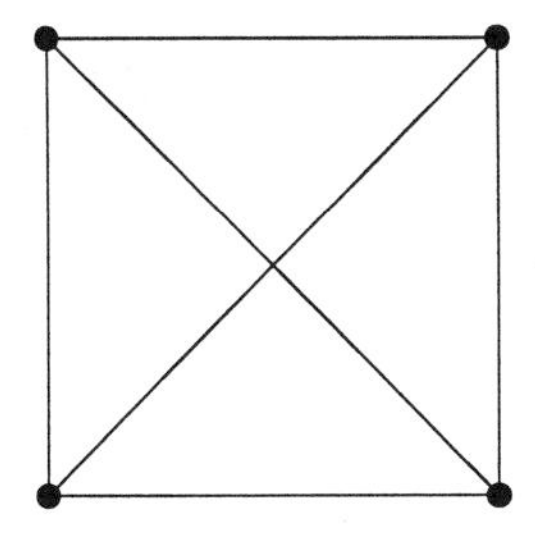

FIGURE 110

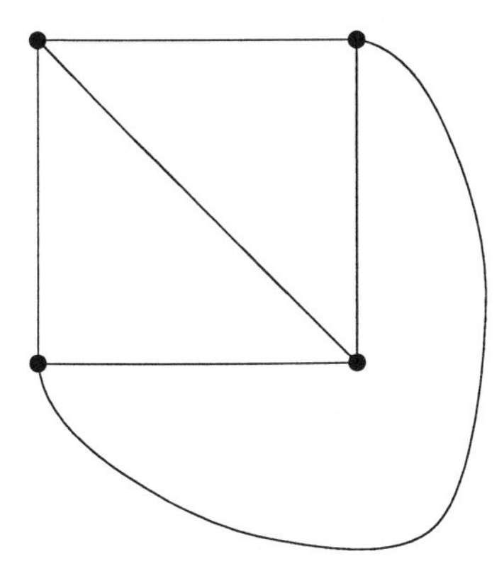

FIGURE 111

We will say that a planar graph is *properly depicted* by a figure if its edges (as shown on the figure) do not intersect at their interior points.

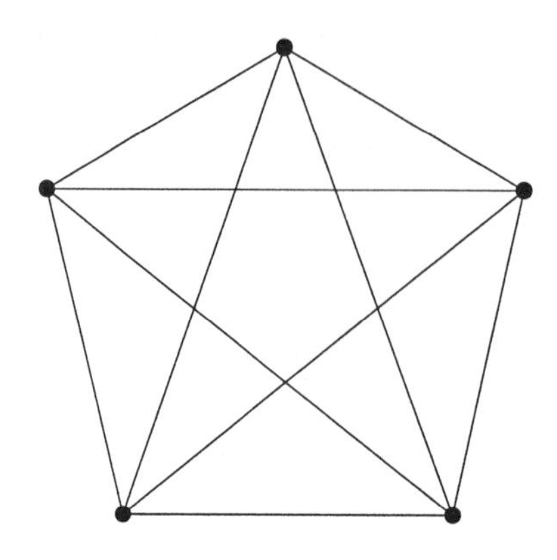

FIGURE 112

For teachers. Perhaps you have already noticed that we are not quite accurate in using the concept of a graph—we do not distinguish between different, but isomorphic, graphs. This inaccuracy is especially clearly seen in the definition of a planar graph. It is important that students understand that a graph may be planar even if some of its edges intersect in a given picture (see Figure 110).

If a graph is depicted properly, then it divides the plane into several regions called *faces*. Let us denote the number of faces by F, the number of the vertices by V, and the number of the edges of the graph by E. For the graph in Figure 111 we have $V = 4$, $E = 6$, $F = 4$ (the outer, infinite, region of the plane is counted as a face).

The following fact is then true.

Euler's theorem. For a properly depicted connected planar graph the equality $V - E + F = 2$ always holds true.

Proof. We repeat the reasoning used in the solution to the problem about the volleyball net: we delete the edges until we get a tree, keeping the graph connected. Look at the behavior of the quantities V, E, and F under such an operation. It is evident that the number of vertices does not change, while the number of edges decreases by 1. The number of faces also decreases by one: Figure 113 shows how two faces adjacent to the deleted edge merge into one new face. Thus the quantity $V - E + F$ does not change under this operation (we can say that the quantity $V - E + F$ is invariant with respect to this operation—see the chapter "Invariants"!).

Since for the resulting tree we have $V - E = 1$ (by the theorem from the previous section) and $F = 1$, for this tree we have $V - E + F = 2$, and therefore, the same equality is true for the original graph.

The equality $V - E + F = 2$ is called *Euler's formula*.

Euler's theorem is a very strong result and we can derive a lot of beautiful and interesting corollaries from it. Let us begin with a simple problem.

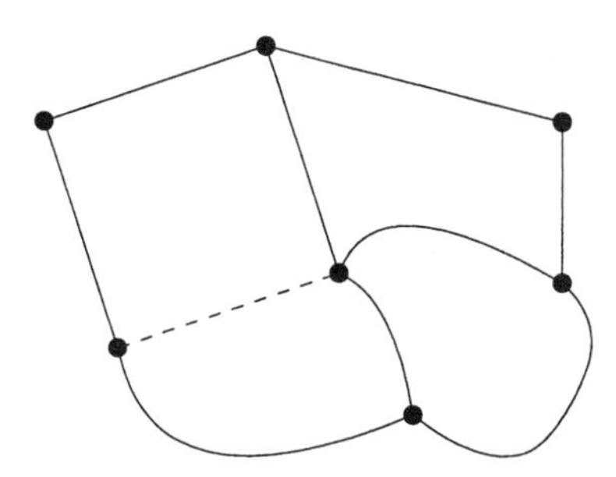

FIGURE 113

Problem 17. There are 7 lakes in Lakeland. They are connected by 10 canals so that one can swim through the canals from any lake to any other. How many islands are there in Lakeland?

The next problem is more difficult.

Problem 18. There are 20 points inside a square. They are connected by non-intersecting segments with each other and with the vertices of the square, in such a way that the square is dissected into triangles. How many triangles do we have?

Solution. We will consider the points and the vertices of the square as the vertices, and the segments and the sides of the square as the edges of a planar graph. For each region (among those into which the graph divides the plane) we calculate the number of edges on its border. Then we add up all these numbers. Since any edge separates exactly two different faces from one another, the total must be simply double the number of edges. Since all the faces are triangles, except for the outer one, which is surrounded by four edges, we get $3(F-1)+4=2E$; that is, $E=3(F-1)/2+2$. Since the number of vertices equals 24, using Euler's formula, we have

$$24-\left(\frac{3(F-1)}{2}+2\right)+F=2\,.$$

Thus $F=43$ (counting the "outside face"). So, the number of triangles our square is divided into is equal to 42.

Problem 19. Prove that for a planar graph $2E\geq 3F$.

We continue with some classical corollaries of Euler's theorem. Let us begin with an inequality.

Problem 20. Prove that for a planar connected graph $E\leq 3V-6$.

Solution. The previous problem gives $2E\geq 3F$. Substituting into Euler's formula we have $V-E+2E/3\geq 2$. Therefore $E\leq 3V-6$, as required.

Problem 21. Prove that for any planar graph (even if it is not connected) $E\leq 3V-6$.

Hint. The inequality is just the result of summing up the corresponding inequalities for each connected component.

It is remarkable that the last inequality allows us to prove the fact claimed at the beginning of this section.

Problem 22. The graph with 5 vertices, each of which is connected by an edge to every other, is not planar.

Hint. The inequality $E \leq 3V - 6$ does not hold true.

A graph in which each vertex is connected by an edge to every other vertex is called *complete*. In Figure 114 you can see the complete graph with 6 vertices.

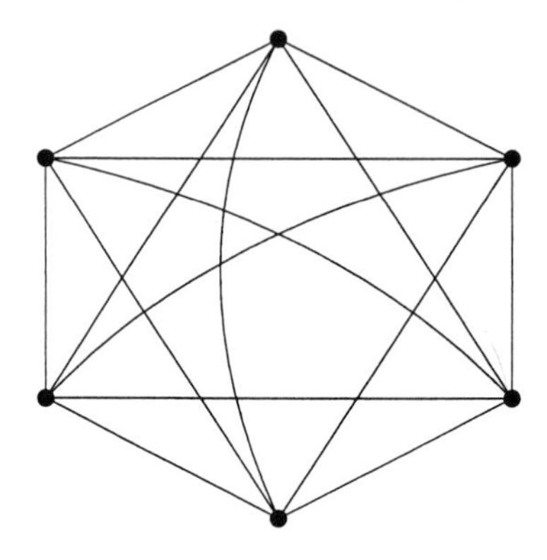

FIGURE 114

The result of Problem 22 means that a complete graph with more than 4 vertices is not planar.

Problem 23. Is it possible to build three houses and three wells, then connect each house with each well by nine paths, no two of which intersect except at their endpoints?

Hint. For the graph given in this problem the inequality $2E \geq 3F$ can be strengthened. Indeed, any cycle in this graph must be of even length, since houses and wells alternate. Assuming this graph is planar and can be properly drawn, we derive that each face in this (presumably) planar representation must have at least 4 edges on its border. Thus the same calculation as in the solution to Problem 19 brings us to the inequality $E \geq 2F$. But this inequality does not hold true, so the answer to our question is no.

Problem 24. Prove that if the degree of each of the 10 vertices of a graph is equal to 5, then the graph is not planar.

The inequality $E \leq 3V - 6$ can be used to prove the following three elegant facts.

Problem 25. Prove that in any planar graph there exists a vertex with degree no more than 5.

Problem 26. Each edge of the complete graph with 11 vertices is colored either red or blue. We then look at the graph consisting of all the red edges, and the graph consisting of all the blue edges. Prove that at least one of these two graphs is not planar.

Problem 27*. A heptagon is dissected into convex pentagons and hexagons so that each of its vertices belongs to at least two smaller polygons. Prove that the number of polygons in the tessellation is no less than 13.

For teachers. Our experience shows that the material covered in this section is quite important for graph theory. A separate session should probably be devoted to this subject.

§4. Miscellaneous problems

In this section we have gathered a few problems from various parts of graph theory. In solving them, one must combine the methods described in this chapter and in the chapter "Graphs–1" with other significant ideas. Therefore, these problems are quite difficult.

Problem 28. Prove that any connected graph having no more than two "odd" vertices (see the chapter "Graphs–1") can be drawn without lifting the pencil off the paper and so that each edge is drawn exactly once.

Sketch of the proof. Assume the graph does not have any "odd" vertices at all. We prove the fact by induction on the number of edges. The base (the graph without edges) is obvious. To prove the inductive step we consider an arbitrary connected graph all of whose vertices are "even". Since this graph has no pendant vertices, it cannot be a tree, and thus, it must contain a cycle. Now we can temporarily delete all the edges belonging to the cycle. After this, the graph splits into several connected components which have common vertices with that "temporarily deleted" cycle and satisfy the condition of the theorem (see Figure 115). By the inductive assumption each of these components can be drawn in the required way. It is clear now how to draw the original graph: we go along the cycle and, coming into a vertex belonging to a connected component, we draw the component starting at this vertex (and, certainly, finishing at the same vertex, which is important!) then continue our movement along the cycle.

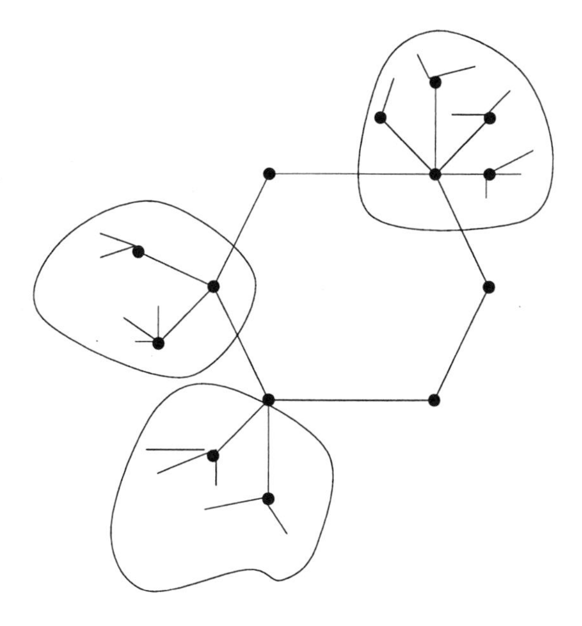

FIGURE 115

The proof for the case when our graph contains two vertices with odd degree is quite similar—we temporarily delete a path connecting these two vertices and apply the same technique.

A graph which can be drawn without lifting the pencil off the paper, so that each edge is drawn exactly once, is called an *Euler graph* or *unicursal*.

In the chapter "Graphs–1" we already proved that an Euler graph cannot have more than two "odd" vertices. The last problem allows us to combine all our results into one theorem.

Theorem. A graph is an Euler graph if and only if it is connected and has no more than two "odd" vertices. Note that we have already proved the "if" part of the theorem.

Here are three more problems.

Problem 29. Is it possible to form the grid shown in Figure 116

a) from 5 broken lines of length 8 each?

b) from 8 broken lines of length 5 each? (The length of the grid's segments is 1.)

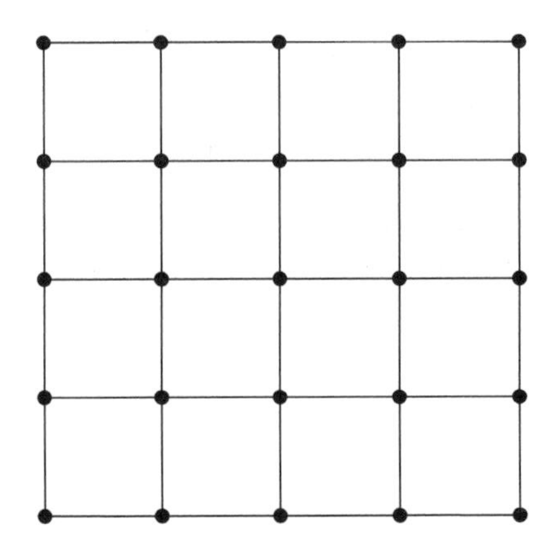

FIGURE 116

Problem 30. There are 100 circles forming a connected figure on the plane. Prove that this figure can be drawn without lifting the pencil off the paper or drawing any part of any circle twice.

Problem 31. Prove that a connected graph with $2n$ "odd" vertices can be drawn without drawing any edge more than once and in such a way that the pencil will be lifted off the paper exactly $n - 1$ times.

* * *

Problem 32. There are 50 scientists at a conference and each of them is acquainted with at least 25 of the others. Prove that there are four of them who can be seated at a round table so that each of them has two acquaintances for neighbors.

Problem 33. Each of 102 students in a school is acquainted with at least 68 other students. Prove that there are four students who have the same number of acquaintances.

Problem 34*. Let us call the length of any simple path connecting two vertices in a tree the *distance between those vertices.* Let us call the sum of all the distances between the vertex and all the other vertices of the graph the *remoteness of a vertex.* Prove that a tree containing two vertices whose remotenesses differ by 1 has oddly many vertices.

Problem 35. Alice drew 7 trees on a blackboard, each having 6 vertices. Prove that some pair of them is isomorphic.

Problem 36. In a certain country any two towns are connected either by an airline route or by railroad. Prove that

a) it is possible to choose one type of transportation so that you can reach each town from any other using only the chosen type of transportation;

b) there is a town and a type of transportation such that you can reach any other town from that one with no more than one transfer, using only the chosen type of transportation;

c) any town possesses the property indicated in b);

d) it is possible to choose a type of transportation so that you can reach each town from any other using only the chosen type of transportation and with at most two transfers on the way.

Problem 37. Each of the edges of a complete graph with 6 vertices is colored either black or white. Prove that there are three vertices such that all the edges connecting them are of the same color.

Problem 38. Each of the edges of a complete graph with 17 vertices is colored either red, blue, or green. Prove that there are three vertices such that all the edges connecting them are the same color.

Problem 39*. Each of the edges of a complete graph with 9 vertices is colored either blue or red. Prove that either there exist four vertices with all the edges connecting them blue, or three vertices with all the edges connecting them red.

Problem 40*. Each of the edges of a complete graph with 10 vertices is colored either black or white. Prove that there are four vertices such that all the edges connecting them are of the same color.

For teachers. Euler's theorem is the most important and essential fact in this section. It demands very thorough discussion and careful proof. Other problems can be used in different ways. The most difficult of them (marked with asterisks) are naturally intended for homework.

§5. Oriented graphs

The main subject of this section is the so-called *oriented graph*; that is, a graph whose edges are supplied with arrows. We will not prove any fundamental theorems about such graphs; however, the concept of an oriented graph is an important element of general mathematical culture and these graphs are common objects in mathematical problems.

Problem 41. After coming back from Fibland, Dmitri told his friends that there are several lakes connected with rivers. He also told them that three rivers flow out of every lake and four rivers flow into every lake in Fibland. Prove that he was wrong.

Solution. Every river has two ends (lakes). It flows out of one and into the other. Therefore, the sum of the number of rivers "flowing in" must be equal to the sum of the number of rivers "flowing out". But if there are n lakes in Fibland then the sum of the numbers of rivers "flowing in" is $4n$, and the sum of the numbers of rivers "flowing out" is $3n$. This contradiction completes the proof.

Problem 42. There is a capital and 100 towns in a country. Some of the towns (including the capital) are connected by one-way roads. Exactly 20 roads lead out from, and exactly 21 roads lead into, every town other than the capital. Prove that it is impossible to drive from any town to the capital and still obey the driving regulations.

* * *

In each of the following two problems the reader is requested to place arrows on the edges of a non-oriented graph to satisfy some conditions.

Problem 43. In some country each town is connected with every other town by a road. An insane king decided to impose one-way traffic on all the roads so that after you drive from any town you cannot return to it. Is this possible?

Problem 44. Prove that it is possible to place arrows on the edges of an arbitrary connected non-oriented graph and choose one vertex in such a way that one can reach any vertex from the chosen one.

* * *

The concept of an Euler graph and its main properties are used in Problems 45 and 46.

Problem 45. The degrees of all the vertices of a connected graph are even. Prove that one can place arrows on the edges of the graph so that the following conditions will be satisfied:

a) it is possible to reach each vertex from any other, going along the arrows;

b) for each vertex the numbers of "incoming" and "outgoing" edges are equal.

Problem 46. Arrows are placed on the edges of a connected graph so that for any vertex the numbers of "incoming" and "outgoing" edges are equal. Prove that one can reach each vertex from any other by moving along the arrows.

* * *

If you are familiar with the method of mathematical induction, then you can use it in solving the problems from the following set.

Problem 47. In a certain country each town is connected with every other by a one-way road. Prove that there is a town from which you can drive to any other.

Solution. We proceed by induction on the number of towns. The base of the induction is obvious. To prove the inductive step we remove one of the towns. For the rest of them, using the inductive step, we can find a town A possessing the required property. Now we replace the removed town B. If there is at least one road going to B, then A is the required town for the original problem. If all the roads lead from B, then B is the town we need.

Problem 48. Several teams played a tournament such that each team played every other team exactly once. We say that team A is stronger than team B, if either A defeated B or there exists some team C such that A defeated C while C defeated B.

a) Prove that there is a team which is stronger than any other team;

b) Prove that the team which won the tournament is stronger than any other.

Problem 49. There are 100 towns in a country. Each of them is connected with every other town by a one-way road. Prove that it is possible to change the direction of traffic on one of the roads such that after this operation each town can still be reached from any other.

Problem 50. Twenty teams played a volleyball tournament in which each team played every other team exactly once. Prove that the teams can be numbered 1 through 20 in such a way that team 1 defeated team 2, team 2 defeated team 3, ..., team 19 defeated team 20.

* * *

Finally, the last three problems of this section.

Problem 51. Some pair of teams showed equal results in a volleyball tournament in which each team played every other team exactly once. Prove that there are teams A, B, and C such that A defeated B, B defeated C, and C defeated A.

Problem 52. There are 101 towns in a country.

a) Each town is connected with every other town by a one-way road, and there are exactly 50 roads going into and 50 roads leaving each town. Prove that you can reach each town from any other, driving along at most two roads.

b) Some pairs of towns are connected by one-way roads, and there are exactly 40 roads going into and 40 roads leaving each town. Prove that you can reach each town from any other, driving along at most three roads.

Problem 53*. In Orientalia all the roads are one-way roads, and you can reach each town from any other by driving along no more than two roads. One of the roads is closed for repair, but it is still possible to drive from each town to any other. Prove that now this can be done by driving along at most three roads.

CHAPTER 14

Geometry

We often hear the question "Who needs plane geometry as it is studied in the school curriculum? This is a science for its own sake—it has no real extensions in higher mathematics and sometimes is too complicated and tricky."

One answer (which is not complete, of course) is that "school geometry" is a wonderful playground for developing logical and consistent thinking. This "science for its own sake" may be regarded as a game played by axiomatic rules created by the ancient Greeks. Euclid and his predecessors (as well as his disciples) were quite convinced that these rules adequately reflected the laws of the real world around them.

As a game, though, geometry can be compared perhaps only to chess in its complexity and elegance. Nowadays probably no one can boast about knowing all the secrets of either of these two great games of mankind. This fact (together with the limits on the size of the chapter) explains why we discuss here only some opening moves of the game.

However, we must remember that geometry also is an inalienable part of mathematics and has various links to other areas of "the queen of sciences": a good teacher will find great opportunities here to demonstrate the integrity of mathematics.

* * *

We don't want to link our explanation of the subject to any existing textbooks—we'd rather that teachers choose for themselves topics for a session depending on the level of their students. We advise teachers to draw upon the school curriculum, but not follow it blindly.

Do not be alarmed that most of the problems have virtually the same look as problems from textbooks. Indeed, it would be strange to demand "olympiad" questions in "school geometry"—the word "olympiad" itself implies an extracurricular atmosphere, stretching the limits of the curriculum. Geometry is, in fact, elaborated on quite well in numerous textbooks. Thus we often will refer the reader to other books for appropriate problem material.

§1. Two inequalities

High school geometry usually deals with precise statements like:

"Points A, B, and C lie on the same straight line."

"Altitudes of a triangle meet at one point."

"The sum of the angles of a triangle equals 180 degrees."

But for earlier study the main tools are undoubtedly the following two inequalities:

Inequality № 1. For any three points A, B, and C on the plane we have $AB + BC \geq AC$, and equality holds if and only if point B belongs to segment AC.

Inequality № 2. In a triangle the larger of any two sides is the side opposite the larger angle. That is, if in triangle ABC we have $AB > AC$, then $\angle C > \angle B$, and vice versa.

In this section we recall these inequalities once more and present a few applications.

Problem 1. Prove that if $b+c > a$, $a+c > b$, and $a+b > c$, where a, b, and c are positive numbers, then there exists a triangle with sides a, b, and c.

Problem 2. Prove that the length of median AM in triangle ABC is greater than $(AB + AC - BC)/2$.

Problem 3. Prove that you can form a triangle from segments with length a, b, and c if and only if there are positive numbers x, y, z, such that $a = x+y$, $b = y+z$, $c = x + z$.

Problem 4. Using Inequality № 2 above prove that if $AB = AC$, then angles ABC and ACB are equal.

Problem 5. In triangle ABC the median AM is longer than half of BC. Prove that angle BAC is acute.

For teachers. Problems 1–5 may be very simple for some students, especially if they have been exposed to the same topic in the school curriculum. After discussing the solutions to these easier problems, such students can tackle the following more difficult ones.

Problem 6. Prove that if you can form a triangle from segments with lengths a, b, and c, then you can do this also with segments with lengths $\sqrt{a}$, $\sqrt{b}$, $\sqrt{c}$.

Problem 7. $ABCD$ is a convex quadrilateral and $AB + BD < AC + CD$. Prove that $AB < AC$.

Problem 8. The centers of three non-intersecting circles lie on the same straight line. Prove that if a fourth circle touches all three given circles, then its radius is greater than that of at least one of the given three.

Problem 9. Let $ABCD$ and $A_1B_1C_1D_1$ be two convex quadrilaterals whose corresponding sides are equal. Prove that if $\angle A > \angle A_1$, then $\angle B < \angle B_1$, $\angle C > \angle C_1$, and $\angle D < \angle D_1$.

Problem 10. Prove that the median of a triangle which lies between two of its unequal sides forms a greater angle with the smaller of those sides.

Problem 11. Is it possible for some five-pointed star $ABCDEFGHIK$ (see Figure 117) to satisfy the inequalities: $AB > BC$, $CD > DE$, $EF > FG$, $GH > HI$, $IK > KA$?

Methodological remark. The problems of this set are a bit more difficult than Problems 1–5, but they, too, are not extremely hard to solve. Other problems using this material can be found in the chapter "The Triangle Inequality" of the present book. See also [**65**] and [**42**].

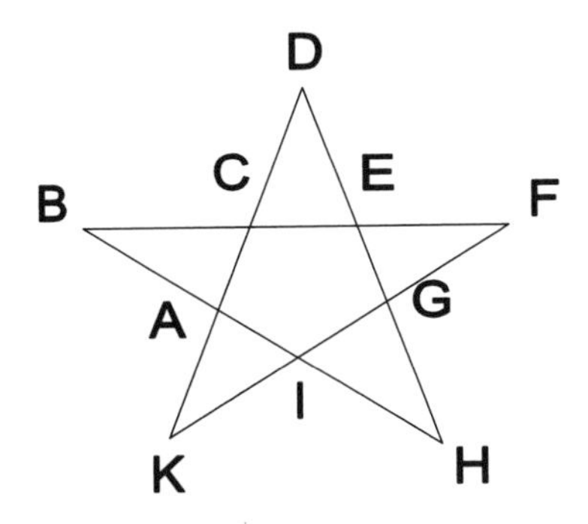

FIGURE 117

For teachers. We don't recommend that you devote an entire session to this section, but believe that it will be useful to give your students 2–3 of these problems in the course of several sessions. The goal is to implant the triangle inequalities into students' minds not as another "problem solving pattern" but as something more basic, which should be used almost unconsciously.

Let's go to the solution to Problem 8. This problem is remarkable because it can be solved using Inequality $\mathcal{N}^{\underline{o}}\,1$ or using Inequality $\mathcal{N}^{\underline{o}}\,2$.

Solution $\mathcal{N}^{\underline{o}}\,1$. We can assume that all circles touch each other externally (see Figure 118) (otherwise the radius of the fourth circle is larger than that of a circle it touches internally). Thus, if we denote the centers of the circles as A, B, C, and D, and their radii, respectively, as r_1, r_2, r_3, and R, then the triangle inequality implies that $AD + DC > AC$; that is

$$R + r_1 + R + r_3 > AC > r_1 + r_3 + 2r_2$$

and, therefore, $R > r_2$.

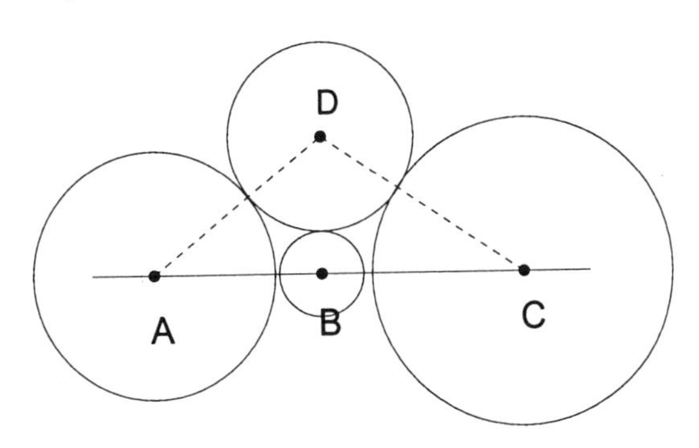

FIGURE 118

Solution $\mathcal{N}^{\underline{o}}\,2$. One of the angles DBA and DBC is non-acute and, therefore, it is the biggest in the triangle DBA or DBC. Without loss of generality we can assume this is angle DBA. Then it follows by inequality $\mathcal{N}^{\underline{o}}\,2$ that $DA > AB$; that is, $R + r_1 > AB > r_1 + r_2$, or $R > r_2$.

*   *   *

To conclude this section we list several problems whose solutions require an auxiliary idea together with the triangle inequalities.

Problem 12. Given isosceles triangle ABC with vertex angle B equal to 20 degrees, prove that a) $AB < 3AC$; b) $AB > 2AC$.

Problem 13. The perimeter of a five-pointed star whose vertices coincide with those of a given pentagon F, the perimeter of F itself, and the perimeter of the inner pentagon of the star are prime integers. Prove that their sum is no less than 20.

Remark. Don't be surprised that this is classified as a geometric problem!

Problem 14. A point is selected on each side of a square. Prove that the perimeter of the quadrilateral formed by these points is no less than twice the length of the square's diagonal.

§2. Rigid motions of the plane and congruence

This theme is very rich in interesting facts and connections with higher mathematics. Students can be led to an understanding of the role of symmetry in mathematics and to the concept of a group, which is central to much of higher mathematics. A crystallographic group, algebraic properties of the group of rigid motions of the plane, and Lobachevsky's geometry are all linked to this important subject.

For teachers. 1. We assume that students are familiar with the basic congruence theorems (see any appropriate textbook or school curriculum).

2. To begin the study of rigid motions of the plane, let the students list all the types of rigid motions they know. The definition of isometries (or rigid motions) is very simple: these are the transformations of the plane which preserve distance. It turns out there are just a few types: translations, rotations, line reflections, and glide reflections (compositions of a line reflection and a translation).

Solutions to the following problems should be carefully discussed, as they are important in later work.

Problem 15. Prove that given any two triangles, each with the same sides a, b, and c, we can make them coincide by moving one of them across the plane (and perhaps reflecting in a line). In other words, they are congruent.

Problem 16. a) If rigid motion T leaves all the vertices of triangle ABC in place, then T is the identity transformation.

b) If two rigid motions T and T' send the vertices of triangle ABC to the same points A', B', C', then T and T' are the same transformation (i.e., any point has the same image under T as it does under T').

Problem 17. a) What is the composition of two translations?

b)* Prove that any translation can be represented as a composition of two symmetries with respect to two points M and N.

c)* Consider the rigid motion which is the composition of a symmetry in line m and a translation with unit distance in a direction parallel to line m. Prove that this rigid motion is neither a rotation, nor a translation, nor a line reflection.

<u>**For teachers.**</u> A geometric solution to the last problem should be thoroughly discussed, making sure that the students learn the concept of composition. This problem is probably more appropriate for homework than for solving in the session.

Problem 18. Two equal circles are given. Is it always possible to map one of them to another by a rotation?

Problem 19. Is it possible for a rotation to map a half-plane onto itself? What about a line symmetry?

Problem 20. It is known that some figure on the plane coincides with itself after a rotation of 48 degrees about point 0. Is it necessarily true that it coincides with itself after a rotation of 72 degrees about the same point?

<u>**For teachers.**</u> 1. The study of rigid motions and their compositions presents an excellent opportunity for a more general discussion of mappings and composition, with illustrations from both algebra and geometry.

2. At some moment the students should be asked whether they know how to "construct" (with compass and straight edge only) the rigid motions of the plane. Can they, for instance, construct the image of a circle under a given line reflection?

The next topic is the use of rigid motions for the solution of geometric problems. It deserves its own (very heavy) book. We will try to give some examples and introduce some basic ideas of the subject.

Problem 21. Point A is given inside a triangle. Draw a line segment with end-points on the perimeter of the triangle so that the point divides the segment in half.

Problem 22. A sheet of paper is given, with two lines. The lines are not parallel, but intersect at a point off of the sheet of paper. Construct an angle which is twice as large as that made by the two lines.

Problem 23. Inscribe a pentagon in a given circle so that its sides are parallel to five given straight lines.

Problem 24. In trapezoid $ABCD$ ($AD \parallel BC$) M and N are the midpoints of the bases, and line MN forms equal angles with lines AB and CD. Prove that the trapezoid is isosceles.

Problem 25. Points P, Q, R, and S are taken on sides AB, BC, CD, and DA of square $ABCD$ respectively so that $AP : PB = BQ : QC = CR : RD = DS : SA$. Prove that $PQRS$ is a square.

Problem 26. Point P and two parallel lines are given on the plane. Construct an equilateral triangle with one of its vertices coinciding with P and two others lying on the given lines.

Problem 27. On a given line find a point M such that a) the sum of the distances from M to two given points is a minimum; b) the difference between these two distances is a maximum.

We repeat that it is virtually impossible to exhaust this beautiful geometric topic, and refer the reader to the books [**42, 65, 71**] and [**68**], where he or she can find dozens of interesting and rather difficult problems in this area. Below are just a few more examples involving the properties of rigid motions and symmetries of figures on the plane.

Problem 28. Prove that if a triangle has two axes of symmetry, then it has at least three axes of symmetry.

Problem 29. Which letters of the English alphabet have an axis of symmetry? A center of symmetry?

Problem 30. Does there exist a pentagon with exactly two axes of symmetry?

Problem 31. Find the set of all points X on the plane such that a given rotation sends X to X', and the straight line XX' passes through a given point S.

* * *

Let us discuss now the proof of the rather difficult Problem 23.

Instead of the five given lines $L_1, L_2, \ldots, L_5$ we consider lines $K_1, K_2, \ldots, K_5$, perpendicular to them and passing through the center of the circle. Then it is clear that lines L_i and AB (for any two points A and B on the circle) are parallel if and only if A and B are symmetric with respect to line K_i. It remains to find a point M on the circle which will remain fixed after reflection in all five lines $K_1, K_2, \ldots, K_5$. Since the composition of five line reflections is a line reflection again (with its axis passing through the center of the circle!), such a point must exist and can be found as one of the points where the axis of symmetry and the circle meet.

Question. There is a small gap in the solution above. Find and fix it.

§3. Calculating angles

What do you need to know to calculate the angles of geometric figures? We give only the most basic facts here:

(1) the sum of the angles in any triangle is 180 degrees;

(2) a pair of vertical angles are equal;

(3) angles lying along a straight line add up to 180 degrees (see Figure 119);

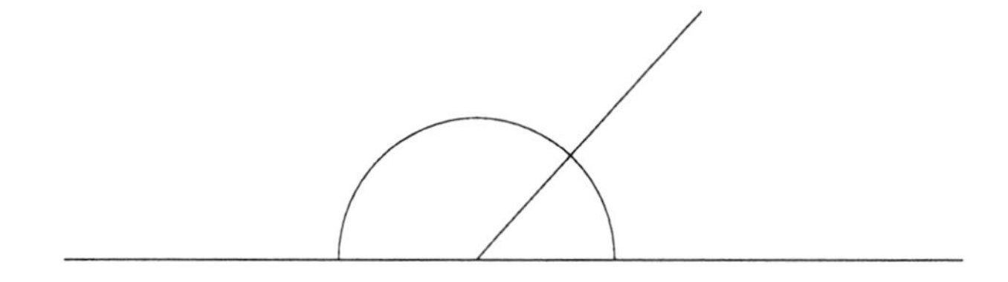

FIGURE 119

(4) an inscribed angle equals half the central angle which intercepts the same arc of a circle, and as a corollary, we have that

(5) two inscribed angles intercepting the same arc of a circle are equal;

(6) rigid motions of the plane do not change angle measure.

* * *

Problem 32. Angle bisector BK is drawn in isosceles triangle ABC, with angle A equal to 36 degrees. Prove that $BK = BC$.

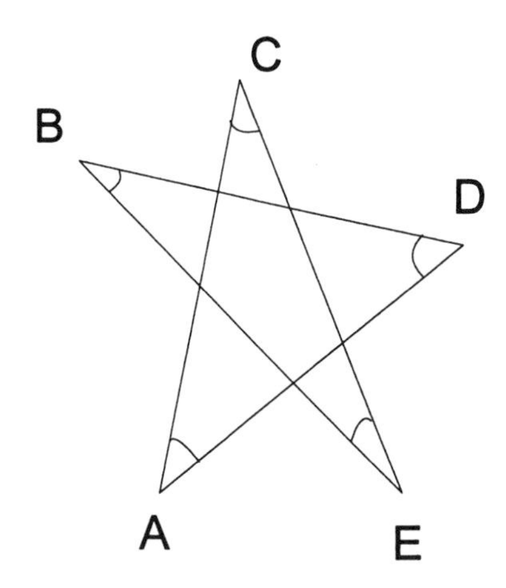

FIGURE 120

Problem 33. Prove that the sum of the angles at the vertices of a five-pointed star (see Figure 120) equals 180 degrees.

Problem 34. Can two angle bisectors in a triangle be perpendicular?

Solution to Problem 32. Since $\angle C = 72°$ and $\angle B = 72°$, we have $\angle KBC = 36°$ and, therefore, $CKB = 72°$. Thus triangle KBC is isosceles and $BK = BC$.

Solution to Problem 33. Clearly, $180° = \angle EBD + \angle BED + \angle BDE = \angle E + \angle B + \angle D + \angle FED + \angle FDE$. Since $\angle FED + \angle FDE = 180° - \angle EFD = 180° - \angle CFA = \angle A + \angle C$, we have $180° = \angle E + \angle B + \angle D + \angle A + \angle C$.

So, we can see that the method is as follows: we denote certain angles as α, β, γ, δ, ... , then we express all the other angles in terms of these. Using facts (1)–(6) we eventually come to the required result.

<u>**Methodological remark.**</u> Here we have a subtle alternative. On the one hand, if we denote just one or two angles by letters, we might not be able to express the remaining angles and parameters of the problem as functions of the variables introduced. On the other hand, introducing too many angles as unknown variables will make our drawing messy and could make our goal unattainable (since possible correlations between the given angles can become obscured).

<u>**For teachers.**</u> Usually, the choice of the "starting" angles (and of their number) constitutes one of the most important parts of the solution. To learn how to introduce the starting variables (in our case, the angles) is one of the crucial ingredients of mathematical culture on the "olympiad" level. Only vast experience or fully developed methodological thinking can help the students to find the right choice.

Here are five more problems (see also [**65, 70**], and any school textbooks).

Problem 35. Chords AB and CD in circle S are parallel. Prove that $AC = BD$.

Problem 36. The ratio of three consecutive angles in an inscribed quadrilateral is $2 : 3 : 4$. Find their values.

Problem 37. In triangle ABC $\angle A = 90°$. Median AM, angle bisector AK, and altitude AH are drawn. Prove that $\angle MAK = \angle KAH$.

Problem 38. Square $ABCD$ is given. A circle with radius AB and center A is drawn. This circle intersects the perpendicular bisector of BC in two points, of which O is the closest to C. Find the value of angle AOC.

Problem 39. Two circles intersect at points A and B. AC is a diameter of the first circle, and AD is a diameter of the second. Prove that points B, C, and D lie on the same straight line.

The solution of Problem 37 is rather standard.

Let us denote angle BCA as α (see Figure 121). Then, since $AM = MC$, we get $\angle MAC = \alpha$ and, therefore, $\angle MAK = 45° - \alpha$ (and we see now that α must be an angle which is not greater than 45 degrees). Further, $\angle ABC = 90° - \alpha$, which implies $\angle BAH = \alpha$. Thus $\angle KAH = 45° - \alpha = \angle MAK$.

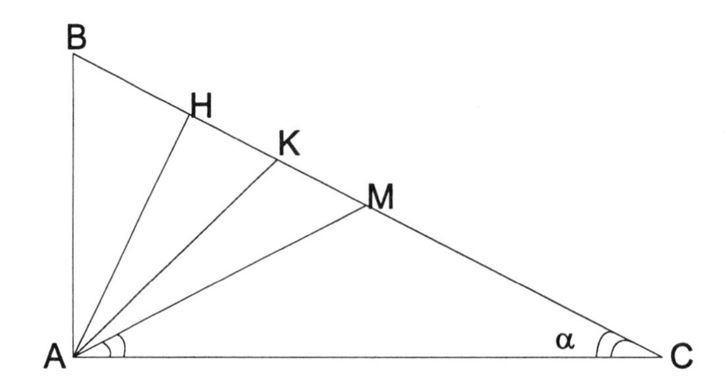

FIGURE 121

For teachers. 1. There are many more difficult problems on calculating angles. In fact, most problems in school geometry involve calculating angles, so it is a good habit to write down neatly the values of the angles on a geometric drawing.

2. We believe that this is not a theme for a separate session. It would suffice to make up a few series of problems and submit them for solution at several different sessions.

§4. Area

This topic is as extensive as many other geometric topics, so we will concentrate more on methods. What are the main principles used in solving problems involving area?

We will name just the most basic facts:

a) The main properties of area: it is invariant under rigid motions of the plane, and if a figure is split into two disjoint figures, then its area equals the sum of the smaller areas.

b) The main formulas: $S = ah/2$ (where S is the area of a triangle, a is its side, and h is the altitude perpendicular to that side); $S = rp$ (where S is the area of a triangle, p is half its perimeter, and r is the radius of the inscribed circle), et cetera.

c) Basic inequalities such as $S \leq ab/2$ (where S is the area, and a and b are two sides of a triangle); see Problem 40.

d) If there are expressions like ab or $a^2 + b^2$ present in the statement of a geometric problem (that is, expressions of degree 2), then it is likely you should try to use area to solve the problem.

e) If there are expressions present in the statement of a problem which can be naturally linked with each other via some area formula, then write this formula down and examine it—this will never do any harm.

From now on, we will frequently denote the area of figure F by $|F|$.

<u>**For teachers.**</u> Items a), b), and c) give you a wonderful opportunity to find out the extent of your students' knowledge in this specific field of geometry.

We do not advise that you turn the entire session into a discussion of this topic, although the basic concept of this theme must be learned quite thoroughly. This cannot be done during a few sessions. For example, only after a year of successful seminars should you investigate more theoretical parts of this branch of geometry (for example, axioms of area).

Problem 40. The lengths of the sides of a convex quadrilateral are a, b, c, and d (listed clockwise). Prove that the area of the quadrilateral does not exceed a) $(ab + cd)/2$; b) $(a + b)(c + d)/4$.

Problem 41. Is it possible that the ratio of the three altitudes of a triangle is $1 : 2 : 3$?

Problem 42. A triangle of area 1 has sides of lengths a, b, and c where $a \geq b \geq c$. Prove that $b \geq \sqrt{2}$.

Problem 43. If all the sides of a triangle are longer than 1000 inches, can its area be less than 1 square inch?

Solution to Problem 41. Let S be the area of a triangle, and a, b, and c the lengths of its sides. Then the altitudes are equal to $2S/a$, $2S/b$, and $2S/c$, and $a : b : c = 1 : 1/2 : 1/3$, which contradicts the triangle inequality.

How did we think of introducing the area and the sides of the triangle? See item e) in the beginning of this section.

The three preceding problems are good representatives of the subtopic "area and inequalities". Below are a few problems dealing with more "exact" calculations.

Problem 44. Points K, L, M, and N are the midpoints of the sides of quadrilateral $ABCD$. Prove that $2|KLMN| = |ABCD|$.

Problem 45. Find the area of convex quadrilateral $ABCD$, if line AC is perpendicular to line BD, $AC = 3$, and $BD = 8$.

Problem 46. Triangle ABC is given. Point A_1 lies on segment BC extended beyond point C, and $BC = CA_1$. Points B_1 and C_1 are constructed in the same way (see Figure 122). Find $|A_1B_1C_1|$, if $|ABC| = 1$.

Problem 47. Point M lies within triangle ABC. Prove that areas of triangles ABM and BCM are equal if and only if M lies on median BK.

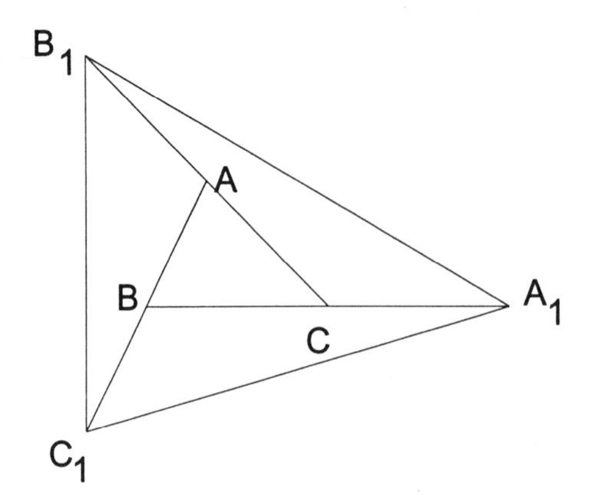

FIGURE 122

Finally, a series of problems whose solutions require not only calculations but also proof.

Problem 48. Prove that if two convex quadrilaterals have the same midpoints for all their sides, then their areas are equal.

Problem 49. The diagonals of trapezoid $ABCD$ (with $BC \parallel AD$) meet at point O. Prove that triangles AOB and COD have equal areas.

Problem 50. Prove that the sum of the distances from a point inside an equilateral triangle to its sides does not depend on the position of the point.

For teachers. You can see from the solutions to Problems 44–47 that even if a problem is about area, standard geometric ideas may apply as well: congruent triangles, similarity, Thales's theorem.[1] This is quite natural. Using such examples, you can make the students see that the solution to a problem usually includes several ideas. It is rare that an "olympiad" problem can be solved in one move. This is a very general principle of problem solving, applicable to higher mathematics as often as to olympiads and other contests.

§5. Miscellaneous

This section consists of three sets of problems on topics not discussed in the present chapter. These problems are intended mostly for homework and can be considered as exercises to accompany more detailed study (see also the other books on geometry in the list of references).

Set 1. Constructions

Problem 51. Construct a triangle if you know

a) its base, altitude, and one of the angles adjacent to the base;
b) the three midpoints of its sides;
c) the lengths of two of its sides and the median to the third side;
d) two straight lines which contain its angle bisectors, and its third vertex.

[1]Thales's theorem, one of the oldest geometric results on record, states that a line parallel to one side of a triangle divides the other two sides in proportion.

Problem 52*. Find the midpoint of a segment

a) using a compass only;

b) using only a two-sided ruler (with two parallel sides), whose width is less than the length of the segment;

c) using only a two-sided ruler, whose width is greater than the length of the segment.

Problem 53. Segment AB is given in the plane. An arbitrary point M is chosen on the segment, and isosceles right triangles AMC and BMD are constructed on segments AM and MB (as their hypotenuses) so that points C and D are on the same side of AB. Find the set of midpoints of all such segments CD.

Problem 54. A certain tool for geometric construction can be used to

a) draw a straight line through two given points;

b) erect a perpendicular to a given line at a point lying on the line.

Show how to use this tool to drop a perpendicular from any given point to any given line.

Problem 55. Peter claims that the set of points on the plane which are equidistant from a given line and a given point is a circle. Is he right?

Set 2. Calculations

Problem 56. Find an error in the following "proof" of the fact that in a right triangle, the hypotenuse has the same length as a leg (see Figure 123). Point M is the intersection of the bisector of angle C and the perpendicular bisector of segment AB. Points K, L, and N are the feet of the perpendiculars dropped from M to the sides of the triangle. Triangles AMK and MKB are congruent since they have equal hypotenuses and equal legs. Thus, $AM = MB$, and triangles ALM and MNB are congruent for the same reason. Therefore, $AL = NB$ and $AC = AL + LC = NB + CN = BC$.

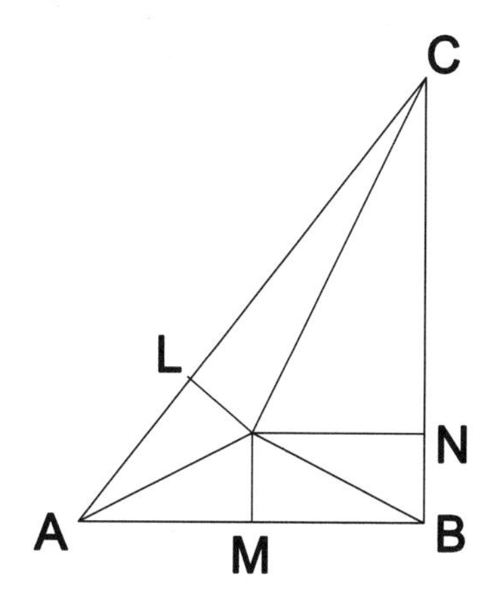

FIGURE 123

Problem 57. $ABCD$ is a quadrilateral such that $BC = AD$, and M and N are the midpoints of AD and BC respectively. The perpendicular bisectors of segments AB and CD meet at point P. Prove that P also lies on the perpendicular bisector of segment MN.

Problem 58. A right triangle with an acute angle of 30° is given. Prove that the length of that part of the perpendicular bisector of the hypotenuse which lies inside the triangle equals one third of the greater leg of the triangle.

Problem 59. Altitudes AA_1, BB_1, CC_1, and medians AA_2, BB_2, CC_2 are drawn in triangle ABC. Prove that the length of the broken line $A_1B_2C_1A_2B_1C_2A_1$ equals the perimeter of triangle ABC.

Set 3. Similarity

Problem 60. One of the diagonals of an inscribed quadrilateral is a diameter of its circumcircle. Prove that the projections of any two opposite sides of the quadrilateral onto the other diagonal are equal.

Problem 61. Arc AB, measuring 60°, is given on a circle with center O, and point M is chosen on the arc. Prove that the straight line that passes through the midpoints of segments MA and OB is perpendicular to the line passing through midpoints of segments MB and OA.

Problem 62. Angle bisector AD is drawn in triangle ABC. Prove that $CD/DB = CA/AB$.

Problem 63. In isosceles triangle ABC perpendicular HE is dropped from the midpoint H of BC to side AC. Prove that if O is the midpoint of HE, then lines AO and BE are perpendicular.

Epilogue

1. The topic "Geometric inequalities", which we merely touched on in §1, can be developed much further. Many very beautiful pearls of geometry, such as isoperimetric inequalities (see [**42**], Chapters 15 and 16), follow from the simple triangle inequalities.

2. Don't forget that in the present chapter we not only give some particular topics and problems but also indicate possible directions of how a session can proceed. We hope you will find the methodological remarks in this chapter useful.

3. "Calculating angles" is only one theme representing the entire realm of "computational geometry", which can also be studied.

4. Unlike other topics in this chapter, "Area" begins with rather difficult problems rather than with mere exercises. You can find many simpler problems in textbooks.

5. **For teachers.** If a problem cannot be "captured" with a quick attack, then, perhaps, it may fall to a long and careful siege (which may take days or weeks). Such a siege is conducted by combining different methods and various ideas, using calculations, or through a gradual accumulation of facts.

6. Many beautiful themes in plane geometry, which are quite accessible to students, are not described in the present chapter. These include similarity and its applications, inscribed and circumscribed polygons, interesting points in a triangle, numerical relations, et cetera. They deserve to be studied, but we cannot include them all here since this would turn this chapter into a huge and dull reference. We have just tried to outline the basic aspects of this peculiar game and science, dropping only similarity from the most important topics. In spite of the diversity of the material covered here, we believe that it is very important to show the students

its integrity, and also the links which connect its various areas with each other and with the other branches of mathematics and science.

7. Quick reference guide:

a) A large number of problems from the usual geometry curriculum can be found in books [**42, 44, 34, 65, 70, 64**] and others. However, the first two of them ascend too quickly to the heights and are not recommended for most students.

b) Books [**67**] and [**71**] are excellent for those who like to see mathematics as algebra rather than geometry. They also provide very good problems on transformations of the plane.

c) Books [**68**] and [**69**] are two of those rare mathematical books which are simply pleasant to read. However, we face the question "for whom are these books written?" It seems that the books were intended for people who already know their contents. These books can be read again and again—the vividness of their explanations will attract all readers interested in the subject. Also instructive are the geometric chapters (and others, too) in Martin Gardner's books [**5, 6**] and [**7**].

CHAPTER 15

Number Bases

§1. What are they?

Any student can say that "2653" stands for the number "two thousand six hundred fifty three", whatever that may mean. How do they know this? We are all accustomed to the following way of writing numbers: the last digit denotes the number of units in the given number, the next-to-last—the number of tens, the third last—the number of hundreds, and so on (though this is a bit ambiguous, since the number of units in 2653 is, in a way, not 3, but 2653!). This way of writing numbers (and interpreting strings of digits) is called in brief a *number base system*. Thus, writing "2653", we think of the number $\mathbf{2}\cdot 1000+\mathbf{6}\cdot 100+\mathbf{5}\cdot 10+\mathbf{3}\cdot 1$, or shortly, $\mathbf{2}\cdot 10^3+\mathbf{6}\cdot 10^2+\mathbf{5}\cdot 10^1+\mathbf{3}\cdot 10^0$. We print the digits of the number in boldface to make it easier to distinguish them from other numbers.

We can easily see that the number ten plays a special part in this representation: any other number is written as a sum of different powers of ten with coefficients taking values 0 through 9. This is why this system is called "decimal" (from the latin word for "ten"). To write a number we use the ten special symbols 0, 1, 2, 3, 4, 5, 6, 7, 8, and 9, called digits. They denote the numbers from zero to nine. The next number, that is, ten, is regarded as a unit of the next level and is written with two digits: 10, which, roughly speaking, means "add up one times ten and zero times one".

Now, what if we used some other number, say, six? Analogously, we would need six symbols as digits. We can take the six familiar symbols 0, 1, 2, 3, 4, and 5, which will denote the numbers from zero to five. The number six will be the unit of the next level, and, therefore, it will be written as 10. Proceeding with this analogy, we can represent each natural number as the sum of different powers of six with coefficients from 0 to 5. For instance (all numbers are written in the decimal system):

$$
\begin{aligned}
7 &= \mathbf{1}\cdot 6^1+\mathbf{1}\cdot 6^0,\\
12 &= \mathbf{2}\cdot 6^1+\mathbf{0}\cdot 6^0,\\
35 &= \mathbf{5}\cdot 6^1+\mathbf{5}\cdot 6^0,\\
45 &= \mathbf{1}\cdot 6^2+\mathbf{1}\cdot 6^1+\mathbf{3}\cdot 6^0.
\end{aligned}
$$

Thus in our new number system (which is called the "base six system") we write the number 7 as "11", the number 12 as "20", 35 as "55", and 45 as "113".

It is easy to see that we can write any natural number in the base six system. We show how to do this for the number 450 (in this example, as earlier, all the given numbers are written in the decimal system unless enclosed in quotes).

The largest power of six that does not exceed 450 is 216. Dividing 450 by 216, we have a quotient of 2 (and a remainder of 18). Thus the first digit of the numeral 450 for the base six system is **2**. Now we take the remainder 18 and divide it by the next smaller power of six—at the previous stage we divided by $6^3 = 216$, and now we divide by $6^2 = 36$. The quotient is 0, hence the second digit is **0**. The remainder is 18 and we divide this by the next smaller power of six; that is, by $6^1 = 6$. Now we see that the next digit is **3** (the remainder is 0). Therefore, the last digit (the quotient after the division by $6^0 = 1$) is 0. Finally, the base six representation of 450 is "2030".

While building our new system, we have not used any particular properties of the number 6, whatever they may be. Similarly, starting with any natural number n greater than 1, we can build a base n number system, in which the digits of a number are connected with its representation as a sum of powers of n. In this system, the number n is called the *base*. To avoid ambiguity, we will write the base of the system as a subscript (in decimal notation) at the right end of the numeral. Using this notation, we can rewrite the equalities indicated earlier as:

$$7_{10} = 11_6, 12_{10} = 20_6, 35_{10} = 55_6, 45_{10} = 113_6 \ .$$

Exercise 1. How many digit symbols do we need for a
a) binary (that is, base 2) system;
b) base n number system?

To write a number in the base n system, we must represent it in the following form:

$$a_k n^k + a_{k-1} n^{k-1} + \ldots + a_2 n^2 + a_1 n^1 + a_0 n^0 \ ,$$

where each a_i takes values from 0 to $n-1$, and a_k is not equal to zero (although the last restriction is, strictly speaking, not necessary).

Exercise 2. Write in decimal notation the numbers 10101_2, 10101_3, 211_4, 126_7, and 158_{11}.

Exercise 3. Write the number 100_{10} in the systems with bases 2, 3, 4, 5, 6, 7, 8, and 9.

Exercise 4. In a system whose base is greater than 10, we need more than ten digit symbols, so we must invent some. For example, in the base 11 system, we might use "A" to represent the "digit" 10. So, for example, 21_{10} could be written as 1A. Using this convention, write the number 111_{10} in the base eleven notation.

Let us learn how to add and multiply numbers written in an arbitrary system. We can do this in exactly the same way as in the decimal system, but we must remember that a "carry" occurs each time the result of adding up digits in a column exceeds or equals the base of the given number system.

Below is an example of the addition of the two numbers 124_{10} and 417_{10} in the base 3 system. First, we rewrite the numbers in the base 3 system: $124_{10} = 11121_3$, $417_{10} = 120110_3$. Then we write them one under another, lining up their rightmost digits. "Carries" are given in the upper row in small print.

$$\begin{array}{rcccccc} & & (1) & & (1) & (1) & \\ & & 1 & 1 & 1 & 2 & 1 \\ + & 1 & 2 & 0 & 1 & 1 & 0 \\ \hline & 2 & 0 & 2 & 0 & 0 & 1 \end{array}$$

To perform these operations successfully one must know the addition and multiplication tables for numbers less than the base of the system—that is, for one-digit numbers. For the decimal system, we have learned it early and well.

Exercise 5. Write down these tables for systems with bases 2, 3, 4, and 5.

Exercise 6. Calculate a) $1100_2 + 1101_2$; b) $201_3 \cdot 102_3$.

For teachers. We explained here very briefly how to add and multiply the numbers in any number base system. In a real session this would take more time. Of course, the goal of this work is not speed or accuracy in computations written in another number base system. An examination of and some practice in the addition and multiplication algorithms written in systems other than base 10 can lead to a deeper understanding of these algorithms.

Now we describe an effective algorithm for converting from one number system to another. It differs from the one we already know, because now the representation of a number will appear digit by digit from right to left rather than from left to right. The last digit is just the remainder when the number is divided by the base of the new system. The second digit can be found as follows: we take the quotient from the previous calculation and find the remainder when the quotient is divided by the base of the new system. Then we proceed in exactly the same way until we complete the representation.

Example. Let us convert the number 250_{10} to the base 8 ("octal") system:

$$\begin{aligned} 250 &= 31 \cdot 8 + \mathbf{2}, \\ 31 &= 3 \cdot 8 + \mathbf{7}, \\ 3 &= 0 \cdot 8 + \mathbf{3}\,. \end{aligned}$$

Thus, $250_{10} = 372_8$.

Exercise 7. Convert to the base 7 system the numbers a) 1000_{10}; b) 532_8.

In conclusion we submit a few more interesting problems.

Problem 1. A teacher sees on the blackboard the example $3 \cdot 4 = 10$. About to wipe it away, she checks if perhaps it is written in another number base system. Could this thought have been right?

Problem 2. Does there exist a number system where the following equalities are true simultaneously:

a) $3 + 4 = 10$ and $3 \cdot 4 = 15$;

b) $2 + 3 = 5$ and $2 \cdot 3 = 11$?

Problem 3. State and prove a condition (involving the representation of a number) which allows us to determine whether the number is odd or even

a) in the base 3 system;

b) in the base n system.

Problem 4. A blackboard bears a half-erased mathematical calculation exercise:

$$\begin{array}{rccccc} & 2 & 3 & ? & 5 & ? \\ + & & & & & \\ & 1 & ? & 6 & 4 & 2 \\ \hline & 4 & 2 & 4 & 2 & 3 \end{array}$$

Find out which number system the calculation was performed in and what the summands were.

Problem 5. A teacher said that there were 100 students in his class, 24 of whom were boys and 32 of whom were girls. Which number system did the teacher use in this statement?

For teachers. The material of this section can be discussed during two or three successive sessions. Students must learn:

—the concept of a number system;

—how to convert numbers from one system to another;

—how to add and multiply in an arbitrary number system.

To make this rather technical material less boring, we recommend using problems similar to 1–5 above.

§2. Divisibility tests

In the previous section we learned how to add and multiply numbers in an arbitrary number base system. The reverse operations—subtraction and division—are performed in the same manner as in the decimal system. However, these operations (like "long division", for example) are a bit more difficult, even in our usual decimal system.

Thus, it is often convenient to use divisibility tests to find out if one number is divisible by another, without actually performing the operation. The tests for the decimal system are discussed in the chapter "Divisibility–2". In non-decimal systems the situation is more unusual and difficult—try, for example, to find out if 123456654321_7 is divisible by 6.

Let us start simply. How do we know that a number with its last digit equal to zero is divisible by 10? The point is that in the decimal representation of any number

$$a_k 10^k + a_{k-1} 10^{k-1} + \ldots + a_2 10^2 + a_1 10^1 + a_0 10^0$$

all the summands are divisible by 10, except, perhaps, the last one. In our case, however, the last summand is zero and, therefore, the whole sum is divisible by 10. We can prove the converse statement similarly: if a natural number is divisible by 10, then its last digit is zero.

Consider now an arbitrary number system. The same ideas allow us to prove the following divisibility test:

In a base n system the representation of a number ends with zero if and only if this number is divisible by n.

Problem 6. State and prove the divisibility test for

a) a power of the base of a system (similar to divisibility tests for 100, 1000, ... in the decimal system);

b) a divisor of the base of a system (similar to divisibility tests for 2 and 5);

c) a power of a divisor of the base of a system.

Methodological remark. We would like to emphasize once again that different number systems are just different ways to write numbers. Thus the divisibility of one number by another does not depend on the particular system in which they are written.

At the same time, in each system there are some tricks to determine divisibility by certain specific numbers. These are the divisibility tests.

Let us investigate now other, less trivial, divisibility tests. Perhaps the most well-known of these are the tests for divisibility by 3 and 9. We will try to generalize these tests for any number base system. First, we must understand the proof of that test in the decimal system (see the chapter "Divisibility and remainders"). The only significant fact used is that $9 = 10 - 1$, and therefore that $10 \equiv 1 \pmod 9$.

Let us formulate and prove the analogous test for divisibility by $n-1$ in the base n system. Indeed, $n \equiv 1 \pmod{n-1}$. Hence, $n^s \equiv 1 \pmod{n-1}$ for any natural number s. Therefore

$$a_k n^k + a_{k-1} n^{k-1} + \ldots + a_1 n^1 + a_0 n^0 \equiv a_k + a_{k-1} + \ldots + a_1 + a_0 \pmod{n-1}.$$

Thus the sum of the digits of a number written in the base n system is divisible by $n-1$ if and only if the number itself is divisible by $n-1$.

Let us recall the question we asked at the beginning of this section: is $12345665432 1_7$ divisible by 6? Now we can answer this question easily: since the sum of the digits (which is 42_{10}) is divisible by 6, the number itself is also divisible by 6.

Problem 7. State and prove the test for divisibility by

a) a divisor of the number $n-1$ in the base n system (similar to the divisibility test for 3 in the decimal system);

b) the number $n+1$ in base n system (similar to the divisibility test for 11);

c) a divisor of the number $n+1$ in the base n system (there is no analog in base 10).

For teachers. We recommend devoting a whole session to the topic of this section. It would be wonderful if during this session students (using hints from the teacher) could formulate and prove various new divisibility tests.

§3. Miscellaneous problems

Up to now we have been interested in number systems for themselves. Now we are going to discuss a few problems which seem to have nothing to do with number systems. However, non-decimal number systems arise quite naturally when we try to solve these problems.

Problem 8. What is the minimum number of weights which enable us to weigh any integer number of grams of gold from 1 to 100 on a standard balance with two pans? Weights may be placed only on the left pan.

Solution. Every natural number can be written in the binary system. Thus, to weigh any number of grams of gold from 1 through 100 it is sufficient to have seven weights: 1, 2, 4, 8, 16, 32, 64. On the other hand, six weights are insufficient since we can obtain no more than $2^6 - 1 = 63$ different weights with them (each weight is either placed or not placed on the left pan).

Remark. Note that we did not assume that the weights must be integers. This assumption would not make the solution simpler.

Problem 9*. The same question as in the previous problem, but the weights can be placed on either pan of the balance.

Solution. To explain the solution to this problem we need the following interesting property of the base 3 system:

Every natural number can be represented as the difference of two numbers whose base 3 representations contain only 0's and 1's.

We can prove this property by writing the original number in the base 3 system and constructing the required numbers digit by digit from right to left. This is a good exercise, and is left to the reader.

Now it is clear that it suffices to have only 5 weights that weigh 1, 3, 9, 27, 81 (can you see why we don't need a weight of 243 grams?).

Four weights are insufficient since we cannot weigh more than $3^4 - 1 = 80$ different weights on them (each weight is either placed on the left pan, or on the right, or not placed on the balance at all).

Problem 10. An evil king wrote three secret two-digit numbers a, b, and c. A handsome prince must name three numbers X, Y, and Z, after which the king will tell him the sum $aX + bY + cZ$. The prince must then name all three of the King's numbers. Otherwise he will be executed. Help him out of this dangerous situation.

Solution. The prince can name the numbers 1, 100, and $100^2 = 10000$. The numbers a, b, and c will then just be the digits of the sum $aX + bY + cZ$ in the base 100 system.

Problem 11*. Prove that from the set 0, 1, 2, ... , 3^{k-1} one can choose 2^k numbers so that none of them can be represented as the arithmetic mean of some pair of the chosen numbers.

Solution. We will use the base 3 system. Let us assume that the base 3 representation of any of the given numbers contains exactly k digits—if there are fewer than k, then we just fill in the rest of the places with zeros. Now we choose those numbers whose base 3 representations contain only 0's and 1's. There are exactly 2^k of them. We show that this can serve as the required subset. Suppose that there were three different numbers in the subset—say, x, y, and z—satisfying the equality $x + y = 2z$. Since the numbers x and y must differ in at least one digit, we could then find the rightmost such digit. The corresponding digit of their sum $x + y$ would be 1. But the base 3 representation of $2z$ contains only 0's and 2's. This is a contradiction.

Problem 12. Prove that a subset of 2^k numbers with the same property can be chosen from the numbers 0, 1, 2, ... , $(3^k - 1)/2$.

For teachers. The problems from this section can be given at sessions or used in various mathematical contests.

§4. The game of Nim

Here we will talk about one form of the famous game of Nim (see Martin Gardner's books [**5–7**]). Its rules are simple. There are three heaps of stones (the initial number of stones in each heap may vary). Two players make their moves in turn by taking several stones from the heaps. It is allowed to take any number of stones but only from one heap at a move. A player who takes the last stone wins the game.

It is remarkable that the winning strategy for this game can be expressed using the binary system. We will discuss this strategy in a more generalized situation:

for an arbitrary number of heaps. We also note that in the case of two heaps this game can be turned into a game with a (chess) rook on a rectangular board (see Problems 10, 16, 22 from the chapter "Games").

As usual, to "solve" the game it is sufficient to determine the set of winning positions (see §3 of the chapter indicated above). Let us write the binary representations for the numbers of stones in each heap one under another, in such a way that the units digits are in the same column, the tens digits are also in the same column, et cetera. Then we calculate the parity of the number of 1's in each column (E denotes "even" and O denotes "odd"). For example, suppose there are three heaps, with 101, 60, and 47 stones. Then we write:

$$\begin{array}{lclclccccccc} n_1 & = & 101 & = & 1 & 1 & 0 & 0 & 1 & 0 & 1 \\ n_2 & = & 60 & = & & 1 & 1 & 1 & 1 & 0 & 0 \\ n_3 & = & 47 & = & & 1 & 0 & 1 & 1 & 1 & 1 \end{array}$$

$$\begin{array}{ccccccc} 1 & 1 & 0 & 0 & 1 & 0 & 1 \\ & 1 & 1 & 1 & 1 & 0 & 0 \\ & 1 & 0 & 1 & 1 & 1 & 1 \\ \hline O & O & O & E & O & O & E \end{array}$$

We claim that a position is winning if and only if the number of 1's in any column is always even; that is, all the letters in the bottom row are E's (so that the position shown above is, presumably, a losing one). We will call such positions "even" and any other position "odd".

To prove that a position is winning if and only if it is even, we must show that:

1. The final position of the game is even.
2. Any move from an even position leads to an odd position.
3. From any odd position you can shift to some even position in one move.

Part 1 is easy. The game ends when there are 0 stones in each heap, and 0 is even.

To prove part 2 we notice that after each move the number of stones in some heap changes, and, therefore, some digit in its binary representation changes. This means that the number of 1's in the corresponding column changes by 1. Since no other row can change (stones can be removed from only one heap at a time), the parity at that column changes, too.

We now show how to move from an arbitrary odd position to an even one. We must take several stones from one heap so that the parity of the numbers of 1's in the columns changes for all the columns with an odd number of 1's in them (and only for these columns!). Consider the leftmost column with an odd number of 1's in it and choose a heap which has 1 as its digit in this column (why does there exist such a heap?). This heap is the one we will take the stones from.

It is easy to understand how many stones must be left in this heap—the binary representation of the number of stones in the heap must change in those digits which correspond to the columns with an odd number of 1's in them. We must take away exactly as many stones as will make this happen. Since the leftmost of these digits will change from 1 to 0, the number of stones in the heap will, in fact, decrease.

Problem 13. Solve the following games:

a) There are eight white pawns in the first row of a chessboard, and eight black pawns in the eighth row. Each of two players, in turn, can move one of his or her pawns towards the other end of the board in a vertical direction for any number of boxes. It is not allowed to jump over a pawn of the opposite color. The player who cannot make a move loses.

b) The same game, except that it is allowed to move the pawns not only forwards but backwards too.

For teachers. It makes sense to discuss the game of Nim only with students of sufficiently high mathematical skill. Students should play each other and submit their own conjectures and strategies. It is rather difficult to devise the correct winning strategy; however, exact and timely hints can facilitate this process and allow students to solve as much as they can of the problem independently.

CHAPTER 16

Inequalities

§1. What's greater?

This question is, perhaps, one of the most common children ask. Children are very curious, and you can hear from them questions like these:

—Who is stronger: my dad or the arm wrestling champion?

—Which is higher: our house or the World Trade Center?

—Do more people live in Chicago than in Atlanta?

In mathematics such "naïve" questions do not make much sense, but they help students learn how to calculate better and more precisely, and to deal with "really large" numbers. Of course, we are not talking here about using calculators, which are not helpful if we want to be absolutely rigorous in our proofs.

For teachers. The technique of calculation and estimation is one of the most valuable aspects of mathematical culture. Students have to master not just "blind usage" of various methods of computation and approximation; they must understand their essence. It is impossible to remember all the technical tricks of mathematics. But it is possible and necessary to teach students how to do things "with their bare hands". The skill of fast and precise estimation can be achieved by solving problems with specific numerical data, like those we consider in this section.

Here is a typical problem of this series.

Problem 1. Which number is greater: 31^{11} or 17^{14}?

Certainly, you can calculate both numbers "manually"—they have no more than 20 digits. However, this way of dealing with the problem is time consuming and will yield nothing in other, more intricate, problems. Let us try another way.

$$31^{11} < 32^{11} = (2^5)^{11} = 2^{55} < 2^{56} = (2^4)^{14} = 16^{14} < 17^{14}.$$

This chain of inequalities shows that 31^{11} is less than 17^{14}. The only thing we needed for this solution was to observe that the numbers 31 and 17 are not far away from powers of 2.

Problem 2. Which number is greater:

a) 2^{300} or 3^{200}?

b) 2^{40} or 3^{28}?

c) 5^{44} or 4^{53}?

Problem 3. Prove that $2^{100} + 3^{100} < 4^{100}$.

Solution. Clearly, $2^{100} < 3^{100}$. So it is enough to prove that $2 \cdot 3^{100} < 4^{100}$, or, equivalently, $\left(\frac{4}{3}\right)^{100} > 2$. But even $\left(\frac{4}{3}\right)^3 = \frac{64}{27}$ is greater than 2.

Problem 4. Which number is greater: 7^{92} or 8^{91}?

To further investigate the situation in Problem 3, let us try to find a natural number n such that $4^n < 2^{100} + 3^{100} < 4^{n+1}$. On our way to the solution we encounter the following problem.

Problem 5.* Prove that $4^{79} < 2^{100} + 3^{100} < 4^{80}$.

Solution. Since $2 \cdot 3^{100} > 2^{100} + 3^{100}$, it is enough to show that $4^{80} > 2 \cdot 3^{100}$; that is, $\left(\frac{4^4}{3^5}\right)^{20} = \left(\frac{256}{243}\right)^{20} > 2$.

We recall Bernoulli's inequality: $(1+x)^n \geq 1 + nx$ for $x \geq -1$, $n \geq 1$ (see Problem 56 or the chapter "Induction"). This fact motivates the question: is it true that $\frac{256}{243} > 1 + \frac{1}{20}$? The answer is yes (check it yourself!). Hence, $\left(\frac{256}{243}\right)^{20} > (1 + \frac{1}{20})^{20} \geq 2$.

Now we prove that $2^{100} + 3^{100} > 4^{79}$. We will show that $3^{100} > 4^{79}$, or $4^{80}/3^{100} < 4$, by building the following chain of inequalities:

$$\begin{aligned}\frac{4^{80}}{3^{100}} &= \left(\frac{256}{243}\right)^{20} < \left(\frac{19}{18}\right)^{20} = \left(\frac{361}{324}\right)^{10} \\ &< \left(\frac{9}{8}\right)^{10} = \left(\frac{81}{64}\right)^{5} < \left(\frac{9}{7}\right)^{5} = \frac{59049}{16807} < 4.\end{aligned}$$

Comment. As you see, the method consists of gradual simplification of an ugly and complicated expression such as $4^{80}/3^{100}$.

For teachers. To learn this method well, it is necessary for the students to be familiar with various small powers of natural numbers. They should also not be afraid to perform any specific (though, perhaps, long) calculation as long as they understand its goal. This can be helped by solving problems similar to the next one.

Problem 6. Find all powers of the natural numbers 2, 3, 4, 5, 6, 7, 8, 9, 11, 12, which are no greater than 10000, and arrange them in increasing order. Find all pairs of these powers with differences no greater than 10.

Comment. This is an excellent problem for homework, using a calculator.

The solutions to the following three problems are based on a very different idea. So far we proved inequalities between specific natural numbers simply by transforming (simplifying) them and calculating the results of the transformations. The new idea is to change some specific number to a variable.

Problem 7. Which number is greater: $1234567 \cdot 1234569$ or 1234568^2?

Solution. We denote the number 1234568 by x. Then the left-hand expression is $(x-1)(x+1) = x^2 - 1 < x^2$. Thus, we avoid the necessity of multiplying seven-digit numbers or raising them to a power.

Problem 8. We are given the two fractions

$$\frac{10\ldots01}{10\ldots001} \quad \text{and} \quad \frac{10\ldots01}{10\ldots001}.$$

In each fraction the number in the denominator has one zero more than the one in the numerator. If the numerator in the left fraction has 1984 zeros, and the numerator in the right fraction has 1985 zeros, which of them is greater?

Problem 9. Which number is greater: $1234567/7654321$ or $1234568/7654322$?

Here are a few more problems using estimates and approximations.

Problem 10. Which number is greater: 100^{100} or $50^{50} \cdot 150^{50}$?

Problem 11. Which number is greater: $(1.01)^{1000}$ or 1000?

Problem 12. Prove that

$$\frac{1}{2} - \frac{1}{3} + \frac{1}{4} - \ldots - \frac{1}{99} + \frac{1}{100} > \frac{1}{5}.$$

Problem 13. (This is almost a joke.) If you apply the factorial function 99 times to the number 100, then you get the number A. If you apply the factorial function 100 times to the number 99, then you get the number B. Which of them is greater?

Let us discuss the solution to Problem 12. We split the summands into pairs

$$\left(\frac{1}{2} - \frac{1}{3}\right) + \left(\frac{1}{4} - \frac{1}{5}\right) + \ldots + \left(\frac{1}{98} - \frac{1}{99}\right) + \frac{1}{100}.$$

Now we have a sum of positive numbers, which begins with $1/6 + 1/20 + 1/42 + \ldots$. Since $1/6 + 1/20 > 1/5$, we immediately have the required result.

Miscellaneous

14. How many digits does the number 2^{1000} have?

15. Find the largest of the numbers 5^{100}, 6^{91}, 7^{90}, 8^{85}.

16*. Prove that the number $\frac{1}{2} \cdot \frac{3}{4} \cdot \frac{5}{6} \ldots \frac{99}{100}$ is
a) less than $1/10$;
b) less than $1/12$;
c) greater than $1/15$.

__For teachers.__ We can create a lot of questions similar to Problems 1–10. For example, let us take two small natural numbers—say, 5 and 7. Then consider a large power of 5—let it be 5^{73}. Now we seek two relatively small powers of 5 and 7, which differ "not by much" (this depends on your sense of measure). For instance, $5^4 = 625$ and $7^3 = 343$ will do (the exponents must be different!). Since $5^4 > 7^3$, we have that $5^{72} > 7^{54}$. Therefore, $5^{73} > 5^{72} > 7^{54}$. After this, we can "fix" the exponent 54 a bit and we get another exercise: prove that $5^{73} > 7^{53}$.

Comment. This class of inequalities, which are the result of combinations of several simple and "rough" inequalities, has a special name in Russian olympiad folklore. They are called "inequalities à la Leningrad".

__For teachers.__ Solving these numerical inequalities helps develop computational skills and approximation technique. However, some students, who may be gifted in logical and combinatorial mathematics, may have a sort of "allergy" to computational problems like these.

§2. The main inequality

The main (and, in some sense, the only!) inequality in the field of real numbers is the inequality $x^2 \geq 0$—its truth is beyond doubt. Other, well-known and useful, inequalities follow from this one. The first among them is, certainly, the inequality of means (or the A.M.–G.M. inequality):

$$\frac{a+b}{2} \geq \sqrt{ab} \quad \text{for any } a, b \geq 0$$

—the numbers $(a+b)/2$ and $\sqrt{ab}$ are called the *arithmetic mean* and *geometric mean* of the numbers a and b.

This is very easy to prove:

$$\frac{a+b}{2} - \sqrt{ab} = \frac{a+b-2\sqrt{ab}}{2} = \frac{1}{2}(\sqrt{a}-\sqrt{b})^2 \geq 0$$

—which implies not only the truth of the A.M.–G.M. inequality, but also the fact that this inequality turns into an equality if and only if $a = b$.

Problem 17. Prove that $1 + x \geq 2\sqrt{x}$, if $x \geq 0$.

Problem 18. Prove that $x + 1/x \geq 2$, if $x > 0$.

Problem 19. Prove that $(x^2 + y^2)/2 \geq xy$ for any x and y.

Problem 20. Prove that $2(x^2 + y^2) \geq (x+y)^2$ for any x and y.

Problem 21. Prove that $1/x + 1/y \geq 4/(x+y)$, if $x > 0$, $y > 0$.

Solution to Problem 18. $(x + 1/x) - 2 = (\sqrt{x} - \sqrt{1/x})^2 \geq 0$.

Remark. Generally speaking, solutions to any of these problems can be reduced either to an appropriate application of the A.M.–G.M. inequality, or, after some "appropriate" transformations, to an application of the main inequality $x^2 \geq 0$.

More complicated inequalities, however, are usually solved either by multiple application of the A.M.–G.M. inequality, or by combining several different ideas. Here is a typical example:

Problem 22. Prove that $x^2 + y^2 + z^2 \geq xy + yz + zx$ for any x, y, and z.

To prove this fact we will use the result of Problem 19, and write three inequalities:

$$\frac{1}{2}(x^2+y^2) \geq xy, \quad \frac{1}{2}(x^2+z^2) \geq xz, \quad \frac{1}{2}(y^2+z^2) \geq yz\,.$$

Adding them up, we are done.

Problem 23. If a, b, $c \geq 0$, prove that $(a+b)(a+c)(b+c) \geq 8abc$.

Problem 24. If a, b, $c \geq 0$, prove that $ab + bc + ca \geq a\sqrt{bc} + b\sqrt{ac} + c\sqrt{ab}$.

Problem 25. Prove that $x^2 + y^2 + 1 \geq xy + x + y$ for any x and y.

Problem 26. Prove that for any a, b, and c, the inequality $a^4 + b^4 + c^4 \geq abc(a+b+c)$ holds true.

Solution. We will use the inequality from Problem 22—twice!

$$\begin{aligned} a^4 + b^4 + c^4 &= (a^2)^2 + (b^2)^2 + (c^2)^2 \geq a^2b^2 + b^2c^2 + c^2a^2 \\ &= (ab)^2 + (bc)^2 + (ca)^2 \geq (ab)(bc) + (bc)(ca) + (ca)(ab) \\ &= abc(a+b+c)\,. \end{aligned}$$

The A.M.–G.M. inequality is remarkable in two ways. First, it allows us to estimate the sum of two positive numbers in terms of their product. Second, it can be generalized for more than two numbers. Here, for example, is the A.M.–G.M. inequality for four positive numbers:

$$\frac{a+b+c+d}{4} \geq \sqrt[4]{abcd} \quad \text{for any } a, b, c, d \geq 0\,,$$

where, as usual, the left and the right sides of the inequality are called the *arithmetic* and the *geometric mean* of the four given numbers respectively.

This version of the A.M.–G.M. inequality can be proved as follows:

$$\frac{a+b+c+d}{4} = \frac{1}{2}\left(\frac{a+b}{2} + \frac{c+d}{2}\right) \geq \frac{1}{2}(\sqrt{ab} + \sqrt{cd}) \geq \sqrt{\sqrt{ab}\sqrt{cd}} = \sqrt[4]{abcd}$$

—we just apply the A.M.–G.M. inequality for two numbers twice.

Problem 27. Prove that $x^4 + y^4 + 8 \geq 8xy$ for any x and y.

Problem 28. If a, b, c, and d are positive numbers, prove that

$$(a+b+c+d)\left(\frac{1}{a} + \frac{1}{b} + \frac{1}{c} + \frac{1}{d}\right) \geq 16\,.$$

It is, however, not so easy to prove the A.M.–G.M. inequality for three positive numbers:

$$\frac{a+b+c}{3} \geq \sqrt[3]{abc} \quad \text{for any } a, b, c \geq 0\,.$$

Let us consider four numbers: a, b, c, and $m = \sqrt[3]{abc}$. Then

$$\frac{a+b+c+m}{4} \geq \sqrt[4]{abcm} = \sqrt[4]{m^3 \cdot m} = m\,.$$

Hence, $(a+b+c)/4 \geq 3m/4$, and, therefore, $a+b+c \geq 3m$, $(a+b+c)/3 \geq \sqrt[3]{abc}$.

Problem 29. If a, b, and c are positive numbers, prove that

$$\frac{a}{b} + \frac{b}{c} + \frac{c}{a} \geq 3\,.$$

Problem 30. Prove that if $x \geq 0$, then $3x^3 - 6x^2 + 4 \geq 0$.

Solution. We show that $3x^3 + 4 \geq 6x^2$. Since $3x^3 + 4 = 2x^3 + x^3 + 4$, we can apply the A.M.–G.M. inequality and get

$$2x^3 + x^3 + 4 \geq 3\sqrt[3]{2x^3 \cdot x^3 \cdot 4} = 3 \cdot 2x^2 = 6x^2.$$

Miscellaneous (for homework)

31. Prove that if a, b, $c > 0$, then $1/a + 1/b + 1/c \geq 1/\sqrt{ab} + 1/\sqrt{bc} + 1/\sqrt{ac}$.

32. Prove that if a, b, $c > 0$, then $ab/c + ac/b + bc/a \geq a + b + c$.

33. Prove that if a, b, $c \geq 0$, then $((a+b+c)/3)^2 \geq (ab+bc+ca)/3$.

34. Prove that if a, b, $c \geq 0$, then $(ab+bc+ca)^2 \geq 3abc(a+b+c)$.

35. The sum of three positive numbers is six. Prove that the sum of their squares is no less than 12.

36. Prove that if $x \geq 0$, then $2x + 3/8 \geq 4\sqrt{x}$.

37. The sum of two non-negative numbers is 10. What is the maximum and the minimum possible value of the sum of their squares?

38*. Prove the A.M.–G.M. inequality for five non-negative numbers; that is, prove that if a, b, c, d, and $e \geq 0$, then

$$\frac{a+b+c+d+e}{5} \geq \sqrt[5]{abcde}\,.$$

Hint. First, prove the A.M.–G.M. inequality for eight numbers, and then use our idea from the proof of the A.M.–G.M. inequality for three numbers.

§3. Transformations

Sometimes a lucky transformation can help solve a problem or prove an inequality immediately. Here is an example: let us return to the inequality $x^2+y^2+z^2 \geq xy+yz+zx$ from Problem 22. It can be proved as follows. We rewrite the difference between parts of the inequality

$$x^2+y^2+z^2-xy-yz-zx = ((x-y)^2+(y-z)^2+(z-x)^2)/2 \geq 0.$$

Actually, the trick is a refinement of the technique of "completing the square", often used to solve simple quadratic equations. Here is another example.

Problem 39. Solve the equation $a^2+b^2+c^2+d^2-ab-bc-cd-d+2/5=0$.

You might say "This is not an inequality!" Certainly, but, first, we can use the same method, and, second, there is an inequality present:

$$\begin{aligned} &a^2+b^2+c^2+d^2-ab-bc-cd-d+2/5 \\ &\qquad = \left(a-\frac{b}{2}\right)^2+\frac{3}{4}\left(b-\frac{2c}{3}\right)^2+\frac{2}{3}\left(c-\frac{3d}{4}\right)^2+\frac{5}{8}\left(d-\frac{4}{5}\right)^2, \end{aligned}$$

and the solution follows. Indeed, a sum of squares can be equal to zero if and only if all the summands are zero. Therefore, we have our answer: $d=4/5$, $c=3d/4=3/5$, $b=2c/3=2/5$, $a=b/2=1/5$.

Problem 40. $a+b=1$. What is the maximum possible value of the product ab?

Hint. $a(1-a)=1/4-(1/2-a)^2$.

Problem 41. Prove the inequality $(a^2/4)+b^2+c^2 \geq ab-ac+2bc$ for all a, b, and c.

Problem 42. Suppose k, l, and m are natural numbers. Prove that

$$2^{k+l}+2^{k+m}+2^{l+m} \leq 2^{k+l+m+1}+1\,.$$

Problem 43. If $a+b+c=0$, prove that $ab+bc+ca \leq 0$.

Let us solve Problem 41. We carry everything over to the left side and rewrite: $(a^2/4)+b^2+c^2-ab+ac-2bc=(a/2-b+c)^2 \geq 0$. This is true because of our "main" inequality.

* * *

Now we discuss another excellent idea which has proven quite useful in dealing with inequalities which show some symmetry (it is also connected with factorization).

Lemma. If $a \geq b$ and $x \geq y$, then $ax + by \geq ay + bx$.

Proof. Indeed, $ax + by - ay - bx = (a - b)(x - y) \geq 0$.

Comment. If, for instance, f is some increasing function, then

$$(a - b)(f(a) - f(b)) \geq 0$$

for any numbers a and b—this is just a reformulation of the definition of an increasing function.

This idea can be applied as follows:

Problem 44. Prove that $x^6/y^2 + y^6/x^2 \geq x^4 + y^4$ for any x and y.

Solution. We denote x^2 by a, and y^2 by b. Then

$$\frac{x^6}{y^2} + \frac{y^6}{x^2} - x^4 - y^4 = a^3/b + b^3/a - a^2 - b^2$$
$$= \frac{(a-b)a^2}{b} + \frac{(b-a)b^2}{a} = (a-b)\left(\frac{a^2}{b} - \frac{b^2}{a}\right) = \frac{(a-b)(a^3-b^3)}{ab} \geq 0.$$

We emphasize once again that the numbers $a - b$ and $a^3 - b^3$ have the same sign.

Problem 45. If x, $y > 0$, prove that $\sqrt{x^2/y} + \sqrt{y^2/x} \geq \sqrt{x} + \sqrt{y}$.

Problem 46. If a, b, $c \geq 0$, prove that

$$2(a^3 + b^3 + c^3) \geq a^2b + ab^2 + a^2c + ac^2 + b^2c + bc^2 .$$

Solution. Carry all the terms over to one side and split them into quadruples: $[a^3 + b^3 - a^2b - ab^2] + [b^3 + c^3 - b^2c - bc^2] + [a^3 + c^3 - a^2c - ac^2]$.

Inside each quadruple the expression can be factored in the following way: $a^3 + b^3 - a^2b - ab^2 = (a - b)(a^2 - b^2) \geq 0$. This completes the proof.

Problem 47*. If $a_1 \leq a_2 \leq \ldots \leq a_n$ and $b_1 \leq b_2 \leq \ldots \leq b_n$, prove that

$$a_1b_1 + a_2b_2 + \ldots + a_nb_n \geq a_1c_1 + a_2c_2 + \ldots + a_nc_n ,$$

where c_1, c_2, $\ldots$, c_n is an arbitrary permutation of numbers b_1, b_2, $\ldots$, b_n.

Miscellaneous

48. Prove that for any x

$$x(x+1)(x+2)(x+3) \geq -1 .$$

49. Prove that for any x, y, and z

$$x^4 + y^4 + z^2 + 1 \geq 2x(xy^2 - x + z + 1) .$$

50*. Prove that

$$\frac{1}{1+\sqrt{2}}+\frac{1}{\sqrt{3}+\sqrt{4}}+\ldots+\frac{1}{\sqrt{99}+\sqrt{100}}>\frac{9}{2}.$$

51. If $x,\, y \geq 0$, prove that $(\sqrt{x}+\sqrt{y})^8 \geq 64xy(x+y)^2$.

For teachers. There exist a number of more difficult inequalities using the idea from our Lemma, as well as the inequality from Problem 47 (see [**7, 8**]). If problems of this level are not appropriate for your sessions, then we recommend giving only the general idea and illustrating it with two or three examples.

§4. Induction and inequalities

Often inequalities contain a variable which takes positive integral values, or one of the numbers involved is just a disguise covering such a variable (see, for instance, Problem 7). By the way, the ability to recognize this kind of variable is an important skill to develop, and not only for inequalities. Such "positive integral" inequalities can often be proved by induction.

Problem 52. Prove that if $n \geq 3$, then

$$\frac{1}{n+1}+\frac{1}{n+2}+\ldots+\frac{1}{2n}>\frac{3}{5}\,.$$

Solution. Base: $n=3$. We have $1/4+1/5+1/6=37/60>3/5$.

Let us prove the inductive step, from $n=k$ to $n=k+1$:

$$\begin{aligned}&\left(\frac{1}{k+2}+\frac{1}{k+3}+\ldots+\frac{1}{2k+2}\right)\\ &\quad=\left(\frac{1}{k+1}+\frac{1}{k+2}+\ldots+\frac{1}{2k}+\frac{1}{2k+1}+\frac{1}{2k+2}-\frac{1}{k+1}\right)\\ &\quad>\frac{1}{k+1}+\frac{1}{k+2}+\ldots+\frac{1}{2k}>\frac{3}{5}.\end{aligned}$$

One of the standard schemes in inductive proofs for inequalities is explained below. Suppose we are given two series of numbers $a_1, a_2, \ldots, a_n, \ldots$ and $b_1, b_2, \ldots, b_n, \ldots$, and suppose it is known also that

a) $a_1 \geq b_1$ (Base);

b) $a_k - a_{k-1} \geq b_k - b_{k-1}$ for any $k \leq n$ (Inductive step).

Then $a_n \geq b_n$.

Problem 53. If n is a natural number, prove that

$$1+\frac{1}{\sqrt{2}}+\frac{1}{\sqrt{3}}+\ldots+\frac{1}{\sqrt{n}}<2\sqrt{n}\,.$$

Problem 54. If n is a natural number, prove that

$$1+\frac{1}{\sqrt{2}}+\frac{1}{\sqrt{3}}+\ldots+\frac{1}{\sqrt{n}}>2(\sqrt{n+1}-1)\,.$$

Problem 55. If n is a natural number, prove that

$$\frac{1}{2^2}+\frac{1}{3^2}+\frac{1}{4^2}+\ldots+\frac{1}{n^2}<1\,.$$

Problem 55 cannot be solved according to the scheme above. The series $a_n = 1/2^2 + 1/3^2 + \ldots + 1/n^2$ increases monotonically, while the series $b_n = 1$ is constant. Thus item b) is not true and we cannot use the scheme.

What can we do then? Here one of the most amazing qualities of the method of mathematical induction reveals itself. It appears that it is sometimes easier to prove a stronger statement, or, as in this case, a more exact inequality. So,

Solution. We will choose another series b_n, namely, $b_n = (1 - 1/n)$.

Base. If $n = 2$, we see that $1/2^2 < 1 - 1/2$.

Inductive step (according to the scheme!): $a_k - a_{k-1} = 1/k^2$, $b_k - b_{k-1} = 1/k(k-1)$; that is, $a_k - a_{k-1} < b_k - b_{k-1}$. Thus, for any natural number n we have $a_n < b_n = 1 - 1/n < 1$.

* * *

Problem 56. (Bernoulli's inequality) If $x \geq 0$ and n is a natural number, prove that $(1+x)^n \geq 1 + nx$.

Hint. This problem has a non-inductive solution, which is clear for those acquainted with Newton's binomial theorem (see the chapter "Combinatorics–2"). Indeed,

$$(1+x)^n = 1 + \binom{n}{1}x + \binom{n}{2}x^2 + \ldots + \binom{n}{n}x^n\,,$$

and all the summands after the first two are non-negative. Therefore, $(1+x)^n \geq 1 + \binom{n}{1}x = 1 + nx$.

On the other hand, there is another general scheme, which deals with problems of this type. It is explained below (using the same notation as in the previous scheme): If

a) $a_1 \geq b_1$ (Base);

b) $a_k/a_{k-1} \geq b_k/b_{k-1}$ for all $k \leq n$ (Inductive step),

then $a_n \geq b_n$. This is another corollary of the MMI.

Problem 57. If n is a natural number, prove that $n^n > (n+1)^{n-1}$.

Problem 58. If n is a natural number greater than or equal to 4, prove that $n! \geq 2^n$.

Problem 59. If n is a natural number, prove that $2^n \geq 2n$.

Problem 60. Find all natural numbers n such that $2^n \geq n^3$.

Let us discuss the solution to Problem 57. We set $a_n = n^n$, $b_n = (n+1)^{n-1}$. The statement is true for the values $n = 1$ and $n = 2$, so the base is proved. To prove the inductive step, it suffices to show that

$$\frac{a_k}{a_{k-1}} = \frac{k^k}{(k-1)^{k-1}} \geq \frac{b_k}{b_{k-1}} = \frac{(k+1)^{k-1}}{k^{k-2}}$$

or, equivalently, $k^{2k-2} \geq (k^2-1)^{k-1}$; that is, $(k^2)^{k-1} \geq (k^2-1)^{k-1}$.

Miscellaneous

61. Prove that for any natural number n the inequality $3^n > n \cdot 2^n$ holds true.

62. Which of the two numbers

$$2^{2^{\cdot^{\cdot^{2}}}} \quad \text{(ten 2's)} \quad \text{or} \quad 3^{3^{\cdot^{\cdot^{3}}}} \quad \text{(nine 3's)}$$

is greater? What if there were eight 3's?

63. The product of the positive numbers $a_1, a_2, \ldots, a_n$ is equal to 1. Prove that

$$(1+a_1)(1+a_2)\ldots(1+a_n) \geq 2^n .$$

Remark. This problem has other, non-inductive solutions.

64. Prove Bernoulli's inequality $(1+x)^n \geq 1+nx$, if $x \geq -1$ and $n \geq 1$.

65. The sum of the positive numbers $x_1, x_2, \ldots, x_n$ equals 1/2. Prove that

$$\frac{1-x_1}{1+x_1} \cdot \frac{1-x_2}{1+x_2} \cdot \ldots \cdot \frac{1-x_n}{1+x_n} \geq \frac{1}{3} .$$

For teachers. In this section two themes are merging, each of which is important enough to take up several sessions. These are "Induction" and "Inequalities". It must be noticed that in the theme "Induction" the subtopic "Applying Induction to Inequalities" is usually understood by students more easily than other (more abstract) ones. We must not make excessive use of this fact, although we can create many problems similar to Problems 52–61.

It is important to give your students more questions which require non-standard thinking. Also, you should allow them to seek non-inductive solutions to the problems.

§5. Inequalities for everyone

The problems in this section are listed without any sorting by method of solution. However, we have tried to arrange them in progressive order of difficulty.

66. A string was stretched along the equator without gaps. Then it was lengthened by 1 centimeter (0.4 inches) and was stretched again along the equator by pulling it off the ground in one place. Is it possible for a man to go through the gap created?

67. Imagine that the Earth is made of dough which is then rolled into a thin "sausage" so that it will reach the Sun. What is the thickness of that "sausage"? Try to deviate from the right answer by no more than 1000%.

68. Is it possible to pack the entire population of the Earth and everything that was created by humankind inside a cube with an edge 2 miles long?

69. Imagine that you are standing on the western bank of the Hudson River. Is it possible, using only things at hand and common sense, to make a good estimate of the length of the radio antenna on one of the buildings of the World Trade Center on the other side of the river?

70. Prove that $100! < 50^{100}$.

71. If n is a natural number, prove that $\sqrt{n+1} - \sqrt{n-1} > 1/\sqrt{n}$.

72. If $1 > x > y > 0$, prove that $(x-y)/(1-xy) < 1$.

73. If a, b, c, $d \geq 0$ and $c+d \leq a$, $c+d \leq b$, prove that $ad+bc \leq ab$.

74. Does there exist a set of numbers whose sum is 1, and the sum of whose squares is less than 0.01?

75. Suppose a, b, $c > 0$ and $abc = 1$. It is known that $a+b+c > 1/a+1/b+1/c$. Prove that exactly one of the numbers a, b, and c is greater than 1.

76. The numbers x, y belong to the segment [0,1]. Prove that

$$\frac{x}{1+y} + \frac{y}{1+x} \leq 1 .$$

77. Suppose a, b, and c are natural numbers such that $1/a+1/b+1/c < 1$. Prove that $1/a+1/b+1/c \leq 41/42$.

78. If x, y, and z are positive numbers, prove that

$$\frac{x}{x+y} + \frac{y}{y+z} + \frac{z}{z+x} \leq 2 .$$

79. Prove that

$$\left(1-\frac{1}{4}\right)\left(1-\frac{1}{9}\right)\left(1-\frac{1}{16}\right)\cdots\left(1-\frac{1}{n^2}\right) > \frac{1}{2} .$$

80. Prove that for any x the inequality $x^4 - x^3 + 3x^2 - 2x + 2 \geq 0$ holds.

81. The numbers a, b, c, and d belong to the segment [0,1]. Prove that

$$(a+b+c+d+1)^2 \geq 4(a^2+b^2+c^2+d^2) .$$

82. Suppose x and y are greater than 0. We denote the minimum of x, $1/y$, and $y+1/x$ by S. What is the maximum possible value of S?

83.* Suppose a, b, c, and d are positive numbers. Prove that at least one of the inequalities

1) $a+b < c+d$;
2) $(a+b)cd < ab(c+d)$;
3) $(a+b)(c+d) < ab+cd$

is not true.

84.* Prove that the three inequalities

$$\frac{a_1b_2}{a_1+b_2} < \frac{a_2b_1}{a_2+b_1}, \quad \frac{a_2b_3}{a_2+b_3} < \frac{a_3b_2}{a_3+b_2}, \quad \frac{a_3b_1}{a_3+b_1} < \frac{a_1b_3}{a_1+b_3}$$

cannot be true simultaneously, if the numbers a_1, a_2, a_3, b_1, b_2, b_3 are positive.

85.* Prove that if $x+y+z \geq xyz$, then $x^2+y^2+z^2 \geq xyz$.

CHAPTER 17

Problems for the Second Year

You probably have already noticed that some topics from the first year were developed in the second part; others (like "Parity", "Pigeon Hole Principle", "Games") —were not. However, some problems pertaining to these themes were not included in the chapters of the first part because they are more difficult. We begin this chapter by filling this gap.

§1. Parity

1. Prove that the equality $1/a + 1/b + 1/c + 1/d + 1/e + 1/f = 1$ has no solutions in odd natural numbers.

2. Eight rooks are placed on a chessboard so that none of them attacks another. Prove that the number of rooks standing on black squares is even.

3. Is it possible to place 20 red and blue pawns around a circle in such a way that a blue pawn is standing on the point opposite to any red pawn, and no two of the blue pawns are neighbors?

4. Points A and B are chosen on a straight line. Then 1001 other points are chosen outside segment AB, and these are colored red and blue. Prove that the sum of the distances from A to the red points and from B to the blue points is not equal to the sum of the distances from B to the red points and from A to the blue points.

5. There are ten pairs of cards with the numbers 0, 0, 1, 1, ... , 8, 8, 9, 9 written on them. Prove that they cannot be laid in a row so that there are exactly n cards between any two cards with equal numbers n on them (for all $n = 0, 1, \ldots, 9$).

6. Twenty points, which form a regular 20-gon, are chosen on a circle. Then they are split into ten pairs, and the points in each pair are connected by a chord. Prove that some pair of these chords have the same length.

7. A 6×6 square is covered by 1×2 dominoes without overlapping. Prove that the square can be cut along a single line parallel to its sides so that none of the dominoes will be damaged.

8. A snail began crawling about a plane, starting from point O with constant speed, making a 60° turn every half hour. Prove that it can return to point O only after a whole number of hours.

9*. We have an audio tape recorder and n reels with tapes which have red leaders on the outside and green leaders on the inside. Find all n such that using only one empty reel we can reach the situation when all the tapes return to their initial reels with their green leaders on the outside.

§2. The Pigeon Hole Principle

10. A line is colored with 11 colors. Prove that one can find two identically colored points an integer number of inches apart.

11. There are seven lines on the plane. Prove that some pair of them forms an angle less than $26°$.

12. Each box of a 5×41 table is colored white or black. Prove that we can choose three columns and three rows such that all nine boxes on their intersections are the same color.

13*. Each point on the plane is colored using a) 2; b) 3; c) 100 colors. Prove that we can find a rectangle with all its vertices the same color.

14. Six friends decided to visit seven movie theaters during the weekend. Shows started at 9 am, 10 am, 11 am, ..., 7 pm. Every hour two of them went to some theater, and all the others to another theater. By the end of the day each of them had visited all seven theaters. Prove that for every theater there was a show which was not attended by any of the friends.

15. What is the largest number of spiders which can amicably share the edges of a cube with an edge equal to 1 meter? A spider will tolerate a neighbor only at a distance of a) 1 meter; b) 1.1 meter (traveling along the edges).

16. Seven vertices are chosen in each of two congruent regular 16-gons. Prove that these polygons can be placed one atop another in such a way that at least four chosen vertices of one polygon coincide with some of the chosen vertices of the other one.

17. A set of ten two-digit numbers is given. Prove that one can choose two disjoint subsets of these numbers with equal sums.

18. Twenty-five points are given on the plane. It is known that among any three of them one can choose two less than 1 inch apart. Prove that there are 13 points among them which lie inside a circle of radius 1.

19. Six points are chosen inside a 3×4 rectangle. Prove that a pair of them can be chosen such that the distance between them is at most $\sqrt{5}$.

20. Set A contains natural numbers, and it is known that there is an element of A among any 100 consecutive natural numbers. Prove that one can find four different numbers a, b, c, and d in set A such that $a + b = c + d$.

21. Ninety-nine 2×2 squares were cut from a sheet of graph paper with dimensions 29×29. Prove that it is possible to cut out another small 2×2 square.

22. A 10×10 table is covered by fifty-five 2×2 squares. Prove that one of them can be removed so that all the others will still cover the table.

23*. A chess grand master plays at least one game per day but no more than 12 games per week (Sunday through Saturday). Prove that one can find several consecutive days during the year when the maestro played exactly 20 games.

24. Ten disjoint segments colored red are given within a segment 10 inches long. It is known that no two red points are exactly 1 inch apart. Prove that the sum of the lengths of the red segments is no more than 5 inches.

25. Inside a 1×1 square, 101 points are given. Prove that some three of them form a triangle with area no more than 0.01.

26. Several circles, with the sum of their radii equal to $3/5$, are placed within a 1×1 square. Prove that there exists a line parallel to a side of the square, which intersects at least two of these circles.

27. Several chords are drawn in a circle of radius 1, and each diameter of the circle intersects no more than four of them. Prove that the sum of their lengths does not exceed 13.

28* There are 100 old (non-digital) watches in an antique shop, all running but not necessarily on time. Prove that at some moment of time the sum of the distances from the center of the shop to the centers of the watches will be less than the sum of the distances from the center of the shop to the ends of the hour hands of the watches. What if some of them are running fast or slow?

§3. Games

In all the problems in this section it is supposed that there are two players making their moves in turn, one after another. Unless otherwise specified, you must determine who wins (the player who makes the first move, or the other one).

29. A pawn is placed in each of the three leftmost squares of a 1×20 table. A move consists of shifting one pawn to any of the free squares to the right of it, without jumping over any other pawns. The player who cannot make a move loses.

30. A pawn is placed in each of the three leftmost squares of a 1×20 table. A move consists of shifting any of the pawns to the neighboring square on the right if it is free. If this neighboring square is occupied but the next one to the right is not, the player can shift the pawn to that free square. The player who cannot make a move loses.

31. The number 1234 is written on a blackboard. A move consists of subtracting some non-zero digit of the written number from this number (the difference is to be written on the blackboard instead of the old number). The player who writes the number zero wins.

32. The numbers 1 through 100 are written in a row. A move consists of inserting one of the signs "+", "−", or "×" in a free space between any two neighboring numbers. The first player wins if the final result is odd, and loses otherwise.

33. The numbers $1, 2, 3, \ldots, 20, 21$ are written in a row on a blackboard. A move consists of crossing out one of the numbers not yet crossed out. The game ends when there are only two numbers on the blackboard. If the sum of these numbers is divisible by 5, then the first player wins; otherwise, the first player loses.

34. A table with dimensions a) 10×10; b) 9×9 is given. A move consists of filling any free box with a plus or minus sign (a player may choose either sign at each turn). The player whose move creates three consecutive identical signs on a straight line (horizontal, vertical, or diagonal) wins the game.

35. There are two heaps of candies on a table: 22 candies in one of them, and 23 in the other. A move consists of either eating two candies from one heap, or of moving one candy from one heap to the other. The player who cannot make a move wins the game.

36. The first player writes on a blackboard one of the digits 6, 7, 8, 9. Each next move consists of writing one of the same digits to the right of the number on the blackboard. The game ends after a) the 10th; b) the 12th move (note that the 10th move, for instance, is the 5th move made by the second player). The first player wins if the resulting number is not divisible by 9, and loses otherwise.

37. Two players play a game on an infinite sheet of graph paper. The first player puts a cross in some square. In each of his next moves he must put a cross in any free square which shares a common side with a square which already has a cross. The second player puts three crosses in any free squares. Prove that no matter how the first player plays, the second player can "stalemate" him—that is, create a position where the first player has no permitted moves.

38. There are 1001 matches in a pile. A move consists of throwing away p^n matches from the pile, where p is any prime number, and $n = 0, 1, 2, \ldots$. The player who takes the last match wins the game.

39. There are 1991 nails in a plank. A move consists of connecting two of them, which are not yet connected with each other, by a wire. If after a move there is a circuit, then the player who made that move a) wins the game; b) loses the game.

40. A move consists of coloring black one or more boxes which form a square in a given table with dimensions a) 19×91; b) 19×92. It is not allowed to color any box twice. The player who colors the last box wins the game.

41*. A 30×45 sheet of graph paper is given. A move consists of making a cut along a line connecting two neighboring nodes of the lattice. The first player begins by cutting from the edge of the paper. Any cut must begin where the last cut ended. The player after whose move the sheet falls apart wins the game.

42*. A king, when it is his turn, can put two crosses into any two free squares on an infinite sheet of graph paper. His secretary, when it is his move, can put a naught into any free square. Is it possible for the king to get 100 crosses in a row (vertically or horizontally)?

§4. Construction problems

43. A traveler checked into a hotel having only a gold chain with 7 links. The owner requires rent for the room of one golden link per day. What is the minimum number of links of the chain which must be opened so that the traveler will be able to pay exactly the required rent every day?

44. Is it possible to write 10 numbers in a row so that the sum of any five consecutive numbers is positive, and the sum of any seven consecutive numbers is negative?

45. Find a ten-digit number such that the first digit is equal to the number of zeros in its decimal representation, the second digit is equal to the number of ones in the representation, et cetera, and the tenth digit is equal to the number of nines in the representation.

46. Ali-Baba wants to get into Sesame cave. There is a barrel in front of the entrance. It has four holes with a jar inside each of them, and there is a herring in every jar. A herring can sit in a jar with its head up or down. Ali-Baba can stick his hands into any two holes, and, after examining the positions of the herrings, change their positions in an arbitrary manner. After this operation the barrel begins to rotate, and after it stops Ali-Baba cannot tell one hole from another. The Sesame cave will open if and only if all four herrings are in the same position. How must Ali-Baba act to get into the cave?

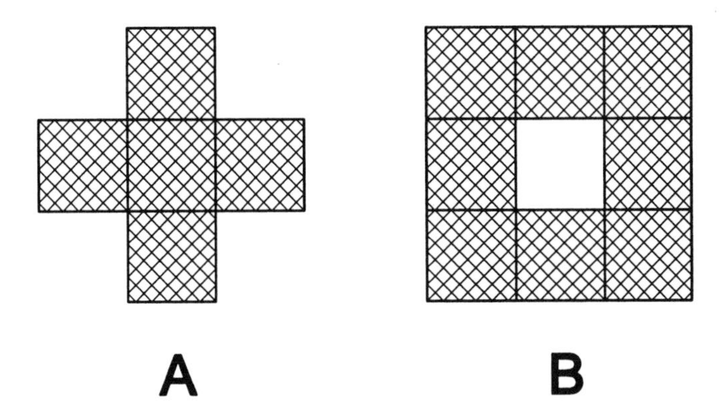

FIGURE 124

47. Find a coloring of a sheet of graph paper using 5 colors such that the boxes in any figure of type A (see Figure 124) are always colored differently, but such that this is not true for any figure of type B.

* * *

48. Draw a figure that cannot cover a semicircle of radius 1, but such that it is possible to cover a circle of radius 1 with two copies of it (the copies may overlap).

49. Choose 6 points on the plane in such a way that any 3 of them lie at the vertices of an isosceles triangle.

50. Draw 11 disjoint squares (that is, not having common interior points) on the plane in such a way that they cannot be properly colored using 3 colors. A coloring of a set of figures is called *proper* if any two figures having more than one point of their boundary in common are colored differently.

51. Cover the plane with non-overlapping squares such that only two of them are the same size.

52. Draw a polygon and a point on the plane in such a way that none of the polygon's sides is completely visible from the point (that is, for any side of the polygon some part of it is obscured by other sides of the polygon if the observer's eye is placed at the chosen point). There are two cases: a) the point is inside the polygon; b) the point is outside the polygon.

53*. Choose 7 points on the plane in such a way that among any three of them there are 2 points 1 inch apart.

§5. Geometry

Set 1. Geometric inequalities

In the chapter "Geometry" we got acquainted with some geometric inequalities and problems using them. In this set we collect several problems on this topic which are more complicated and less standard.

54. Prove that the circles constructed on two sides of a triangle as diameters cover the entire triangle.

55. Prove that the circles constructed on the four sides of a quadrilateral as diameters cover the entire quadrilateral.

56. Prove that a convex polygon cannot have more than three acute angles.

57. Prove that in a convex polygon the sum of any two angles is greater than the difference of any other two angles.

58. A circle of radius 1 and five straight lines intersecting it are given on the plane. Point X is known to be 11.1 inches from the center of the circle. Prove that if X is consecutively reflected in all five lines, then the resulting point cannot lie inside the circle.

59. An astronomer observed 50 stars such that the sum of all the pairwise distances between them is S. A cloud obscured 25 of the stars. Prove that the sum of the pairwise distances between the visible stars is less than $S/2$.

60. A 1×1 square is cut into several rectangles. For each of them we calculate the ratio of the smaller side to the bigger one. Prove that the sum of these ratios does not exceed 1.

61. The vertices of triangle ABC lie on the nodes of a sheet of graph paper (whose boxes have sides of unit length). It is known that $|AB| > |AC|$. Prove that $|AB| - |AC| > 1/p$, where p is the perimeter of triangle ABC.

Set 2. Combinatorial geometry

Here are some problems in so-called "combinatorial geometry". This branch of geometry studies various combinatorial properties of arrangements of geometric figures, like points, lines, polygons, etc., on the plane (and in space). Convexity is usually included in this theme as well.

62. There are 200 points chosen on segment AB in such a way that the set is symmetric with respect to the midpoint of the segment. One hundred points are colored red, the others blue. Prove that the sum of the distances from A to the red points is equal to the sum of the distances from B to the blue points.

63. Five points are given on the plane such that no three of them lie on the same straight line. Prove that some four of them lie on the vertices of a convex quadrilateral.

64. A 2×2 square is cut into several rectangles. Prove that we can color some of them black so that the projection of the colored figures on one of the square's sides has length no greater than 1, while the projection on the other side has length no less than 1.

65. Six dimes lie on a table, forming a closed chain. A seventh dime rolls along the outer side of the chain without "sliding", touching all six dimes on its way. How many full revolutions will it make before it first returns to its initial position?

66. An 8-segment closed broken line whose vertices coincide with the vertices of a cube is given in space. Prove that one of the segments coincides with one of the edges of the cube.

67*. Several segments are given on a straight line, and it is known that any two of them intersect. Prove that some point on the line belongs to all the segments.

68. Several segments cover the segment $[0;1] = \{x : 0 \leq x \leq 1\}$. Prove that you can choose some of them so that they are all disjoint, and the sum of their lengths is at least $1/2$.

69*. There are several segments lying inside the segment $[0;1]$ and covering it. Prove that their left halves cover at least one half of the segment $[0;1]$.

§6. Integers

70. Find all natural numbers equal to the sum of the factorials of their digits.

71. The number $\overline{abc}$ is prime. Prove that $b^2 - 4ac$ cannot be a perfect square.

72. The fourth power of a natural number is written with the digits 0, 1, 4, 6, 7, 9 in some order. Find this number.

73. What is the smallest natural number of the form $111\ldots11$ which is divisible by $333\ldots33$ (one hundred 3's)?

74. Prove that among any 39 consecutive natural numbers one can find a number, the sum of whose digits is divisible by 11.

75. Do there exist two different seven-digit numbers, each written with digits 1, 2, ..., 7 without repetitions, and such that one of them is divisible by the other?

76. The difference of the numbers $\overline{abcdef}$ and $\overline{fdebca}$ is divisible by 271. Prove that $b = d$ and $c = e$.

77. Does there exist a two-digit number whose square ends in the same two digits, but in the reverse order?

78. Integers a, b, and c are given, and it is known that $ax^2 + bx + c$ is divisible by 5 for any integer x. Prove that a, b, and c are themselves divisible by 5.

79. Find all natural numbers n such that the number $n^n + 1$ is prime and has no more than 19 digits.

80*. The natural number y was obtained from the number x by rearranging its digits. It is known that $x + y = 10^{200}$. Prove that x is divisible by 50.

81*. Prove that the number $\overline{a000\ldots00b}$ cannot be a perfect square.

§7. Optimization problems

Set 1. The principle of the extreme

The main idea of the solutions to the problems of this set is a consideration of an "extreme" (in some sense) object: the largest number, a pair of points with the greatest distance between them, the smallest angle, et cetera.

82. a) Numbers are placed on the vertices of a 100-gon in such a way that each of them equals the arithmetic mean of its neighbors. Prove that all the numbers are equal.

b) Prove the same fact if the numbers are placed on the squares of a chessboard and it is known that each of them does not exceed the arithmetic mean of its neighbors.

The solution to Problem 82a) is rather simple. Let us consider the smallest number. It is clear that its neighbors must coincide with it—otherwise their arithmetic mean would be greater than the smallest number. Therefore, their neighbors must be equal to them too, and so on. This shows that all the numbers are equal.

83. Ten points are given on the plane. Prove that one can find 5 disjoint segments whose endpoints coincide with the given points.

84. Is it possible to draw several segments on the plane so that every endpoint of each of them lies within another of the drawn segments?

85. There are several coins lying on the table of a money-changer so that they do not touch each other. An armless thief wants to push one of them, using his nose, to the edge of the table, where he can pick it up. The coin being moved may never touch any other coin (lest the money-changer hears the sound). Is this always possible?

86. a) Several identical coins lie on a table. Prove that one of the coins touches no more than three others.

b) Several coins of different sizes lie on a table. Prove that one of the coins touches no more than five others.

87. There are several airports in a country, and it is known that the distances between them are all different. An airplane takes off from each airport and flies to the closest airport. Prove that there will be no more than five of the planes at each airport after they all have landed.

88. In outer space there are 1991 asteroids, and an astronomer lives on each of them. All the distances between the asteroids are different. Each astronomer is observing the closest asteroid to him or her. Prove that there is an asteroid which is not being observed.

Set 2. Semi-invariant

The idea of a "semi-invariant" is a quite natural extension of the idea of an invariant. We will call some quantity a "semi-invariant" if this quantity changes monotonically during some transformation process. A typical semi-invariant is the age of a person, which unfortunately can only increase as time passes.

89. There are several cities in a certain kingdom. An obnoxious citizen is exiled from city A to city B, which is the farthest city in the kingdom from A. After a while, he is again exiled from city B to the farthest city from it, which happens to be different from A. Prove that, if his exiles continue the same way, he will never return to city A.

90. One hundred coins lie in a row, arranged head, tail, head, tail, ... A move consists of turning over several coins lying in succession. What is the minimum number of moves necessary to reach the situation when all the coins lie with heads up?

91. At midnight a virus was placed into a colony of 1984 bacteria. Each second each virus destroys one bacterium, after which all the bacteria and viruses divide in two. Prove that eventually all the bacteria will be destroyed, and determine the exact time of this event.

92. One real number is written in each box of a rectangle table. It is allowed to change the sign of all the numbers in any row or column. Prove that, using these operations, one can make the sum of the numbers in the table non-negative.

93*. In a given graph with n vertices the degree of each vertex does not exceed five. Prove that the vertices can be colored using three colors in such a way that there are no more than $n/2$ edges with endpoints of the same color.

94*. Real numbers are written along a circle. If some four successive numbers a, b, c, and d satisfy the inequality $(a-d)(b-c)>0$, then it is permitted to transpose the numbers b and c. Prove that it is impossible to perform infinitely many such permutations.

§8. Discrete continuity

Here we gather problems whose solutions are based mostly on one very simple idea, which can be illustrated by the following example. There is a flea jumping along the integers on a number line. It is known that at each jump the flea moves no further than to a neighboring number. If the flea was originally sitting on a negative number, and it ends up on a positive number, then at some moment the flea was on zero.

95. There are 100 points chosen on the plane. Prove that there exists a straight line such that exactly fifty of the points lie on one side of it.

Let us discuss the solution to Problem 95. First, we draw all straight lines passing through pairs of points in our 100-element set. Second, we find a line which is parallel to none of these, and such that all the points are to one side of it. Now we move this line toward the points so it is always parallel to its initial position, and observe how the number of the points behind the line changes. It is quite obvious that we cannot pass through two points simultaneously. Thus we have the same sort of "discrete continuity" as in the flea example. Since we started with zero chosen points to one side of the line and will eventually have all 100 points to that side of the line, then, clearly, at some moment the number of points behind the line will be equal to 50, and we will have the required result.

96. One hundred black balls and one hundred red balls are laid in a row so that the leftmost and the rightmost balls are black. Prove that it is possible to choose several successive balls (but not all of them!) from the left side of the row in such a way that there are equally many red and black balls among them.

97. The soccer match between the New York Centaurs and Liverpool ended with the score 8:5. Prove that there was a moment during the match when the Centaurs were going to score the same number of points as Liverpool already had at that moment.

98. Prove that you can rearrange the digits of any six-digit number in such a way that the difference between the sum of the first three and the sum of the last three is between 0 and 9.

99. The numbers $+1$ and -1 are placed in the squares of an 8×8 board in such a way that the sum of all the numbers equals zero. Prove that the board can be cut into two parts so that the sum of the numbers in each part is zero.

100. The faces of eight unit cubes are colored black and white so that there are equally many black and white faces. Prove that one can form a $2 \times 2 \times 2$ cube from the unit cubes in such a way that there are equally many white and black faces on its surface.

101. The numbers $+1$ and -1 are placed in some boxes of a 50×50 square, and it is known that the absolute value of their sum is not greater than 100. Prove that there is a 25×25 square such that the absolute value of the sum of the numbers within it is not greater than 25.

102.* The sequence (a_n) is such that $a_1 = 1$, and $a_{n+1} - a_n$ is always either 0 or 1. It is also known that $a_n = n/1000$ for some natural n. Prove that $a_m = m/500$ for some natural m.

§9. Power questions

This section is devoted to so-called "power questions". Each of these is just a chain of several simpler facts (lemmas) which add up to the proof of a rather difficult problem. First we state the required result, then the lemmas you must prove in order to solve the problem. Such problems are very useful for homework, and some of them can be discussed at a special session.

103. Equilateral triangles on graph paper. Prove that there is no equilateral triangle with its vertices on the lattice points of graph paper.

a) Prove that for any triangle with vertices on the lattice points of graph paper, its doubled area is an integer.

b) Find the length of the altitude of an equilateral triangle with side a and calculate its area.

c) Prove that the square of the length of any segment with its endpoints on two lattice points is an integer.

d) Prove that if there exists an equilateral triangle with its vertices on three lattice points, then $\sqrt{3}$ can be represented as a ratio of two natural numbers.

e) Prove that if the square of a fraction, whose numerator and denominator are relatively prime, is an integer, then the denominator is 1.

f) Now prove the statement of the problem.

104. Pick's formula. Prove that the area of a polygon with vertices on the lattice points of a piece of graph paper equals $a + b/2 - 1$, where a is the number of lattice points inside the polygon, and b is the number of lattice points on its boundary.

a) Prove Pick's formula for a rectangle with its sides on the lines of the grid.

b) Prove Pick's formula for a right triangle with its legs on the lines of the grid.

c) Prove Pick's formula for a polygon which can be dissected into two polygons satisfying Pick's formula.

d) Assume that we have a polygon for which Pick's formula is true, and this polygon is dissected into two smaller polygons. Prove that if Pick's formula is true for one of these smaller polygons, then it is true for the other one.

e) Prove that Pick's formula is true for any triangle with its vertices on the lattice points of the grid.

f)* Prove that any polygon can be dissected into triangles by a set of its disjoint diagonals (see Problem 31 in the chapter "Induction").

g) Prove Pick's formula in the general case.

105. Bolyai-Gervin's theorem. Two polygons with equal areas are given. Prove that the first polygon can be dissected into several parts which can be rearranged to form the second polygon. We will call two figures that can be so transformed into each other "equiform".

a) Prove that two parallelograms with a common base and equal altitudes are equiform.

b) Prove that if polygons P_1 and P_2 are equiform, and it is also known that P_2 and P_3 are equiform, then P_1 and P_3 are equiform too.

c) Prove that two parallelograms with equal areas are equiform.

d) Prove that every triangle is equiform to some parallelogram.

e) Prove that every triangle is equiform to a rectangle, one of whose sides has length 1.

f) Prove Bolyai-Gervin's theorem.

106. Divisibility of Fibonacci numbers. Prove that if F_n (the nth Fibonacci number) is prime, then either $n = 4$ or n is prime.

a) Prove (by induction on n) that for any natural numbers m and n

$$F_{m+n} = F_{m+1}F_n + F_mF_{n-1} .$$

b) Prove that F_{km} is divisible by F_m (use induction on k).

c) Prove that if n is divisible by $m > 2$, then F_n is composite.

d) Prove that every composite number different from 4 has a proper divisor greater than 2, and complete the solution to the problem.

107. Helly's theorem. There are several convex sets on the plane, and it is known that any three of them have a common point. Prove that all these sets have a common point.

a) Four points marked with the sets of numbers $\{1, 2, 3\}$, $\{1, 2, 4\}$, $\{2, 3, 4\}$, and $\{1, 3, 4\}$ are given on the plane. It is allowed to connect any two points marked with some sets of numbers by a segment, and mark all the points of the segment with the numbers which are common to both ends of this segment. Prove that using these operations one can find a point which is marked with all four numbers 1, 2, 3, and 4.

b) Prove that the intersection of convex sets is convex.

c) Prove Helly's theorem for four convex sets A_1, A_2, A_3, and A_4; remember to assume that any three of these have a common point.

d) Prove the general case of Helly's theorem by induction on the number of convex sets.

§10. Miscellaneous

108. The boxes of a 9×9 table are colored using two colors. It is allowed to choose any 3×1 rectangle and color all its boxes with the color that there had been the most of in this rectangle before. Prove that using these operations one can reach the situation where all the boxes are the same color.

109. Is it possible for the six differences between some set of four numbers to be equal to 2, 2, 3, 4, 5, and 6?

110. There are 1970 people living in a village. Each day some of them exchange a dime for two nickels with other people in the village. Is it possible that during one week each of the people of the village has given away exactly 10 coins in the course of the exchanges?

111. Pete, Paul, and Mary were solving problems from a problem book. Each solved exactly 60 problems, but they solved only 100 problems altogether. We will call a problem "easy" if it was solved by all three of them, and "difficult" if it was solved by only one of them. Prove that the number of difficult problems exceeds the number of easy problems by 20.

112. It is known that the square of the number x has the following decimal representation: $0.999\ldots99\ldots$ (one hundred 9's after the decimal point). Prove that the number x itself has a decimal representation of the same type (but, perhaps, with another number of 9's).

113. Forty-nine rooks are placed on a 100×100 chessboard, and a king is placed in the lower left corner of the board. The king makes moves towards the upper right corner, and after each of its moves one of the rooks makes a move too. Prove that at some moment the king will be under attack by one of the rooks.

114. In the numeral 0.1234567891011... all the natural numbers are written in succession after the decimal point. Is this fraction periodic?

115. Three players play table tennis, and the one of them who does not participate in a particular game plays the winner in the next game. By the end of the day the first player has played 10 games, and the second has played 21. How many games has the third player played?

116. The numbers x and y are such that $\sqrt{x} - \sqrt{y} = 10$. Prove that $x - 2y < 200$.

117. In a 10×10 table all the boxes in the left half of the table are colored black, and the others are colored white. It is permitted to change the color of all the boxes in a row or in a column. Is it possible to achieve the standard chess coloring of the table, using only these operations?

118. A children's game has several pieces of railroad track of two types (see Figure 125) arranged in a closed loop in such a way that the direction of a train's movement coincides with the directions of the arrows. One of the type 1 parts was broken and replaced with a spare part of type 2. Prove that now it is impossible to arrange all the parts in a closed loop.

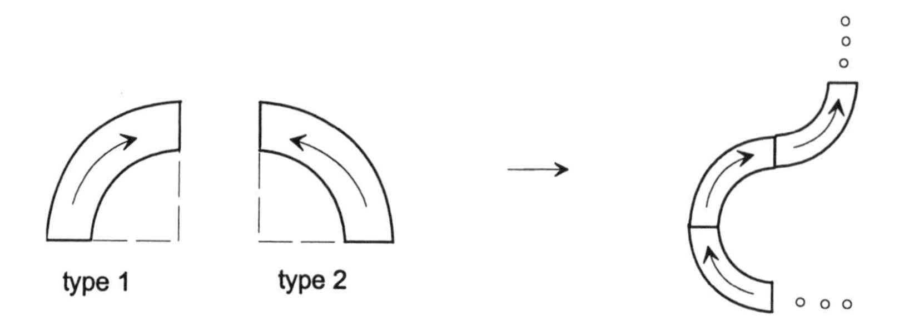

FIGURE 125

119. A town is laid out in the form of a convex polygon, whose diagonals are the streets. Their intersections and the polygon's vertices are the crossings. There are tram routes along some of the streets, and it is known that for every crossing there is at least one route passing through it. Prove that one can reach any crossing from any other, changing routes no more than twice.

120. There are 100, 200, and 300 students respectively in three towns. Where must a school be built to minimize the overall distance the students walk every day?

121. Is it possible to cut a convex 17-gon into 14 triangles?

122. There are three frogs sitting on three vertices of a square. They play leapfrog, jumping one over another. If frog A jumps over frog B, then A lands at point A′ such that B is the midpoint of segment AA′. Is it possible for one of them to jump onto the fourth vertex of the square?

123. Twenty-five elephants stand in a row in a circus arena, and each of them weighs an integer number of kilograms. It is known that if you add the weight of any of them (except the rightmost one) and half the weight of its neighbor to the right, then the result is 6 tons. Find the weights of the elephants.

124. A billiard table with dimensions 101×200 has pockets only at the four corners. A ball rolls out of the table's corner along its angle bisector and moves on, reflecting from the sides of the table. Will it ever roll into a pocket again?

125. The diagonals of a convex 13-gon cut it into several regions. What is the maximum number of sides of such a region?

126. Prove that if the natural number n is greater than 1, then the number $1 + 1/2 + 1/3 + 1/4 + \ldots + 1/n$ cannot be an integer.

127. What is the maximum possible radius of a circle which lies on a chessboard and does not intersect any white squares (except at their corners)?

128. A round socket has 6 holes uniformly spread along its border. We also have a similar plug with 6 pins. The holes in the socket are numbered with the numbers 1 through 6 in some order; the same is done with the pins on the plug. Prove that the plug can be plugged into the socket so that none of the connectors will go into the hole with the same number. Is this true if there are 7 holes and 7 connectors?

129. Two natural numbers m and n have natural divisors $a_1, \ldots, a_p$ and $b_1, \ldots, b_q$ respectively, and it is known that

$$a_1 + \ldots + a_p = b_1 + \ldots + b_q,$$
$$\frac{1}{a_1} + \ldots + \frac{1}{a_p} = \frac{1}{b_1} + \ldots + \frac{1}{b_q}.$$

Prove that $m = n$.

130. In a certain country the following coins are in circulation: 1 cent, 2 cents, 5 cents, 10 cents, 20 cents, 50 cents, and 1 dollar. It is known that you can pay A cents with B coins. Prove that you can pay B dollars with A coins.

131. Is it possible to place one hundred natural numbers around a circle so that the product of any two neighboring numbers is a perfect square?

132. There are n physicists and n chemists sitting at a round table. It is known that some of them always lie, and others always tell the truth. It is also known that the number of physicist liars is equal to the number of chemist liars. Each of these people said: "My neighbor to the right is a chemist." Prove that n is even.

133. The natural numbers a and b are such that a^2+ab+1 is divisible by b^2+ba+1. Prove that $a = b$.

134. Each of two mathematical geniuses is given his or her own secret natural number, and they both know that these numbers differ by 1. They ask each other in turn: "Do you know my number yet?" Prove that eventually one of them will answer the question affirmatively.

135. One hundred integers are written around a circle, and it is known that their sum is 1. We will call a subset of several successive numbers a "chain". Find the number of chains whose members have a positive sum.

136. In a chess tournament each of the players gained exactly half of their points in games with the players occupying the last three places in the tournament. How many players took part in the tournament?

137. The numbers a, b, and c are such that $a+b+c = 7$ and $1/(a+b)+1/(b+c)+1/(c+a) = 7/10$. Find the value of the expression $a/(b+c) + b/(a+c) + c/(a+b)$.

138*. One hundred and nineteen people live in a building with 120 apartments. An apartment is called "overpopulated" if there are more than 15 people living in it. Each day all the tenants of some overpopulated apartment have a quarrel, and they all move to different apartments in the same building. Is it true that eventually none of the apartments will be overpopulated?

139*. There are several cars on a circular track, and there is enough gas in all their tanks taken together to drive one car around the track. Prove that at least one of the cars can drive around the track by taking gas from other cars on its way.

140*. The number sequence 1, 9, 8, 2, ... satisfies the following rule: each element of the sequence, starting from the fifth, is equal to the last digit of the sum of the previous four members. Will we ever meet four successive members equal to 3, 0, 4, 4 in this sequence?

141*. There are 25 stones in a heap. The heap is divided into two parts, then one of the parts is divided in two again, et cetera, until we have 25 separate stones. After each division of one of the heaps into two smaller heaps we write the product of the numbers of stones in these two heaps on a blackboard. Prove that at the end the sum of all the numbers on the blackboard is 300.

142*. For any boy in a certain village all the girls who are acquainted with him are acquainted with each other. Also, for each girl there are more boys than girls among her acquaintances. Prove that the number of boys in the village is greater than or equal to the number of girls.

143*. A snail crawled along a straight line for 6 minutes while several people watched its progress. Each of these people watched the snail for 1 minute, and during this minute the snail crawled exactly 1 foot. It is also known that the snail was always watched by at least one person. What is the maximum possible distance the snail could crawl during these 6 minutes?

APPENDIX A

Mathematical Contests

§1. Introduction

The Sphinx ... lay crouched on the top of a rock, and arrested all travellers who came that way, proposing to them a riddle, with the condition that those who could solve it should pass safe, but those who failed should be killed. Not one had yet succeeded in solving it, and all had been slain.

Oedipus was not daunted by these alarming accounts, but boldly advanced to the trial. The Sphinx asked him, "What animal is that which in the morning goes on four feet, at noon on two, and in the evening upon three?"

From "Mythology" by Thomas Bulfinch. A modern abridgment by Edmund Fuller. Dell Publishing, New York, 1959.

It is very likely that the riddle of the Sphinx was the very first olympiad problem in the full meaning of the word. As everybody knows, the outcome of that ancient contest was quite fortunate for Oedipus.

Contemporary mathematical contests do not require human sacrifice, and many students gladly participate in a variety of them. This unique sort of competition combines mathematics, sports, and a test of psychological endurance. Some contestants allow themselves to get involved to such a degree that they become olympiad "professionals" (which is not always good for their real mathematical education). We, nevertheless, hope that the reader will find the contests discussed in this chapter interesting, useful, and instructive.

*   *   *

For teachers. 1. Remember that students, especially younger ones, like to turn every serious matter into a game, sport, or recreation. At first, this is acceptable, and you can use this as one of the ways to get them acquainted with new areas of mathematics. However, you may want a more serious tone for the main activity of your seminar.

2. Problems for mathematical contests can be found in any chapter of this book and in many others. Use the list of references (try [**44**], for example).

§2. Mathematical battle

This is one of the most popular mathematical contests in Lenigrad (and in Russia). It was invented in the middle of the 1960's by Iosif Verebeichik, then a teacher of mathematics in one of the Leningrad schools. This is a team competition which, remarkably, unites mathematics, sports, team spirit, and theatrical performance.

We will briefly explain the basic rules. Each of two teams receives a list of problems prepared by the jury (the problems are the same for both teams). They are given a certain amount of time (which may vary from 30 minutes to a week) to solve these problems. After the time expires, the teams' members and the jury gather in some auditorium (with a big blackboard and a supply of chalk), and the battle begins.

First, through a "captains' contest", the jury determines which team goes first. The two team captains are given a simple question which must be answered on the spot, on the blackboard, and without consulting the other team members. For example, is the number 7999 prime? Or, there are seven rubber rings in space; is it possible that each of these is linked with exactly three other rings?

As soon as one of the captains gives an answer the captain's contest is over. If the answer is correct, then the team whose captain has come up with it wins. Otherwise, the other team wins.

The victorious team decides which of the teams—say, team A—will "issue a challenge", and after that the "challenge" itself follows. That means that team A declares that they want to hear team B's solution to some problem from the list. Team B can accept the challenge by sending one of its members to the blackboard to act as "storyteller" (that is, to explain the solution). Team B can also reject the challenge.

In the former case, team A sends one of its members to act as an opponent to the storyteller. His or her responsibility is to check the solution, reveal its weak points, or, perhaps, even prove it wrong.

In the latter case team A must prove that they challenged team B in a correct manner. This means that they must delegate a "storyteller" , who gives a correct solution to the problem. As before, team B sends an opponent who tries to disprove the solution or at least find some minor errors or unproved points.

In all these cases but one the next challenge must be made by the other team. The exception occurs if after checking the correctness of the challenge (see the previous paragraph) the jury decides that team A did not provide a correct solution to the problem (the jury always has the right to check the solution, ask questions, et cetera). In this case, team A is to be fined, and they must also repeat the challenge (for another problem, of course).

After the discussion of a solution is finished, the jury distributes the points (each problem is worth 12 points). Points may be granted to the opponent even if the solution of the storyteller was correct; for example, if there were some minor errors which were pointed out by the opponent and then corrected by the storyteller. The jury is also permitted to award points to itself.

If a major error found by the opponent (or by the jury) was not corrected by the storyteller within some standard amount of time (usually 1 minute), the explanation terminates, and the jury can decide to hear the other side. When this extra discussion is finished the jury distributes the points.

If one of the teams runs out of problems they have solved and is not willing to take chances by challenging the other team for an unsolved problem, then they can forfeit their right to challenge. In this case, the other team can give the rest of the solutions they have at that moment. These explanations are carried out as usual, with opponents present.

Many other minor rules and restrictions have been added to the basic code over 30 years of mathematical battles. Among them are, for instance:

1) the setting of a fine for an "incorrect challenge" at 6 points;

2) the restriction that no one of the contestants can appear at the blackboard more than x times (the captain's contest does not count), where x is some natural number announced when the list of problems is given. Usually, $x = 2$ or 3.

3) a rule that only the captain or the temporary deputy (in case the captain acts as storyteller or opponent, or is absent) can talk to the jury.

* * *

The last and the most revered law of the mathematical battle reads that in any uncertain situation the jury rules—it is both the judicial and legislative system of the mathematical battle.

* * *

The organization of a battle requires certain skills and some experience: for example, it is a very standard opening gambit (like the $e2$–$e4$ move in chess) to challenge your adversary to the most difficult problem solved by your team. We also recommend that you organize a few simple training mathematical battles within your circle or school.

We should mention here that the popularity of mathematical battles in Leningrad was sometimes so high that specialized schools even held championships (in the senior grades only). There were also "triple" mathematical battles (with three teams participating) which are far more difficult to organize.

As an illustration, we give here an example of a mathematical battle. Below is the problem list and the record of a battle which was actually held in 1986 between two advanced Leningrad circles for sixth graders (ages 12-13) affiliated with the Leningrad Palace of Pioneers and the Youth Mathematical School.

* * *

1. A "crocodile" is a pawn which can move across an infinite sheet of graph paper in the following way: standing on any square it first moves to a neighboring square (horizontally or vertically) and then moves n boxes in the direction perpendicular to the direction of the first shift. For example, if $n = 2$, the "crocodile" is a chess knight. Find all n such that the "crocodile" can reach any box from any other.

2. The numbers p and $2^p + p^2$ are primes. Find p.

3. Can the cube of a natural number end with 1985 ones?

4. Prove that one can always choose three diagonals from among the diagonals of a convex pentagon, which can form a triangle.

5. The boxes of an $n \times (n+1)$ table are filled with integers. Prove that one can cross out several columns (but not all of them!) so that after this operation all the sums of the numbers in each row will be even.

6. How many ways are there to represent the number 15 as the sum of several natural addends, if we distinguish representations which differ in the order of their addends?

7. Point A is called a "pseudocenter of symmetry" of set M (containing more than one point on the plane) if it is possible to remove a point from M in such a way that A will be the center of symmetry of the resulting set. How many pseudocenters of symmetry can a finite set have?

8. Thirty numbers are placed around a circle so that each of them equals the difference of the two numbers following it in a clockwise direction. Given that the sum of these numbers is 1, find them.

* * *

Captain's contest: is the number 227 prime? (won by LPP team)

The Jury		Y M S		L P P	
	0	Ivanov	—6⟶	Dogolyatsky	12
18	0		| ⟵5—	incorr.challenge	−6
	12	Demchenko	⟵8—	Gurevich	0
	12	Vyskubov	—4⟶ |	Pchelintsev	0
	12	Ivanov	⟵2—	Roginskaya	0
	12	Viro	—3⟶ |	Dogolyatsky	0
	12	Novik	⟵1—	Gurevich	0
18	−6	incorr.challenge	|—7⟶		0

Remark 1. The symbol —6⟶ denotes the challenge for Problem 6. An arrow ending with a vertical line (like —3⟶ |) denotes a rejection of the challenge.

Remark 2. The record shows that the YMS team committed a serious tactical error. What was this?

§3. Mathematical fight

Unlike the mathematical battle, this is an individual event. Therefore, we recommend its use in a circle which is rather homogeneous in mathematical strength.

Some problems are submitted for solution: each of them is supplied with its "price" in winning points. As soon as somebody wants to give a solution, he or she steps forward to the blackboard and explains it. If the solution is correct, then the storyteller collects the price. Otherwise, the price of the problem must be slightly increased—the amount of this increase is determined by the teacher—and the same amount is subtracted from the storyteller's balance.

This contest is risky since it can actually last forever if some of the problems are too difficult. The teacher must take measures against this.

This contest teaches the students the art of self-control: they must double-check their solutions. If they do not they can finish the fight with a negative number of points.

We conclude this section with an example of a real mathematical fight:

9. An "emperor" in checkers is a piece that can move backwards and forwards any number of squares along a diagonal, capturing a piece by passing over it (but it cannot pass over two or more pieces in diagonally adjacent squares). What is the largest number of checker "emperors" that can be placed on a checkers board with dimensions 8×8 so that each of them can be directly taken by some other "emperor"? (5 points)

10. Perpendiculars are drawn from the midpoints of the sides of acute triangle ABC to its sides. Prove that the area of the resulting hexagon is half the area of the triangle. (6 points)

11. Find 2 three-digit numbers x and y such that the sum of all the other three-digit numbers equals $600x$. (6 points)

12. A right-angle tool can be used to draw a line through two given points, and erect a perpendicular to a given line at a given point. Use this tool to drop a perpendicular from a given point to a given line. (10 points)

13. There are 10 girls and 10 boys in a dancing class. It is known that for any $1 \leq k \leq n$ and for any group of k boys the number of girls who are friends with at least one boy from this group is not less than k. Prove that it is possible to split the class into 10 pairs for a dance so that every pair consists of a boy and a girl who are friends. (20 points)

Record list:

(9): 5—Peterson
(10): 6 → 7—Thompson
(11): 6 → 9—Johnson
(12): 10 → 15 → 18—Smithson
(13): 20 → 27—Peterson

Results:

Peterson (+24)
Thompson (+7)
Johnson (+9)
Smithson (+12)
Knickerbocker (−5)

Exercise. In the list above the numbers in parentheses give the order of the problems. The numbers that follow record how each price changed in the course of the contest. Then the name of the solver is given. Try to reconstruct the fight.

Remark. The problems from the problems list can be submitted for solution one at a time or all at once.

§4. Mathematical marathon

This is an oral olympiad (which can also be held in written form). To organize it the teacher might want to recruit some assistants, and to create a rather long list of sufficiently simple problems. Before we describe the details we should say a few words about the system of oral olympiads, which is extremely popular in St. Petersburg but is not well-known beyond the former iron curtain.

All solutions must be submitted orally to one of the jurors, but they do not have to be written in advance. The result of the presentation is recorded (+ for an adequate solution and − for none). Each contestant is allowed three attempts on every problem. Therefore, the record list can contain such grades as "two minuses" (=) or "plus with two minuses" ($\overset{+}{=}$). The latter is usually considered equivalent to +.

An oral olympiad is usually split into two stages: a "preliminary" and a "final" stage (this is how it is done, for example, at the All-City St. Petersburg olympiad, which is oral). At the first stage all contestants are given 4 preliminary problems to solve, and only those who have correctly solved three of them (or sometimes two) are transferred to another auditorium where they get the complete list of 6 or 7 problems (they are already familiar with the first four of them).

The mathematical marathon is the ultimate version of this elimination system. The teacher must have a list of 10-20 problems carefully prepared in advance. The first 5-10 of them must be very simple and quite standard. The level of difficulty should increase slowly with the number of the problems in the list.

At the beginning of the marathon all the students receive problems 1 and 2 from the list. After that they start solving and explaining their solutions to the teacher or assistants. If a problem is solved correctly, then the juror puts a plus sign in the record, and the contestant gets the next problem from the list. For instance, if Mary had problems 3 and 7, then after solving either of them she will get problem 8.

Thus, at any moment of the olympiad all participants have exactly two unsolved problems on their hands.

Experience has shown that during the marathon students manage to solve more problems than they otherwise would in the same time period, probably because their concentration is much higher. Experience also shows that students find marathons more interesting and attractive than other, more well-known olympiads.

For teachers. Using problems from this book, one can create several very simple marathons and some difficult ones. We recommend that you take several of the simplest exercises from the chapters you are interested in and mix them up in an arbitrary order to make up the first half of a marathon. The second part can be made up similarly but using slightly more difficult problems. The level of difficulty must gradually increase as the students move through the list.

Warning. The most dangerous mistake one can make in preparing a marathon is to underestimate the complexity of the first problems on the list. If one of the first 5–6 problems is not solved by most of the students, then the marathon will be a failure.

§5. Mathematical hockey

This interesting mathematical competition is intended for students of ages 10 through 12 or older. The game is played by two teams, each consisting of 5 players: one "goalkeeper", two "defensive players", and two "forwards".

The teacher must have a rather long list of extremely simple problems, preferably of a numerical nature, so that each of them could be solved in five minutes.

At the beginning of the game the puck (imaginary, of course; though you can draw a hockey field with zones and puck on a blackboard) is at the center of the field. The puck is thrown in—this means that all the field players of both teams are given the first problem from the list. If team A finds the solution sooner the puck moves to the zone of defeated team B, where the forwards of team A will play against the defensive players of team B. Their struggle is over the next problem from the list. Depending on the outcome of this combat, the puck moves again either back to the center of the field or to the goal zone of team B. In the latter case, the goalkeeper of team B plays alone against the forwards of team A during the next struggle, as they try to solve the next problem from the list. If the forwards win they score a point, and the next problem will be "thrown in" at the center of the field.

You can think of the field as consisting of the five zones shown in Figure 126. At each moment of the game the puck is in one of these zones. Depending on the outcome of each struggle, the puck moves to the adjacent zone, either to the left or to the right.

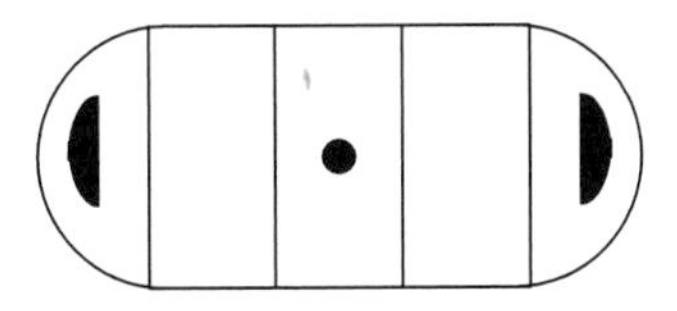

FIGURE 126

Hints. 1. To make this game more dynamic, remove the central part of the field. In this case the initial throwing-in can be done by tossing a coin.

2. If there are too many students in your circle, then you can make three or more teams. They can play a round-robin or elimination tournament. All the teams not playing in the current game act as spectators.

3. You should choose up sides quite carefully. Try to reach an approximate equality in the average strength of the teams (this goes for any other mathematical team contest as well).

§6. Mathematical auction

This contest is very close to gambling, though of course it is not. Our experience shows that students participate in such competitions with great enthusiasm. What is also important is that a mathematical auction can be held as an individual contest as well as in team format.

The rules are as follows. The teacher gives the students one problem of a special type (we will call it a "research problem"), a complete solution to which may be unknown even to the teacher (though this is not recommended). More specifically, a research problem must allow intermediate answers which make possible a gradual approach to the final result. To make the competition more interesting there must be at least 5 or 6 of these problems.

14. What is the maximum number of chess bishops which can be placed on an 8×8 chessboard (other versions: chess knights, rooks; 10×10 board, et cetera) so that no two of them attack each other?

15. What is the maximum number of figures shown in Figure 127 which can be placed without overlapping inside a 10×10 table?

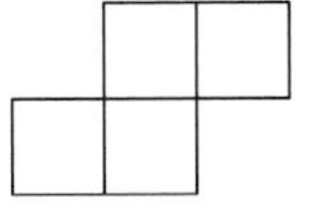

FIGURE 127

16. Find as many solutions as possible to the following alphanumeric puzzle: BACK + BOA = SCAM.

17. Using the digits 1, 9, 8, and 4 in the given order, and four arithmetic operations, write as many consecutive natural numbers as possible starting with 1. Example: $5 = (1+9)/(8/4)$.

18. Write the number 1991 using only 4's. Try to use as few digits as possible. You may use as many arithmetic operations as necessary.

19. Draw 7 lines on the plane (other versions: 8, 9, or 10 lines) in such a way that among the parts they dissect the plane into, the number of triangles is a maximum.

20. Arrange the maximum possible number of "maharajas" on a 10×10 board so that each of the squares of the board is under attack by at least one of them (a maharaja is a "superchess" piece which can move like a queen and also like a knight).

* * *

Important remark. The exact statements of the problems above can be altered to get other problems, simply by changing particular numbers, figures, and so on.

* * *

After a problem is submitted to the students, they are given a certain amount of time to solve it (the event is best managed if more than one problem is submitted at the same time). Then each team is allotted an equal number of units of fictional "currency", such as dinars. For instance, their initial capital may be 1000 dinars.

Finally, the auction itself begins. Each problem is put up for bid by an auctioneer (usually the teacher). The auctioneer announces the beginning of the auction and the value of the first problem.

Suppose Problem 17 (above) is put up for auction, and is valued at 180 dinars. Then the team which solves this problem will get 180 dinars for the solution. But, as we will see, the cash flow may not be straightforward.

* * *

Example. Team A pays 132 dinars for the right to give their results to the problem. They demonstrate how to write all natural numbers 1 through 62 in the required way. However, during this demonstration it is discovered that their representation of the number 51 is incorrect. Therefore, they get credit only for expressing the numbers 1 through 50.

* * *

Then the problem is again put up for auction, but this time the team who buys it is permitted to give only stronger results than those of the previous team.

* * *

Example (cont.). Suppose team B now pays 25 dinars for the right to improve on team A's results. They demonstrate how to express in the required way the numbers 51 through 68 (of course, they need not repeat the representations of the numbers 1 through 50).

* * *

The problem is put up for auction again and again, until no team wants to buy the problem anymore. When this happens, the team with the best result collects the value of the problem. In cases when the winner can give a complete solution (that is, by showing that their result cannot be improved), they are eligible for special prize money (say, another 50 dinars).

* * *

Example (cont.). For example, team C pays 6 dinars for the right to give a better solution than team B. But their representation of the number 69 turns out to be wrong. Their 6 dinars are simply wasted, and the auction trade of Problem 17 is terminated.

Results of part of the auction:

Team A lost 132 dinars.

Team B gained 155 dinars ($180 - 25 = 155$).

Team C lost 6 dinars.

* * *

Now the next problem is put up for auction, and so on.

Here are five more research problems that can be used at auctions.

21. Make the minimum possible number of marks on a wooden plank so that every integer number of inches from 1 through 15 (other versions: from 1 through 20, 30, et cetera) can be measured using this plank; that is, this length can be represented as the distance between some pair of the marks.

22. What is the minimum possible number of straight cuts necessary to split a $5 \times 5 \times 5$ cube into 125 unit cubes, if the pieces can be rearranged arbitrarily between cuts?

23. There are 10 bricks, each 10 inches long. It is permitted to arrange them in a stable stack, and the bricks are not required to lie exactly over each other. What is the maximum possible horizontal distance between the right edges of the top and the bottom bricks in the stack?

24. Dissect a square into the smallest possible number of acute triangles.

25. Find as many solutions of the equation $x^2 + y^2 = z^2$ in natural numbers no greater than 50 as possible (other versions: greater than 40 or 100).

APPENDIX B

Answers, Hints, Solutions

0. CHAPTER ZERO

1. Thinking backwards, we can see that if the glass is full after 60 seconds, it must have been half full one second before. Answer: after 59 seconds.

2. Again, we can think of this backwards. One of the tourists, say Alex, can pay 15 chips for all three of them. Then each of the others owes Alex 5 chips. They can easily pay, for example, by giving Alex a 20-chip coin and receiving a 15-chip coin in return.

3. The key insight here is that the parity of the number of the last page torn out is opposite that of the first page. This is not hard to see if we watch what happens if Jack tears out 1 page, then 2 pages, then another small number of pages. Of the three-digit numbers with digits 1, 3, and 8, only 318 is greater than 183 and has parity opposite that of 318. Now, $318 - 183 + 1 = 136$, which is the answer.

4. The problem allows us only to divide a given set of nails into two parts with equal weights. So, for example, we can get piles of 12, 6, and 3 pounds by continually halving the original pile. Then three 3-pound piles will give 9 pounds of nails.

5. Most people would reason that the caterpillar's net achievement each day and night is 1 inch, so it will take him 75 days. However, on the 70th day the caterpillar will have climbed 70 inches, and the next day's effort will supply him with the "missing" 5 inches. Answer: the caterpillar will be on the top of the pole at the end of the 71st day (before the 71st night begins).

6. The 1st, 8th, 15th, 22nd, and 29th days of any month are the same day of the week. Since January has 31 days, if the month starts on a given day, there will be five of those days, and also five of the next two days of the week. Hence January could not have started on Saturday, Sunday, or Monday (otherwise there would be five Mondays), and not on a Wednesday, Thursday, or Friday (otherwise there would be five Fridays). Hence January 1 was a Tuesday, and it is not difficult to see that January 20th was a Sunday.

7. Without loss of generality we can assume that the given diagonal goes from the upper left corner to the lower right. Then, we will color all the boxes crossed by the diagonal black. In every row of the table, let us mark the black box which is the nearest to the left (vertical) side of the table with the letter R. Similarly, in each column, we will mark the box which is the nearest to the upper (horizontal) side of the table with the letter C. We can prove that each of the black boxes is marked at least once, and that only the box in the upper left corner is marked more than once. Therefore, the number of black boxes is the sum of the number of boxes marked R and the number of boxes marked C minus 1. That is, the answer is $199 + 991 - 1 = 1189$.

Now we try to prove our previous claims. First, why was each of the squares colored at least once? If one of the squares—say, A—is not marked with any letter, then its left and upper neighbors must also be black; that is, they intersect with the diagonal, which is impossible. Second, if a black square is marked with both letters R and C, then there are no black squares in the same row to the left of it and no black squares in the same column up from it. This means that the diagonal passes through the upper left corner of this square, which is also impossible, though the reason is more complicated (it is because 199 and 991 are relatively prime, but we do not think it is appropriate to go into such technicalities in discussing problems from this chapter).

8. We would like as many 5's on the left as possible. We can accomplish this by crossing out the initial sequence 1234, leaving a 5, then crossing out another sequence of 1234. Clearly, had we left any digit other than 5 on the left our number would have been smaller. However, we cannot gain another 5, since we can only cross out two more digits. So we cross out the next two small digits: 1 and 2. It is not hard to see that the result, 553451234512345 is the largest possible.

9. We need to have as much time as possible between the uttering of this sentence and Peter's next birthday. We can manage this if he made his statement on January 1, and he was born on December 31. He will turn 13 at the end of the next calendar year.

10. No, he is not. Indeed, the fact that event A (rain) always causes B (cat's sneeze) does not mean that B causes event A. This is one example of a very common sort of logical error, the confusion of a statement with its converse.

11. There are 12 circles: five of them are on one side of the sheet, and the other seven are on the other side. This is the only possible explanation of what happened in the class.

12. Yes, it is possible, if the professor is a woman.

13. The third turtle lied.

14. He reasoned as follows: "If my face is clean, then one of my colleagues, seeing that the third one is laughing at something, would realize that his face is also black with soot. Since he is still laughing, my face must be black as well."

15. Certainly, the percentage of milk in the tea is the same as the percentage of tea in the milk, since the total amount of milk (or of tea) in both glasses does not change.

16. See Figure 128. This answer is unique up to rotations and reflections.

4	3	8
9	5	1
2	7	6

FIGURE 128

17. There are 95343 "loves" in "there".

18. There is only one answer: $51286+1582 = 52868$. Hints: $L+L < 10$; $S+S \geq 10$; otherwise, the hundreds and the units digits in the number $BASES$ could not be equal ($B \neq E$!).

19. The dollar bills can be distributed as follows: $1+2+4+8+16+32+64$.

20. See Figure 129.

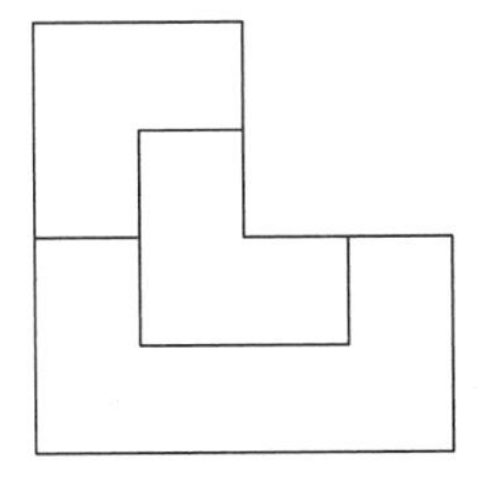

FIGURE 129

21. See Figure 130.

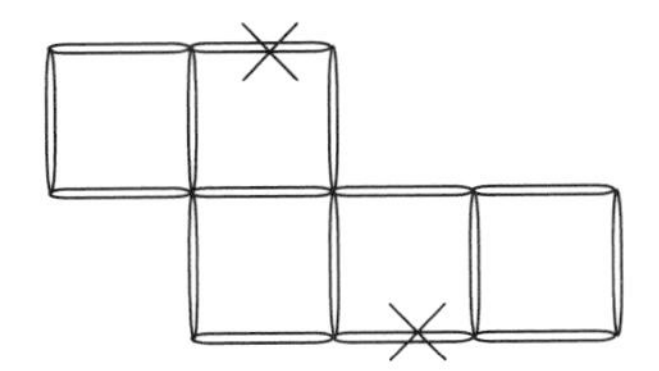

FIGURE 130

22. See Figure 131.

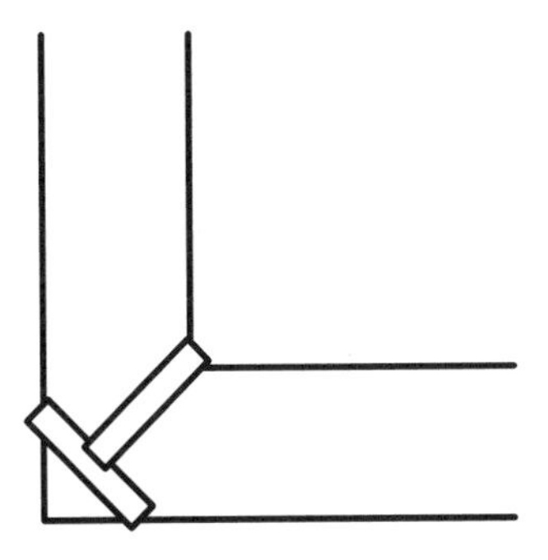

FIGURE 131

23. Take three pencils and arrange them as shown in Figure 132. Another three pencils are to be arranged similarly but "spun" in the opposite direction and put on the top of the first three.

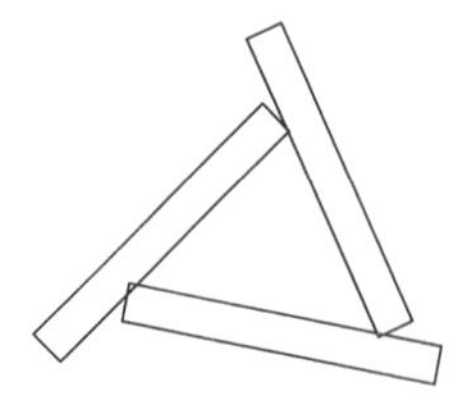

FIGURE 132

24. See Figure 133.

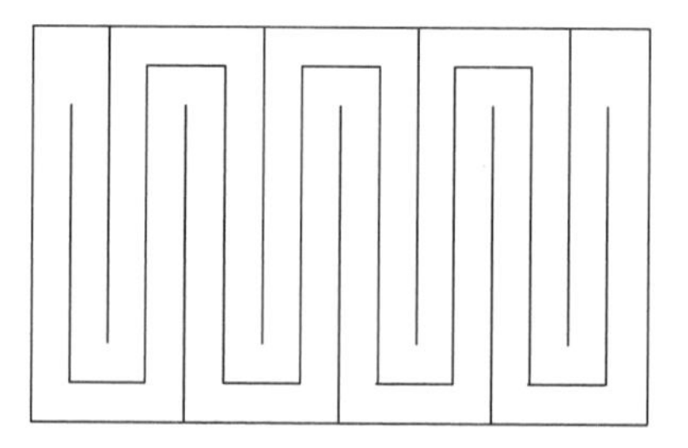

FIGURE 133

25. Remove the four coins shown in Figure 134.

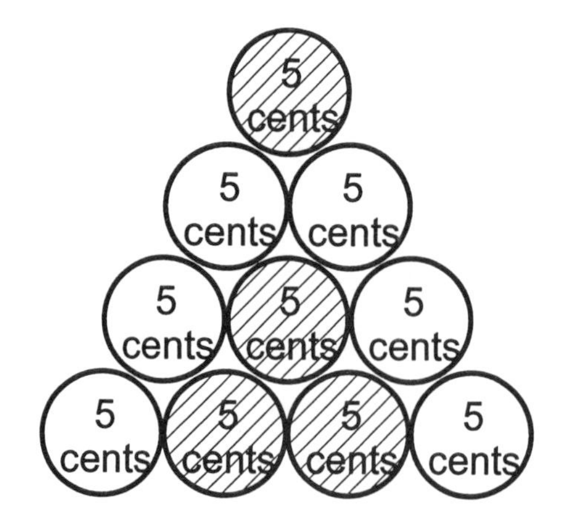

FIGURE 134

1. PARITY

2. A knight always moves from a square of one color to a square of the opposite color. Thus the colors of the squares occupied by the knight alternate between white and black. To get back to a square of the same color as he started on (in particular, the same square), he must make evenly many moves.

4. Answer: No, it cannot. Suppose we did have such a line. If we trace the path, each time we intersect the given line we pass from the half-plane on one side of the line to the half-plane on the other side (any line divides a plane into two half-planes). Since the path is closed, we begin and end on the same side of the line. The sides of the line alternate, so the polygon would have evenly many vertices.

5. Answer: No, he cannot. Let us call a position of the three pucks "correct" if, in tracing triangle ABC from A to B to C (and back to A), we travel clockwise. Let us call a position "incorrect" in the opposite case. It is not hard to see that after each move, the "correctness" of the pucks' positions changes. Hence the original position cannot be recovered.

6. Answer: five. If any of Katya's friends are standing next to children of their own sex, then it is clear that all the children are of the same sex. This means that the boys and girls must alternate, so that there are as many girls as boys.

8. Answer: No. There are 25 squares on the board. Since each domino covers two squares, the dominoes can only cover an even number of squares.

9. If the axis of symmetry did not pass through a vertex, then the 101 vertices could be partitioned into pairs of symmetric vertices. This is impossible, since 101 is odd. However, a regular decagon is an example of a 10-gon with an axis of symmetry which does not pass through any of its vertices.

10. Within the chain of dominoes, each number of spots occurs in a pair (placed end-to-end). Since there are eight 5's in the set of dominoes, the last square must also have five spots on it.

11. Answer: No. We prove this by contradiction. If there were such a chain, then one of the numbers 1, 2, 3 does not occur at the ends of the chain. Suppose the number 3 does not occur. Now inside the chain the 3's occur in pairs, so 3 occurs an even number of times. However, with the "zeros" discarded, there will be seven 3's in the set altogether. This is a contradiction.

12. The answer is no. Suppose we could partition a convex 13-gon into parallelograms. Let us choose one side of the 13-gon, and consider the parallelogram it belongs to (it is clear that there are not two such parallelograms). The opposite side of this parallelogram is also a side of a second parallelogram. This second parallelogram has another side parallel to the first, and we can continue this "chain" of parallelograms until we arrive at a side of the 13-gon. This side is therefore parallel to the side with which we started, and since a convex polygon cannot have three mutually parallel sides it is parallel to no other side of the convex 13-gon.

This argument shows that if we could partition the 13-gon into parallelograms, we would be able to find pairs of parallel sides. Since 13 is an odd number, this is impossible.

14. Suppose that no checker is placed in the center square. We connect all the pairs of checkers which are located symmetrically with respect to one of the diagonals with a thread. We then divide all the checkers into "necklaces": groups connected by threads. Then in each "necklace" there will be either two or four checkers. This means that the total number of checkers will be even, which is a contradiction.

15. It is not hard to see that there must be fifteen 1's in the table, for example, by noticing that there is one 1 in each column. Problem 13, applied to the boxes with 1's in them, shows that there must be at least one 1 along the main diagonal. Reasoning analogously, the main diagonal must contain a 2, a 3, and so on. An example of such a table is shown below.

1	2	3	4	5	6	7	8	9	10	11	12	13	14	15
2	3	4	5	6	7	8	9	10	11	12	13	14	15	1
3	4	5	6	7	8	9	10	11	12	13	14	15	1	2
4	5	6	7	8	9	10	11	12	13	14	15	1	2	3
5	6	7	8	9	10	11	12	13	14	15	1	2	3	4
6	7	8	9	10	11	12	13	14	15	1	2	3	4	5
7	8	9	10	11	12	13	14	15	1	2	3	4	5	6
8	9	10	11	12	13	14	15	1	2	3	4	5	6	7
9	10	11	12	13	14	15	1	2	3	4	5	6	7	8
10	11	12	13	14	15	1	2	3	4	5	6	7	8	9
11	12	13	14	15	1	2	3	4	5	6	7	8	9	10
12	13	14	15	1	2	3	4	5	6	7	8	9	10	11
13	14	15	1	2	3	4	5	6	7	8	9	10	11	12
14	15	1	2	3	4	5	6	7	8	9	10	11	12	13
15	1	2	3	4	5	6	7	8	9	10	11	12	13	14

17. Answer: No. The sum of the pair of numbers on each page is odd, and the sum of 25 odd numbers will also be odd. The number 1990 is even.

18. Clearly, each integer is either $+1$ or -1, and there are an even number of $+1$'s (since their product is positive). If their sum were zero, there would have to be 11 numbers $+1$, which is a contradiction.

19. Answer: No. Among the given numbers, only one (2) is even, and the rest are odd. Therefore, the sum of the numbers in the row containing the 2 is odd, while the sum of any other row is even.

20. Answer: No. The sum of the numbers from 1 through 10 is 55, and changing the sign of any one of them changes this sum by an even number. The sum must thus remain odd.

21. The proof is the same as in Problem 20, since the sum $1 + 2 + 3 + \ldots + 1985$ is odd.

22. Answer: No. It is not hard to see that the given operation does not change the parity of the sum of the numbers on the blackboard. Since this parity is initially odd, the sum can never be 0.

23. Answer: No. Each domino covers one black square and one white square, but if we leave out squares $a1$ and $h8$, there will be two more white squares than black squares remaining.

24. Suppose there were a 17-digit integer whose "reversed sum" contained no even digit. For convenience, we number the columns of digits from right to left, and consider the usual addition algorithm. The ninth digit of our number will be added to itself. This would produce an even digit in the answer, unless there is a "carry" from the 8th column. But if there is such a carry, then there must be one also from the 10th column to the 11th (the 10th column is identical to the 8th except for the order of the digits). Hence the 7th column has digits of the same parity, and requires a carry from the sixth column.

Proceeding similarly, we find that there must be a carry into each odd numbered column. But there cannot be a carry into the first column, so we have a contradiction.

25. Answer: No. Since a given soldier shares each tour of duty with two others, if he shared duty with every other soldier exactly once, the 99 remaining soldiers could be partitioned into pairs with whom he shared his tours of duty. This is a contradiction, since 99 is an odd number.

26. For any point X lying outside segment AB, the difference $AX - BX = \pm AB$. If we assume that the sum of the distances from A and from B are equal, then the expression $\pm AB \pm AB \pm \ldots \pm AB$, in which there are 45 addends, is zero. This is impossible.

27. We can analyze this situation by working backwards. If there are nine 1's in the circle, then there must have been either nine 1's or nine 0's before the operation was applied. Since there are not nine 1's to begin with, nine 1's cannot arise in this way. If there were nine 0's, then the desired situation was achieved in the previous step. But could this have happened? If there are nine 0's, then in the step before this, the 0's and 1's would have alternated. This is impossible, since there are oddly many numbers altogether.

28. Let us number the students, starting with any one of them. We use an indirect method of proof, supposing no student has two neighbors who are boys. Suppose there is a boy in the kth position. Then there is a girl either in position $k-1$ or position $k+1$. If position $k+1$ is a girl, then $k+2$ cannot be a boy (or else that girl would have two boys for neighbors). If position $k+1$ is a boy, then a girl must sit in position $k+2$ (or else that boy would have two boys for neighbors). A similar argument shows that a girl must be seated in position $k-2$ as well.

Continuing the analogous reasoning and taking the numbers "modulo 50", we can show that if there is a girl in the kth position, then there are boys in both the $(k-2)$nd and $(k+2)$nd positions. If we now look only at those 25 students sitting in even places, we find that among them the boys and the girls alternate around the table. But 25 is an odd number, so this is impossible.

Students should be encouraged to complete the analogous reasoning in this problem, rather than simply relying on the symmetry of the situation with respect to boys and girls.

29. Suppose the snail has returned "home" after tracing over N vertical segments. Then it is not hard to see that the snail has also traced over N horizontal segments. Altogether, it has traced $2N$ segments, and spent $30N = (2N)15$ minutes. Since the snail has returned home, N is even (the number of upward segments traced, for example, must equal the number of downward segments, and their sum is N). Hence $2N$ is a multiple of 4, and $(2N)(15$ minutes$)$ is a whole number of hours.

30. Answer: No. Let us name the grasshoppers A, B, and C. We call the positions ABC, BCA, and CAB (left to right) correct, and the positions ACB, BAC, and CBA incorrect. It is easy to see that after each leap, the type of position changes from correct to incorrect or back.

31. Peter must lay the chosen coin aside, divide the remaining coins into two piles of 50 coins each, and weigh these piles against each other. We will show that if the chosen coin is genuine, the difference between the weights must be even, and if the coin is counterfeit, the difference must be odd.

First suppose the chosen coin is genuine. If we knew the total weight of the remaining genuine coins, we would calculate the total weight of the counterfeit coins by adding fifty numbers, each equal to ± 1. This means that if we put the 50 remaining genuine coins in one pan of the balance, and the 50 counterfeit coins in the other, the difference in weights would be even. It is not difficult to see that if we exchange a coin from one side of the balance for a coin from the other, the difference will change by ± 2. We can keep exchanging coins between the pans. At each exchange, if the coins are identical, the difference does not change. If one is genuine, the difference changes by ± 2. If one is too heavy and the other too light, the difference changes by ± 4. In any case, the operation of exchanging coins preserves the parity of this difference. We can achieve any arrangement of the coins on the balance by performing exchanges on the original arrangements. Since the difference was originally zero, the parity of any difference must be even.

Similarly, we can show that if the coin is counterfeit, then the difference is odd. For if we put all the counterfeit coins on one side of the balance (and one genuine one as well), the difference would show an odd number (the sum of 49 differences of $+1$ or -1). Again, the parity of this difference does not change when two coins are exchanged. Hence a counterfeit coin will yield an odd difference.

32. Answer: No. Suppose the numbers were arranged as required. Then number the places in which they stand from 1 through 9 (say, from left to right). If the number 1 is in place N, then it is not hard to see that the place number for 2 differs from N by an even number, and so is of the same parity. The same is true of 2 and 3, of 3 and 4, and so on. This means that the places in which the numbers stand all have the same parity. Since there are nine numbers, and at most 5 places of the same parity (if that parity is odd), this is a contradiction.

2. COMBINATORICS–1

28. Since each of the five envelopes can be independently supplied with any of the four stamps, we must multiply the numbers of choices: $5 \cdot 4 = 20$.

29. There are two different vowels and three different consonants which can be chosen independently. Thus, the answer is $2 \cdot 3 = 6$.

30. Since no choice will restrict any other, we multiply the numbers of choices. Answer: $7 \cdot 5 \cdot 2 = 70$.

31. Since any two stamps can be exchanged, there are $20 \cdot 20$ ways to exchange stamps, and similarly, there are $10 \cdot 10$ ways to exchange postcards. Therefore, the answer is $20 \cdot 20 + 10 \cdot 10 = 500$.

32. There are two cases: a number can have all its digits odd or it can have all its digits even. The first case gives us 5^6 numbers, since each of the six digits can be chosen from the set $\{1, 3, 5, 7, 9\}$ independently. The second case, however, is slightly different since the first digit cannot be zero, which gives us only $4 \cdot 5^5$ numbers. Thus, the answer is $5^6 + 4 \cdot 5^5 = 28125$.

33. Each of the letters can be given out in three different and independent ways. Therefore, to obtain the answer we must multiply six 3's, and the answer is $3^6 = 729$.

34. We must have one card of each suit, and the spade can be chosen in 13 ways. The club cannot be of the same value as the spade, and therefore there are only 12 ways to choose it. The number of choices for the diamond is 11, and for the heart is 10. The order in which we have named the suits does not affect the analysis. Answer: $13 \cdot 12 \cdot 11 \cdot 10$.

35. Consider five cases, depending on how many books the stack consists of. If there is only one book, then it can be chosen in 5 ways. A stack of two books can be chosen in $5 \cdot 4$ ways, since the number of ways to choose the second book in the stack is 4. Similarly, we can calculate the numbers of ways to arrange 3, 4, and 5 books in the stack. The answer is $5+5\cdot4+5\cdot4\cdot3+5\cdot4\cdot3\cdot2+5\cdot4\cdot3\cdot2\cdot1=325$.

36. Exactly one of the rooks must be in each row. Whether some of the rooks attack each other depends only on the choice of the columns the rooks are in. Since the numbers of the columns belong to the set of natural numbers 1 through 8, and two rooks can attack each other if and only if they stand in the same column, we have the familiar problem about the number of ways to arrange eight objects in a row. The answer is $8! = 40320$.

37. This problem is quite similar to the previous one. We can think of the boys as rows, and the girls as columns. Each square then represents a boy-girl pair, and each arrangement of "non-attacking" rooks yields a pairing for the class. Answer: $n!$.

38. See the solution to Problem 23. Here, the players are vertices of an n-gon, and the diagonals represent matches played. Answer: $18 \cdot 17/2 = 153$.

39. The answers are

a) $(28 \cdot 56 + 20 \cdot 54 + 12 \cdot 52 + 4 \cdot 50)/2 = 1736$;

b) $(4 \cdot 61 + 8 \cdot 60 + 20 \cdot 59 + 16 \cdot 57 + 16 \cdot 55)/2 = 1848$;

c) $(28 \cdot 42 + 20 \cdot 40 + 12 \cdot 38 + 4 \cdot 36)/2 = 1288$.

To demonstrate the method, we will prove part a). There are 28 squares on the border of the chessboard, and from one of these the first bishop attacks 8 squares (including the one it stands on). Therefore, there are 56 squares left for the second bishop. Further, there are 20 squares adjacent to the border squares. When positioned on these squares, the first bishop attacks 10 squares, so there are 54 squares on which to place the second bishop. Analogously, there are 12 squares from which the first bishop attacks 12 squares, and, finally, 4 central squares (standing on these, the first bishop attacks 14 squares). After adding up all the variants, we must divide the sum by two, since we counted each arrangement exactly twice (we do not distinguish the bishops).

40. This problem can be restated as the following question: how many ways are there to arrange two apples, three pears, and four oranges in a row? The solution is just the same as in Problems 17–21. Answer: $9!/2!3!4!$.

41. Distributing the students is equivalent to arranging them in a row, since after doing that the first student can be sent to live in a single room, the next two to the double, and the remaining four to the room for four students. However, each of the distributions can be obtained from several arrangements. Indeed, we can permute the students within the pair and within the quadruple (we can do this with the first student too, though nobody will notice our efforts). Since there are $2!$ and $4!$ possible permutations of the pair and the quadruple respectively, we must divide the number of these arrangements (which is equal to $7!$) by the product of $2!$ and $4!$. Thus, the answer is $7!/1!2!4!$.

42. Using the same method as in the previous solution, we obtain the answer $8!/2!2!2!$.

43. The answer can be obtained as the sum of four numbers, each representing the number of words with five letters A and 0, 1, 2, and 3 letters B respectively: $1 + 6!/5!1! + 7!/5!2! + 8!/5!3! = 84$.

44. Hint: calculate the number of ten-digit numbers which do not possess the property described. The answer is: $9 \cdot 10^9 - 9 \cdot 9!$.

45. Since the number of seven-digit numbers without 1 in their decimal representation equals $8 \cdot 9^6$, and $8 \cdot 9^6 < 9 \cdot 10^6 - 8 \cdot 9^6$, we conclude that there are more numbers with 1 in their decimal representation.

46. The number of outcomes without occurrences of six is equal to 5^3. Thus, the answer is $6^3 - 5^3 = 91$.

47. The first pair can be chosen in $\binom{14}{2}$ ways, the second pair can be chosen in $\binom{12}{2}$ ways, et cetera. So we have the product $\binom{14}{2}\binom{12}{2}\dots\binom{2}{2}$. But here each splitting is counted 7! times since every set of 7 pairs can be obtained in 7! ways (depending on the numeration of the pairs in the set). Therefore, the answer is $\binom{14}{2}\binom{12}{2}\dots\binom{2}{2}/7!$ which equals $13 \cdot 11 \cdot 9 \cdot 7 \cdot 5 \cdot 3 \cdot 1$.

48. The first eight digits can be chosen arbitrarily. There are $9 \cdot 10^7$ ways to do this. Then the last digit can always be chosen in exactly 5 ways (if the sum of the previous eight digits is odd, then we must choose an odd digit, otherwise the last digit must be even). Hence the answer is $9 \cdot 10^7 \cdot 5 = 450000000$.

3. DIVISIBILITY AND REMAINDERS

1. The answers are a) 4; b) 6; c) 9; d) $(n+1)(m+1)$. We prove the last result, since it is a generalization of the previous ones. Every divisor of $p^n q^m$ equals $p^i q^j$ for some $0 \le i \le n$ and $0 \le j \le m$. Therefore, the choice of a divisor is equivalent to the choice of two integers satisfying the inequalities above. The first of them, i, can be chosen in $n+1$ ways, and the second, j, in $m+1$ ways. Multiplying the number of choices, we get the answer.

3. Part b) implies part a), so we discuss only the former. There must be a number divisible by 3 among the given five numbers. Analogously, there is a number divisible by 5, and at least two even numbers, one of which is a multiple of 4. Multiplying 3, 5, 2, and 4 gives us 120, and we are done.

4. The answers are a) $p-1$; b) $p^2 - p$. Part a) is not difficult: all natural numbers less than p are relatively prime to p. The second result follows from the observation that the only numbers which are *not* relatively prime to p are the multiples of p, and there are p of these less than or equal to p^2.

5. Since $990 = 2 \cdot 3^2 \cdot 5 \cdot 11$, $n!$ must contain a factor of 11. Since 11 is prime, it must itself be contained in the product, so $n = 11$ is the smallest possible value.

6. If a number has n terminal zeros, then it is divisible by 10^n. So we are asking how many factors of 10 are contained in 100!. But since $10 = 5 \cdot 2$, we are asking how many factors of 5 and 2 there are. Since 2 is smaller than 5, for any factor of 5 there will be enough factors of 2 to make a factor of 10. Thus we need to count only factors of 5.

Since $100 = 20 \cdot 5$, there are 20 multiples of 5 in the product $1 \cdot 2 \cdot 3 \cdot \ldots \cdot 99 \cdot 100$. But there are more factors of 5, since the numbers 25, 50, 75, and 100 contain two factors of five, to make four "extras". There are therefore 24 factors of 5, so 24 factors of 10, so 24 terminal zeros in the product 100!.

Query. How many terminal zeros are there in 1000! ?

7. We find that 24! ends with four zeros and 25! ends in six. It is not difficult to see that as n increases, the number of terminal zeros in $n!$ cannot decrease. Hence the answer to our question is no.

8. Let us split all the divisors of n into pairs of the following type: $(d, n/d)$. The only obstacle to this is that the members of some pair may coincide. However, this can happen if and only if n is a perfect square, which completes the proof.

9. Tom must have erred: the right-hand number is a multiple of 11, but neither of the left-hand numbers are. Since 11 is prime, this is impossible.

Note that the placement of this problem in a set of exercises on divisibility provides a hint to its solution. The problem is more difficult if taken out of this context.

11. Observe that $65(a + b) = 65a + 65b = 65a + 56a = 121a$. Since 65 and 121 are relatively prime, it follows that $a + b$ is divisible by 121, which is a composite number, so $a + b$ is composite as well.

12. Answers: a) $x = 16$, $y = 15$; b) $x = 152$, $y = 151$, or $x = 52$, $y = 49$. To prove part a), let us rewrite the equation as follows: $(x - y)(x + y) = 31$. Since 31 is a prime number, the smaller factor must be 1, and the greater must be 31. Thus, we have the simultaneous equations

$$\begin{cases} x - y = 1, \\ x + y = 31, \end{cases}$$

which gives us the answer shown above.

13. We have $x(x^2 + x + 1) = 3$. Hence either $x = \pm 1$ or $x = \pm 3$. After analyzing all cases we find that $x = 1$.

14. Hint: check that both sides of the equality are divisible by the same powers of any prime number p.

15. a) 0 (the remainders when 1989, 1990, 1991, and 1992^3 when divided by 7 are 1, 2, 3, and 1 respectively); b) 1, since 9 gives remainder 1 when divided by 8.

17. Hint: analyze the remainders when divided by 5.

18. Hint: analyze the remainders when divided by 3.

19. Hint: analyze the remainders when divided by 9.

21. a) Let us prove that the given numbers are divisible by 3 and by 8. We have $p^2 - 1 = (p - 1)(p + 1)$. If $p > 3$ is prime, then p is odd. Hence both $p - 1$ and $p + 1$ are even, and one of them is a multiple of 4. It follows that $p^2 - 1$ is divisible by 8. Also, since $p - 1$, p, $p + 1$ are three consecutive integers, one of them is divisible by 3. It is not the prime p, so it must be either $p - 1$ or $p + 1$. Thus $p^2 - 1$ is divisible by 3 and 8, and therefore by 24.

b) We have $p^2 - q^2 = (p - q)(p + q)$. Proceeding as before, we find that both factors are even. To show that one of them is a multiple of 4, we assume the opposite; that is, that these numbers give remainders 2 when divided by 4. Then their sum must be divisible by 4. On the other hand, their sum is $2p$, which is not a multiple of 4, since p is odd.

Furthermore, the numbers p and q must have either equal or different remainders modulo 3. In the former case, their difference is divisible by 3; in the latter case, their sum is. This proves that $(p - q)(p + q)$ is divisible by 3 and by 8.

22. If neither x nor y is divisible by 3, then x^2 and y^2 have remainders 1 when divided by 3. Therefore, their sum has remainder 2, which is impossible for a perfect square.

23. Hint: check that both a and b are divisible by 3 and by 7.

24. Hint: check that numbers x^3 and x have equal remainders when divided by 6.

25. If d is odd, then one of the numbers p and q is even, which is impossible. If d is not divisible by 3, then one of the numbers p, q, and r is divisible by 3, which again gives us a contradiction.

26. Hint: find all possible remainders given by perfect squares when divided by 8.

27. Possible remainders of perfect squares when divided by 9 are: 0, 1, 4, and 7. Check that if the sum of some triple of them is divisible by 9, then some pair of them are equal.

30. Using the method of Problem 28, we find that the answer is 7.

31. The answer is 1.

32. The answer is 6.

34. The answer is 3. Hint: the units digit of 7^n has a cycle of length 4. We must determine when in this cycle 7^7 occurs; that is, we need the remainder when 7^7 is divided by 4.

35. Hint: One of these numbers is always divisible by 3. a) $p = 3$; b) $p = 3$.

36. The answer is $p = 3$. The method of the previous solution works here as well.

37. Hint: prove that $p = 3$ using the same trick as before.

38. Hint: analyze the remainders when divided by 3.

39, 40. Hint: check that the remainder of the square of an odd number when divided by 4 is always 1, and the remainder of the square of an even number is always 0. The answer to both questions is no.

41. The answer is $p = 5$. Analyze the remainders upon division by 5.

42. The remainder of this number when divided by 9 is 7, and this cannot be true of a perfect cube.

43. Hint: find all possible remainders of the number $a^3 + b^3 + 4$ when divided by 9.

44. Hint: find all possible remainders of the number $6n^3 + 3$ when divided by 7.

45. If neither of the numbers x or y is divisible by 3, then z^2 gives a remainder of 2 when divided by 3, which is impossible. Now notice that the square of an odd number always has remainder 1 when divided by 8; the square of an even number not divisible by 4 always has remainder 4; and the square of a multiple of 4 always has remainder 0. Using this, we can show that either both numbers x and y are even, or one of them is divisible by 4.

46. Hints: a) $4 + 7a = 4(a + 1) + 3a$; b) $a + b = (2 + a) - (35 - b) + 33$.

47. The answer is 0. First, calculate the last digit of $0^2 + 1^2 + 2^2 + \ldots + 9^2$. Second, notice that this last digit is always the same for every set of ten consecutive natural numbers.

48. Prove that any two numbers out of the given seven—say, x and y—have the same remainder when divided by 5. To accomplish this, consider two 6-tuples, the first containing all the numbers but x, the second containing all the numbers but y.

49. Denoting the first one of these numbers by a, we get

$$a + (a+2) + (a+4) + \ldots + (a + 2(n-1)) = na + 2(1 + 2 + 3 + \ldots + (n-1))$$
$$= na + n(n-1) = n(a + n - 1) .$$

50. Note that this number increased by 1 is divisible by 2, 3, 4, 5, and 6. The answer therefore is one less than the LCM of these numbers, or 59.

51. If n is composite and greater than 4, then $(n-1)!$ is divisible by n. Indeed, $n = kl$ where k and l are less than n. If $k \neq l$, then the product $(n-1)!$ contains both these numbers as factors and our point is proved. If $k = l$, that is, $n = k^2$, where $k > 2$, then the product $(n-1)!$ contains factors k and $2k$, and we are done.

55. Using Euclid's algorithm we get $\gcd(30n + 2, 12n + 1) = \gcd(12n + 1, 6n) = \gcd(6n, 1) = 1$.

56–57. Use Euclid's algorithm. The answers are: $2^{20} - 1$ and $111\ldots11$ (twenty 1's) respectively.

4. THE PIGEON HOLE PRINCIPLE

3. The pigeon holes are the remainders when divided by 11. The pigeons are the numbers. (See also the solution to Problem 21.) If two numbers have the same remainder when divided by 11, their difference must be divisible by 11.

4. The pigeon holes here are the numbers of hairs on a person's head (from 1 to 1,000,000). The pigeons are the citizens of Leningrad.

6. Let us sort the football players by team as they come off their airplanes. There will be $10M + 1$ players to sort. The General Pigeon Hole Principle assures us that there will be one team which has 11 players, and this team is complete.

8. There are five possible numbers of acquaintances for any person: 0, 1, 2, 3, or 4. So it would seem that each could have a different number of friends. However, if any person has four acquaintances, then no person may have zero acquaintances. Hence two people must have the same number of acquaintances.

9. If there are k teams, then the number of games played by each team varies from 0 to $k - 1$. However, if any team has played $k - 1$ games, then it has played every other team, and no team has played 0 games. Hence we are fitting k teams into $k - 1$ pigeon holes, which are either the numbers from 0 through $k - 2$ or the numbers 1 through $k - 1$.

10a. The answer is 32. Indeed, suppose that 33 or more squares are colored green. Then, after we have divided the board into sixteen 2×2 squares, the Pigeon Hole Principle guarantees that at least one of these squares contains 3 or more small green squares. These 3 green squares form the "forbidden" tromino in some position, and we have a contradiction. On the other hand, we can color all the black fields (of the usual coloring) green, and this is an example of 32 green squares with the property needed.

10b. The answer is 32 (again!). Indeed, if 31 or fewer squares are colored green, then one of those sixteen 2×2 squares contains 1 or 0 green squares. Then the other 3 or 4 squares are not colored green, and they form the tromino without green squares in it. This contradiction (and the same construction as above) completes the proof.

11. At least $1 + 2 + 3 = 6$ problems were solved by the students mentioned in the problem statement. Therefore, there are 29 problems left to be solved, and 7 students to account for them. If each student had solved only 4 problems, then there would have been only 28 problems solved. Therefore, one student must have solved at least 5 problems.

12. Answer: 12 kings. See the hint to Problem 10a.

13. Divide the cobweb into 4 sectors as shown in Figure 135, each of which can hold no more than one spider.

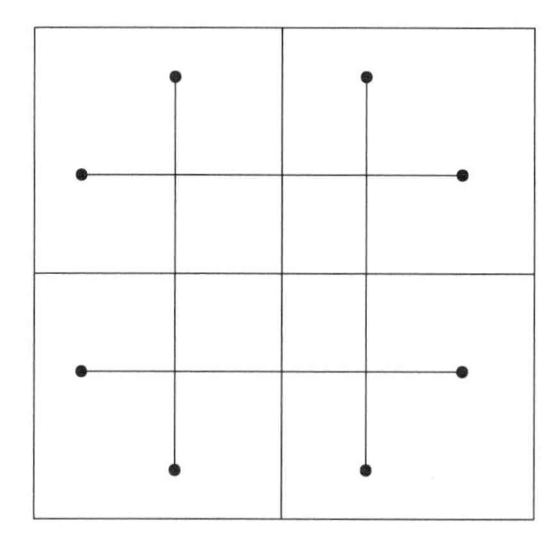

FIGURE 135

14. Each of the smaller triangles can cover only one vertex of the larger triangle.

18. Color all the dry land red, and color each point diametrically opposite dry land green. Then there must be a point which is both red and green. Start the tunnel at this point. Do you see why this is, in a way, a Pigeon Hole Principle?

19. There are only 1987 possible remainders when a number is divided by 1987. If we examine, for example, the first 1988 powers of 2, we find that two of them must have the same remainder when divided by 1987. These two powers then differ by a multiple of 1987.

20. When divided by 100, a perfect square can give only 51 remainders, since the numbers x^2 and $(100 - x)^2$ give the same remainder. Hence of 52 integers, the squares of two must have the same remainder when divided by 100. These two squares differ by a multiple of 100.

22. If 3^m and 3^n (where $m > n$) are two powers of 3 which give the same remainder when divided by 1000, then $3^m - 3^n = 3^n(3^{m-n} - 1)$ is divisible by 1000. Now the prime factors of 1000 are 2 and 5, and neither divides 3^n. It follows that 1000 must divide $3^{m-n} - 1$, which means that 3^{m-n} is a power of 3 ending in the digits 001.

23. This sum can take on only seven values: the numbers from -3 through 3.

24. Divide all the people into 50 pairs who are sitting diametrically opposite each other. Consider these pairs as the pigeon holes. Since there are more than 50 men, one pair must include more than one man.

25. If the conclusion is false, then it is clear that the boys will have gathered at least $0 + 1 + 2 + \ldots + 14 = 105$ nuts, which is a contradiction.

26. The product of the numbers in all the groups is $9! = 362880$. If the product of each group were no greater than 71, the product of all the numbers could only be $71^3 = 357911$. It should be noted here that this method of proof is, in a way, more general than the simple Pigeon Hole Principle.

27. One can move from any square to any other by passing through neighboring cells, and we can always choose a path such that the number of squares visited is less than 19. This means that if a is the smallest number on the board, all the numbers are included between a and $a + 95$. Therefore there can be no more than 96 different numbers among the 100 on the board, and two must be equal.

28. We choose any one person in the group. Let us call him Bob. We sort the others into two pigeon holes: those who know Bob and those who do not. There are at least three of the remaining five people in one of these categories. Suppose Bob has three acquaintances. If two of these know each other, then they, together with Bob, form the required triple. If none of them knows each other, then they themselves form the required triple. A similar argument holds if there are three people whom Bob does not know.

29. Consider the parity (remainder upon division by 2) of the coordinates of the points. There are four possibilities: (odd, odd); (odd, even); (even, odd); (even, even). Since there are five points, we can choose two of them whose coordinates both match in parity. It is not hard to see that the midpoint of the line segment they determine has integer coordinates.

30. There are two categories into which we can fit the three sizes: those sizes for which there are more right boots than left boots, and those sizes for which there are more left boots than right boots (if there happens to be an equal number of right and left boots in one size, we put that size in the second category). It follows that two sizes lie in the same category. Let us say that sizes 41 and 42 have more right boots than left boots (an analogous argument will hold if two sizes have more left boots than right boots).

Now there are 300 left boots in all, and at most 200 left boots in any one size. Therefore, the sum of the left boots in any two sizes is at least 100. We have shown that there are at least 100 left boots in sizes 41 and 42 (taken together), and that each of these sizes contains more right boots than left boots. Hence each left boot has a match, and there are at least 100 good pairs in the warehouse.

31. There are 11 more consonants than vowels in the alphabet. Therefore, if we add the differences between the number of consonants and the number of vowels in each of the six subsets, these differences must sum to 11. It follows that there must be at least one subset in which this difference is less than 2, and the letters of this subset must form a word.

32. Consider the ten sums: x_1, $x_1 + x_2$, $x_1 + x_2 + x_3$, $\ldots$, $x_1 + x_2 + \ldots + x_{10}$. Two of these must have the same remainder when divided by 10. The difference between these two sums gives a set whose sum is divisible by 10.

33. We can divide the numbers from 1 through 20 into ten disjoint sets, such that if a pair of numbers is selected from the same set, one of the pair divides the other: $\{11\}$, $\{13\}$, $\{15\}$, $\{17\}$, $\{19\}$, $\{1, 2, 4, 8, 16\}$, $\{3, 6, 12\}$, $\{5, 10, 20\}$, $\{7, 14\}$,

$\{9, 18\}$. Then, of any eleven numbers not greater than 20, two of them must fit in one of these pigeon holes, and one of these two divides the other.

34. We can number the study groups with the numbers 1 through 5. Then, instead of considering each student him or herself, we can consider the set of numbers belonging to the study groups he or she is part of. Each of these is a subset of the set $\{1, 2, 3, 4, 5\}$. We solve the problem by dividing the 32 subsets of this set into 10 collections such that if two subsets are chosen from the same collection, one of them contains the other (compare this with the solution to Problem 33). The following is such a collection. The subsets in each collection are written as numerals:

$$\begin{aligned}
&\big[\emptyset, \{1\}, \{1,2\}, \{1,2,3\}, \{1,2,3,4\}, \{1,2,3,4,5\}\big],\\
&\big[\{2\}, \{2,5\}, \{1,2,5\}, \{1,2,3,5\}\big],\\
&\big[\{3\}, \{1,3\}, \{1,3,4\}, \{1,3,4,5\}\big],\\
&\big[\{4\}, \{1,4\}, \{1,2,4\}, \{1,2,4,5\}\big],\\
&\big[\{5\}, \{1,5\}, \{1,3,5\}\big],\\
&\big[\{2,4\}, \{2,4,5\}, \{2,3,4,5\}\big],\\
&\big[\{3,4\}, \{3,4,5\}\big],\\
&\big[\{3,5\}, \{2,3,5\}\big],\\
&\big[\{4,5\}, \{1,4,5\}\big],\\
&\big[\{2,3\}, \{2,3,4\}\big].
\end{aligned}$$

5. GRAPHS–1

3. Yes, such a path is possible. See, for example, Figure 136, in which a graph is drawn similar to the one in the solution of Problem 2. An example of a path satisfying the conditions of the problem can then be constructed easily.

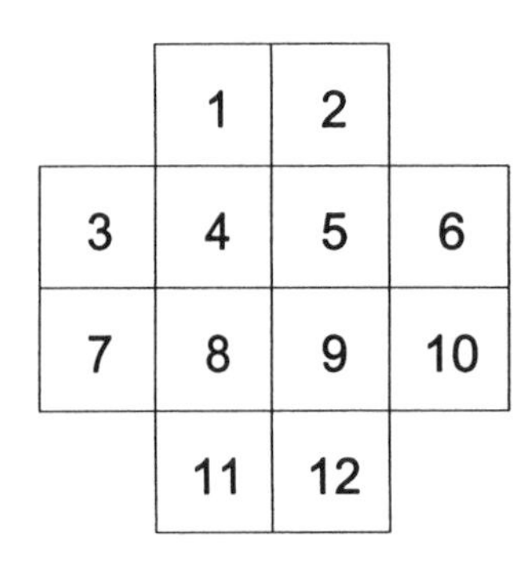

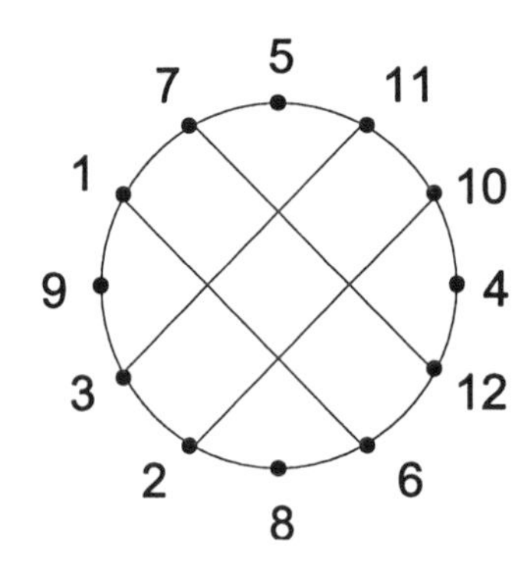

FIGURE 136

4. If the number AB is divisible by 3, then so is the number BA. This means that if a traveler can get from city A to city B directly, she can also get from city B directly to city A. This observation allows us to draw a graph of the connections, such as the one in Figure 137. Clearly, a traveler cannot get from any city to another. For example, she cannot get from city 1 to city 9.

7. Draw a graph where the cities are vertices and the roads are edges. We can then count the edges of this graph using the method illustrated in Problem 5. The

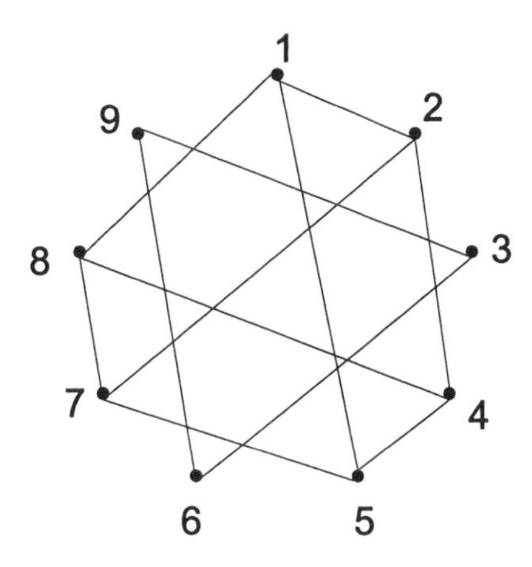

FIGURE 137

problem states that the degree of each vertex is 4, so the total number of roads is equal to $100 \cdot 4/2 = 200$.

9. Both situations (a) and (b) are impossible. In each case, we can think of a graph similar to that in Problem 5, and count the odd vertices. We find that the number of odd vertices is not even, so the graph cannot be drawn.

10. Answer: No. We can imagine a graph in which the vertices represent vassals, and neighboring vassals are connected by edges. A count of the odd vertices shows that there are not evenly many of them, so the graph cannot be drawn.

11. Answer: No. If the kingdom had k towns, then there would be $3k/2$ roads. This number cannot equal 100 if k is an integer.

12. Answer: Yes, it is true. Suppose it were not. Draw the graph in which the vertices represent islands and the edges represent the bridges connecting them. The problem says that each of the seven islands is represented by an odd vertex, so there would be oddly many odd vertices. Since this is impossible, the graph must show at least one edge leading to the shore. Figure 138 shows a graph representing a possible situation such as John described.

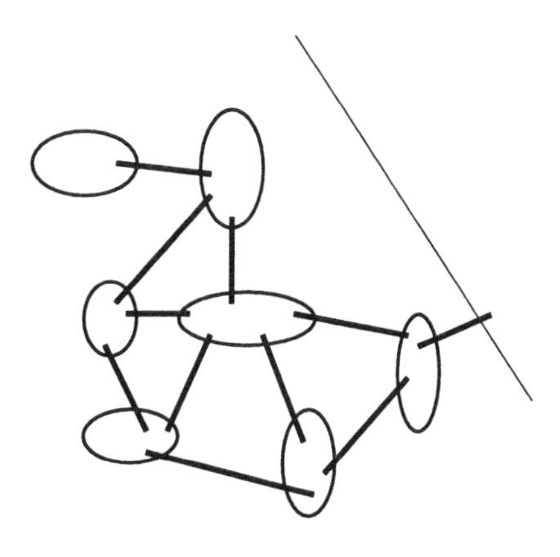

FIGURE 138

13. Imagine a huge graph in which each person who ever lived on earth is represented by a vertex, and each handshake is represented by an edge connecting the vertices corresponding to the two shakers. Then we are counting the odd vertices of this graph, and our theorem assures us that there must be evenly many of them.

14. Answer: No. The difficulty in this problem is to decide how to draw the graph. Taking the line segments themselves as edges of a graph probably won't work (which may confuse some students at first). Instead, we can consider a graph where the actual line segments are represented by vertices (!), and two vertices are connected by an edge if and only if the corresponding line segments intersect. Then this graph has nine vertices of degree 3, which is impossible.

16. We can generalize the solution to Problem 15. Suppose such a graph was not connected. Certainly it could not consist of fewer than two towns (with what could a single town fail to be connected?). Select two towns which, supposedly, cannot be connected by a path. Consider all the towns to which these two are connected. There are at least $2(n-1)/2 = n-1$ of these. As before, these new towns must all be distinct: if two new towns were the same, the two selected towns would be connected by a path through it. Therefore, the graph would have $n-1+2 = n+1$ towns, which is a contradiction. Hence the graph must be connected.

Here again, it is clear that students should attempt to construct the graph in question. They will quickly find that it has "too many" edges not to be connected. This intuition can be the springboard for a formal discussion of the result.

18. If road AB is closed, than it is enough to prove that we can still get from A to B. If this were not true, then in the connected component containing A, all the vertices other than A would be even. This situation of having exactly one odd vertex in a connected component contradicts our theorem about the odd vertices of a graph.

20. Answer: No, such a stroll is not possible. We represent the islands and the shores by vertices of a graph, and bridges by edges. As Figure 139 shows, the graph has 4 odd vertices, which is too many.

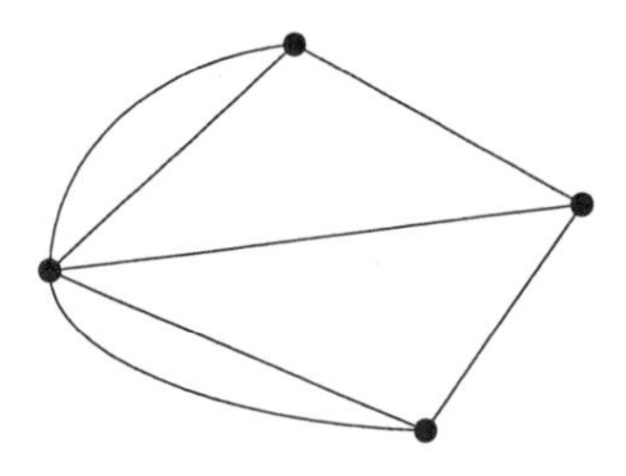

FIGURE 139

21. Answers: (a) six bridges; (b) five bridges; (c) four. Students need simply count the number of bridges used to visit Thrice on each occasion.

22. (a) The required cube is not possible. First note that the wire cannot double back on itself, since the total length of all the edges is 12×10 cm $= 120$ cm (using up the whole length of wire). Let us draw the graph of the cube's edges (Figure

140). If the wire frame could be formed, then we could follow the wire and traverse the graph without lifting the pencil from the paper. But this graph has eight odd vertices, which is too many to allow this. Therefore, the wire cannot be bent as required.

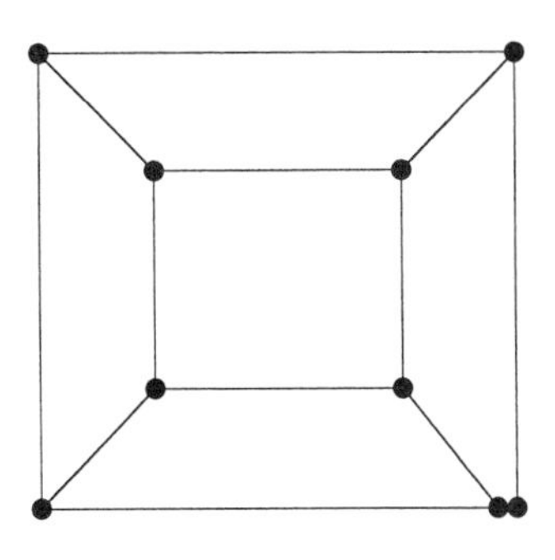

FIGURE 140

(b) Since the graph of the cube has 8 odd vertices, there must be at least four such pieces.

6. THE TRIANGLE INEQUALITY

1. Suppose $AB \geq BC$. If A, B, and C form a triangle, then the triangle inequality assures us that $AC + BC > AB$, which leads to the desired result. If $AB \leq BC$, then we can start with $AB + AC > BC$ (which the triangle inequality also assures us), to get the same result.

Equality holds if and only if A, B, and C are collinear, and B is not between A and C.

2. The length of side BC must be less that $AC + AB = 4.4$. On the other hand, BC must be greater than $|AB - BC|$ (see Problem 1), which is 3.2. The only integer within these bounds is 4.

3. If the sides of the triangle are a, b, and c, then the triangle inequality tells us that $b+c > a$. Adding a to each side, we find that $a+b+c > 2a$, which is equivalent to the required result.

4. Answer: 350 kilometers.

6. We will show that $OB + OC + OD > OA$. Adding the triangle inequalities $AC + OC > OA$ and $OB + OD > BD$, we find $AC + OB + OC + OD > OA + BD$ (see Figure 141). Since $AC = BD$, this gives the required result. Note that the same proof holds even if point O is outside the plane of square $ABCD$.

7. Suppose the diagonals of the quadrilateral intersect at O (Figure 142). Then $AB + BC > AC$, $BC + CD > BD$, $CD + AC > AC$, and $AD + AB > BD$. Adding, we find that $2(AB + BC + CD + DA) > 2(AC + BD)$, which proves the first result. Also, $OA + OB > AB$, $OB + OC > BC$, $OC + OD > CD$, and $OD + OA > AD$. Adding, we find that $2(OA+OB+OC+OD) = 2(AC+BD) > AB+BC+CD+DA$, which proves the second result.

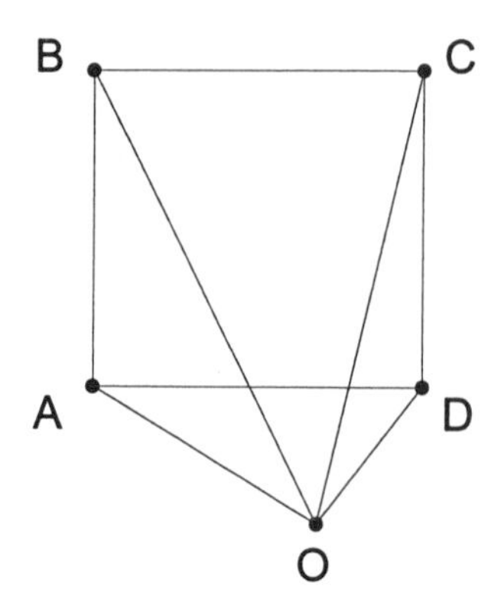

FIGURE 141

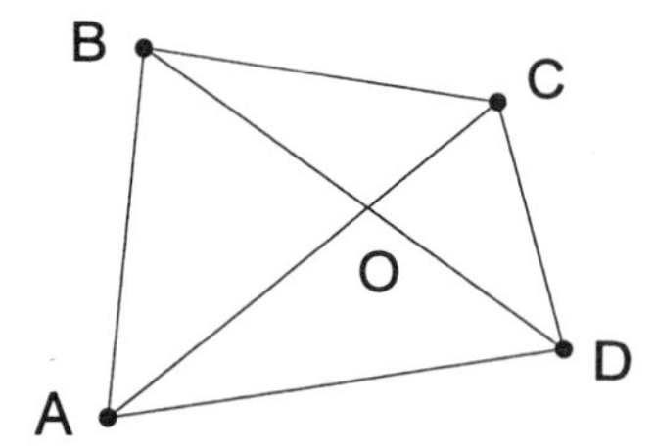

FIGURE 142

8. We have: $AP+PB > AB$; $BQ+QC > BC$; $CR+RD > CD$; $DS+SE > DE$; $ET+TA > EA$ (see Figure 143). Adding these inequalities gives $AP+PB+BQ+QC+CR+RD+SE+ET+TA > AB+BC+CD+DE+EA$. The right side of this inequality is the perimeter of the pentagon, while the left side is less than the sum of the diagonals (it will equal this sum if we add the perimeter of the inner rectangle $PQRST$). This proves the first result.

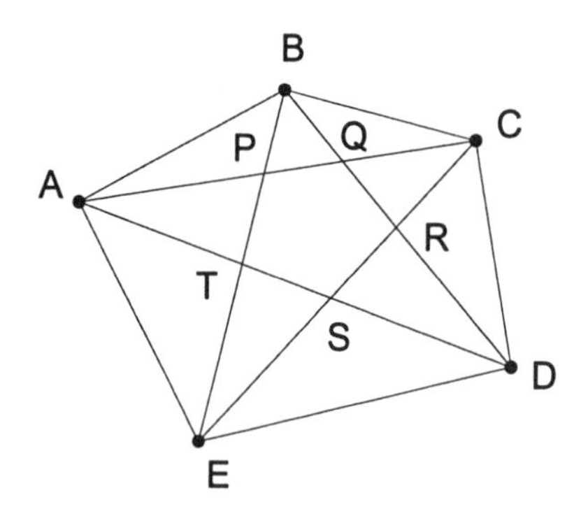

FIGURE 143

To get the second result, add the inequalities $AC < AB+BC$, $BD < BC+CD$, $CE < CD+DE$, $DA < DE+EA$, $EB < EA+AB$.

9. If the internal points are X and Y, we extend the segment connecting them in both directions, until it intersects the sides of the triangle (see Figure 144). Then

$EF < EA + AF$, and $EF < EB + BC + CF$. Adding, we find that EF is less than half the perimeter of the triangle. Since $XY < EF$, XY is also less than half the perimeter.

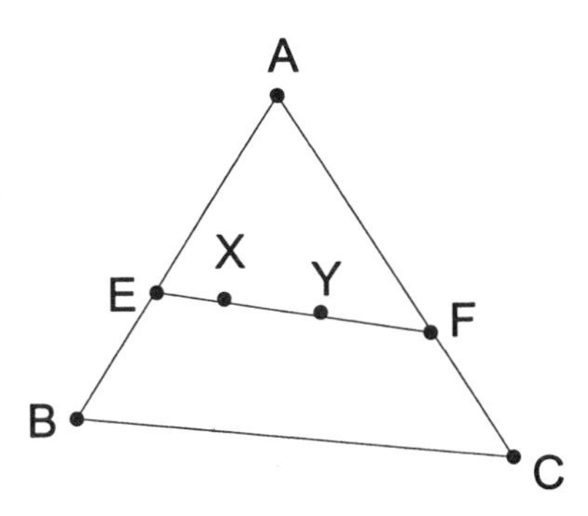

FIGURE 144

11. The solution is the path $ADEA$ as shown in Figure 145. Indeed, any other path will correspond to a path between points B and C (in that diagram) which is not a straight line.

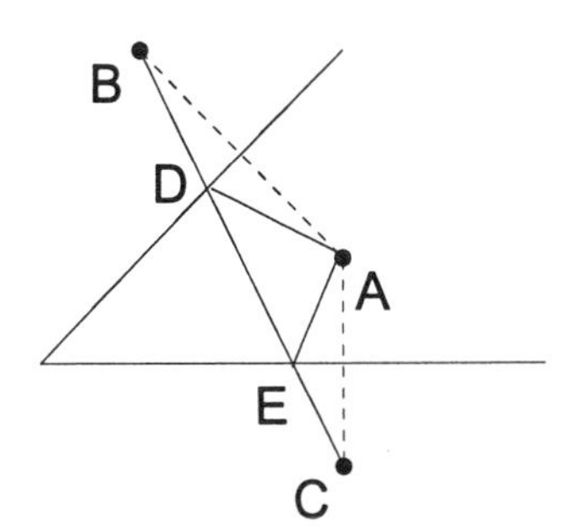

FIGURE 145

12. If we draw AD and AE (Figure 145), then $BC = BD + DE + EC = AD + DE + EA$. Then DE is less than half of the perimeter of triangle ADE (Problem 3), hence less than half of BC.

14. We can unfold the cube to form a diagram such as in Figure 146. Then if the fly is at A, the shortest distance to the opposite vertex B is a straight line. Folding the cube back up gives the answer. Students can make a paper model of this problem. Assigning a numerical value to an edge, they can be asked to find the length of the shortest path.

15. We can "unroll" the surface of the glass to get a rectangle, then "unfold" the front and back of the rectangle to get Figure 147. The shortest path is again given by a straight line.

16. Extend segment AO to intersect with side BC at point D. Then add the triangle inequalities $AB + BD > AD$ and $OD + DC > OC$, and subtract OD from each side of the inequality.

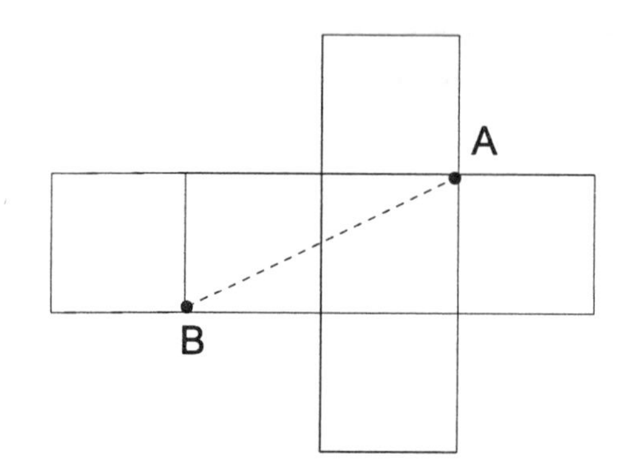

FIGURE 146

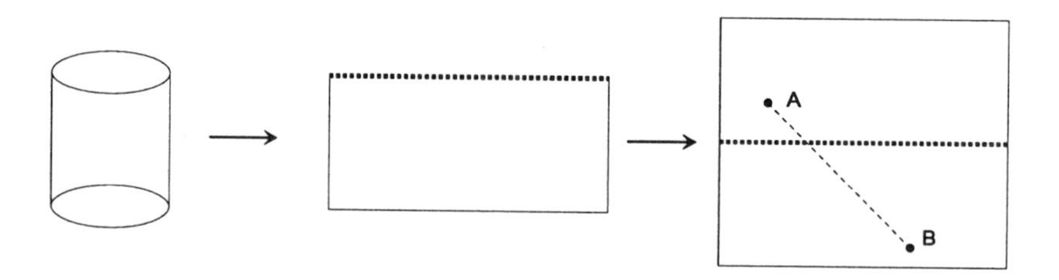

FIGURE 147

18. The woodsman must walk to the vertex of the angle, then back home. If the given point is A, and the given obtuse angle is BOC (see Figure 148), we choose that one of the angles $\angle AOC$ or $\angle AOB$ which is acute (maybe they both are)—say, $\angle AOC$. Then we drop a perpendicular BD from point B on line OC. By the result of Problem 13 we have $AB + BC + CA > 2AD$, and, obviously, we have $AD > AO$, which completes the proof.

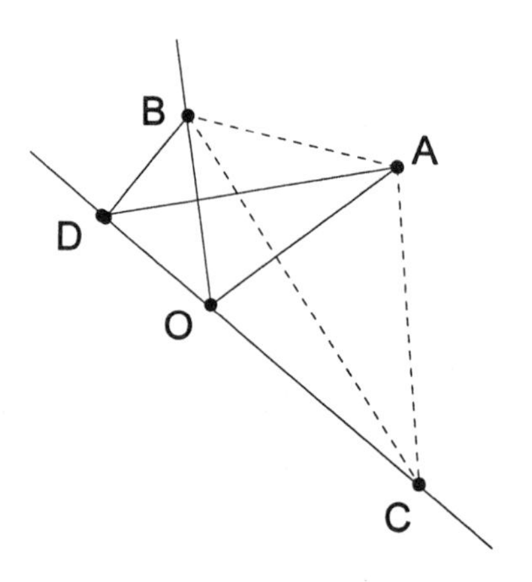

FIGURE 148

19. Construct parallelogram $ABDC$ from triangle ABC (Figure 149). Then, from the triangle inequality, $AB + BD > AD = 2AM$. Since $BD = AC$, this gives the first result. The second follows from writing the corresponding inequalities for each median and adding.

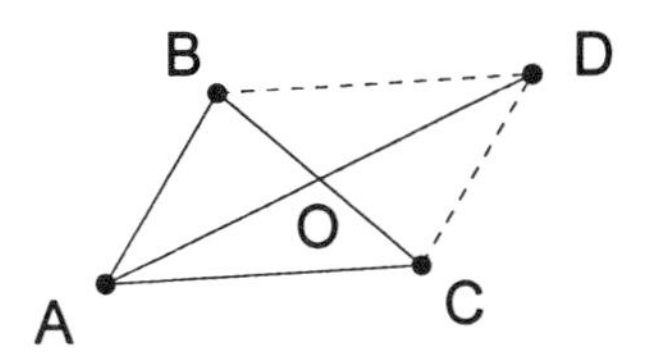

FIGURE 149

20. In finding the perimeter of the folded polygon, we lose the portion of the original perimeter represented by broken line $AXYB$ (in the example of Figure 150), but we add the length of segment AB. Since the sum of all but one side of a polygon is greater than the remaining side, the perimeter must have decreased.

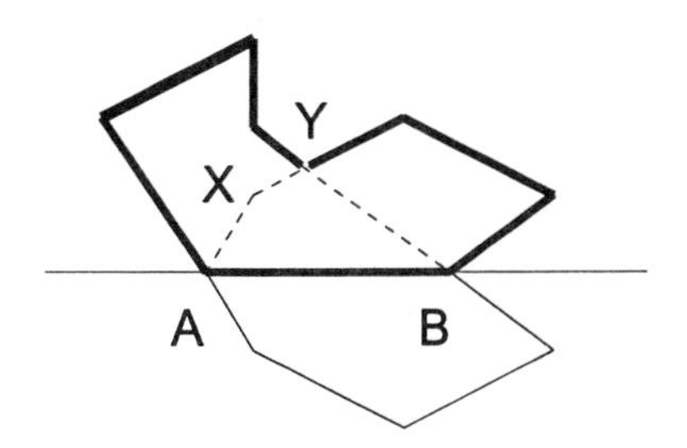

FIGURE 150

21. Consider two of the sides which do not have a common endpoint—say sides AB and CD. Then, on the one hand, $AC + BC < AB + CD$ (since AC and BD are diagonals). On the other hand, if AC and BD intersect at point O, then $OA + OB > AB$ and $OC + OD > CD$. Adding these inequalities we find that $AB + CD < AC + BD$, which is a contradiction.

22. Suppose the medians intersect at point M. Then, adding the inequalities $AM + BM > AB$, $BM + CM > BC$, and $CD + AM < AC$, and noticing that the lengths of AM, BM, and CM are each $2/3$ of a median, we reach the required inequality.

23. If the width of the river is h, and the towns are situated at points A and B, then the ends of the bridge must be placed at the points of intersection of lines $A'B$ and AB' with the banks, where A' and B' are obtained from A and B by a translation of distance h towards the river (Figure 151).

24. Consider the longest diagonal XY of the pentagon. One pair of vertices of the pentagon lies on the same side of this diagonal, so there exist two intersecting diagonals, each of which has X and Y as one endpoint respectively. It is not hard to see that these three diagonals will form a triangle.

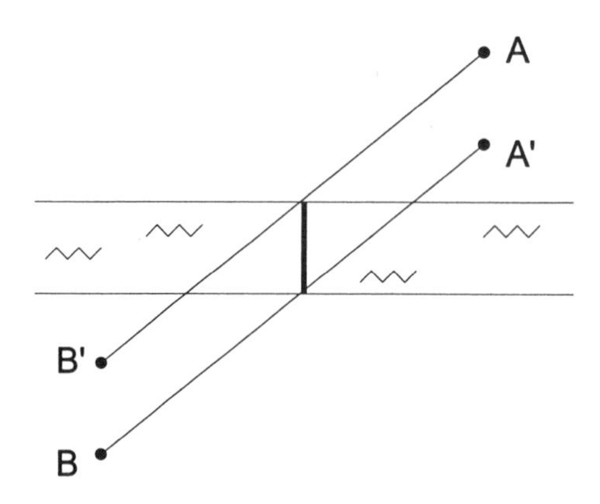

FIGURE 151

7. GAMES

2. After each move, the number of piles increases by 1. At first there are three piles, and at the end of the game there are 45. Therefore, 42 moves are made altogether. The last and winning move is always made by the second player.

3. The parity of the result does not depend on the position of the pluses and minuses, but only on the number of odd integers in the original set of numbers. Since there are 10 odd integers to begin with, and 10 is an even number, the first player will win.

4. After each move, the number of rows in which it is possible to place a rook decreases by 1, as does the number of columns. Therefore, there can only be 8 moves altogether, and the second player will make the last (winning) move.

5. The parity of the number of 1's on the blackboard remains unchanged after each move. Since there are evenly many 1's to begin with, there cannot be a single 1 left at the conclusion of the play (since 1 is an odd number!). The second player will therefore win.

6. In playing the game, the greatest common divisor of the two initial numbers must eventually be written down (compare this game with Euclid's algorithm). Therefore, every multiple of the greatest common divisor, not greater than the original numbers, will also appear. In this case, the greatest common divisor of the original numbers is 1, so that every number from 1 to 36 must appear. Therefore there will be 34 turns, and the second player will win.

7. This game is not entirely a joke, since the player who should win can in fact make a mistake and lose his or her advantage. This mistake consists in moving so that the remaining blank squares are all in one column or all in one row, allowing the opponent to win in the next move. The loser in this game, it turns out, is the player who makes just this fateful move. Notice that after crossing out a row of an $m \times n$ board, we can consider the remaining squares to be an $(m-1) \times n$ board. Analogously, in crossing out a column of an $m \times n$ board, we form an $m \times (n-1)$ board. The unique situation in which each move is "fateful" is the case of a 2×2 board. Therefore, the player who leaves this position for his opponent will win. However, as we have seen, after each turn the sum of the rows and columns decreases by 1. Therefore, the parity of this sum at the beginning of play will determine the winner. In case (a) this is the first player, while in the remaining cases it is the second. Note that in case (b) the second player can follow a strategy of symmetry (see §2).

11. Since a knight always moves from a black to a white square, or vice-versa, the second player can win, using either point or line symmetry.

12. The first player will win, if he moves first to the center of the board, then adopts a symmetric strategy.

13. The second player wins in both cases, using (a) line symmetry; (b) point symmetry. In the former case, the proof is quite simple: the second player just maintains the symmetry by always moving to the square symmetric to the previous move of the first player with respect to the line between the fourth and the fifth rows of the board. Since two squares symmetric in this line always have different colors we cannot encounter the situation when the current move of the first player prohibits the symmetric move of the second player.

The solution for the latter case is more tricky though the idea is similar: the second player uses symmetry with respect to the center of the board. The details are left to the reader.

14. The second player wins, using a point symmetric strategy.

15. The first player wins, if he removes the center checker first, then follows a point symmetric strategy.

16. The first player wins, if he first makes the two piles equal, then adopts the second player's strategy from Problem 10.

17. The first player wins. He must first draw a chord which separates the points into two groups of 9. He then replies symmetrically to each move of his opponent. Note that this strategy does not depend on how the points are arranged on the circle.

18. The second player wins in both cases. No matter how the first player begins, the second player can reply so as to leave two identical rows of petals on the flower. He can then follow a symmetric strategy.

19. In cases (a) and (b), the second player wins, following a strategy of point symmetry.

In case (c), the first player will win. In his first move, he skewers the row consisting of the center cubes of the four 3×3 layers. After this, he plays symmetrically with respect to the center point of the figure.

20. The loser is the player who breaks off a rectangle of width 1. The first player will win, by first breaking the chocolate bar into two 5×5 pieces. After that, he plays symmetrically.

21. The first player will win, if he places his first **x** in the center square, then replies to each of the second player's moves with an **x** placed symmetrically with respect to the center square.

23. The first player wins. We number the rows and columns of the chessboard in the usual order, so that the coordinates of square $a1$ are $(1,1)$, and those of square $h8$ are $(8,8)$. The winning positions are those in which the king occupies a square, both of whose coordinates are even. The first move is to square $b2$.

24. The first player wins. The winning positions are those in which both piles have oddly many pieces of candy. The first move is to eat the pile of 21 candies and divide the pile of 20 candies into any two piles of oddly many candies.

25. The second player wins. The winning positions are those in which the number of unoccupied squares between the checkers is divisible by 3.

26. The first player wins. The winning positions are those in which the box contains $2^n - 1$ matches. The first move is to leave 255 matches in the box.

27. The first player wins. The winning positions are those in which the largest pile of stones contains $2^n - 1$ stones. The first move consists in dividing the first two piles in any way at all, and dividing the third pile into two piles of 63 and 7 stones, respectively.

28. In this game, the player who obtains a 1 will win. This is the first player, if he recognizes that writing an odd number is a winning position.

29. In case (a) the second player wins, and in case (b) the first. The winning positions are those in which each pile contains oddly many matches.

For Problems 32–38, we give answers provided by analysis from the endgame. The reader can supply details.

32. The second player will win. Figure 152 shows the arrangement of pluses and minuses.

-	-	+	+	-	-	+	+
-	-	+	+	-	-	+	+
-	-	-	-	-	-	-	-
-	-	-	-	-	-	-	-
-	-	+	+	-	-	+	+
-	-	+	+	-	-	+	+
-	-	-	-	-	-	-	-
+	-	-	-	-	-	-	-

FIGURE 152

33. We can reformulate both cases (a) and (b) in terms of a chessboard. Game (a) turns out to be equivalent to the game of Problem 23. The arrangements of pluses and minuses in both cases are identical, and are shown in the figure to Problem 23 (Figure 55).

34. The first player wins. The arrangement of pluses and minuses, after a chessboard reformulation, is given in Figure 153.

35. This problem gives an example in which a geometric interpretation is not essential to an analysis of a game from the endgame. Here, it is convenient to mark

+	+	+	+	+	+	+	+	+	+	+	+
-	-	-	-	-	-	-	-	-	-	+	+
-	-	-	-	-	-	-	-	-	-	-	+
+	+	+	+	+	+	+	+	+	-	-	+
-	-	-	-	-	-	-	+	+	-	-	+
-	-	-	-	-	-	-	-	+	-	-	+
+	+	+	+	+	+	-	-	+	-	-	+
-	-	-	-	+	+	-	-	+	-	-	+
-	-	-	-	-	+	-	-	+	-	-	+
+	+	+	-	-	+	-	-	+	-	-	+
-	-	+	-	-	+	-	-	+	-	-	+
-	-	+	-	-	+	-	-	+	-	-	+

FIGURE 153

each number with a plus or a minus. The plus signs belong to those numbers which are multiples of 10. Therefore, the second player will win.

36. The winning positions are the numbers from 56 to 111, or from 4 to 6. Thus the first player wins, by moving to any of the numbers 4, 5, or 6.

37. The winning positions are 500, 250, 125, 62, 31, 15, 7, and 3. The first player wins.

38. The winning positions are the multiples of 3. The first player wins, for instance, by subtracting 1, 4, or 16 on the first move.

9. INDUCTION

8. The base can be either $n = 1$ or $n = 2$. To prove the inductive step let us take $k+1$ points on a circle. The segments connecting all of these points but the $(k+1)$st divide the interior of the circle into $k(k-1)(k-2)(k-3)/24 + k(k-1)/2 + 1$ parts by the inductive assumption. The segment connecting the $(k+1)$st point with the ith one (where i is a positive integer not greater than k) intersects $(i-1)(k-i)$ other segments. Thus, adding this segment would increase the number of the parts by $(i-1)(k-i)+1$. Adding all the segments connecting the $(k+1)$st point with the other k points would increase the number of parts by

$$1 + (1 \cdot (k-2) + 1) + \ldots + ((i-1)(k-i) + 1) + \ldots + ((k-2) \cdot 1 + 1) + 1.$$

This last expression can be rewritten as

$$(k+1)(1 + 2 + \ldots + k) - (1^2 + 2^2 + \ldots + k^2) - k^2 + k.$$

Using the identities

$$1 + 2 + \ldots + k = \frac{k(k+1)}{2}$$

(see the solution to Problem 6) and

$$1^2 + 2^2 + \ldots + k^2 = \frac{k(k+1)(2k+1)}{6}$$

(see Problem 10), we obtain

$$1 + (1 \cdot (k-2) + 1) + \ldots + ((i-1)(k-i) + 1) + \ldots + ((k-2) \cdot 1 + 1) + 1$$
$$= \frac{k(k+1)^2}{2} - \frac{k(k+1)(2k+1)}{6} - k^2 + k.$$

It remains to verify that

$$\left(\frac{k(k-1)(k-2)(k-3)}{24} + \frac{k(k-1)}{2} + 1\right) + \left(\frac{k(k+1)^2}{2} - \frac{k(k+1)(2k+1)}{6} - k^2 + k\right)$$
$$= \frac{(k+1)k(k-1)(k-2)}{24} + \frac{k(k+1)}{2} + 1,$$

which is just an algebraic calculation.

9. The base is $n = 1$. Let us prove the inductive step. By the inductive assumption we have

$$1 + 3 + \ldots + (2k-1) = k^2.$$

Thus,

$$1 + 3 + \ldots + (2k-1) + (2(k+1) - 1) = k^2 + (2(k+1) - 1) = (k+1)^2.$$

10. The base is $n = 1$. Let us show the inductive step. By the assumption

$$1^2 + 2^2 + \ldots + k^2 = \frac{k(k+1)(2k+1)}{6}.$$

Thus,

$$1^2 + 2^2 + \ldots + k^2 + (k+1)^2 = \frac{k(k+1)(2k+1)}{6} + (k+1)^2$$
$$= \frac{(k+1)(k+2)(2(k+1)+1)}{6}.$$

11. The base is $n = 2$ and is clear. By the assumption we have

$$1 \cdot 2 + 2 \cdot 3 + \ldots + (k-1) \cdot k = \frac{(k-1)k(k+1)}{3}.$$

Thus,

$$1 \cdot 2 + 2 \cdot 3 + \ldots + (k-1) \cdot k + k \cdot (k+1) = \frac{(k-1)k(k+1)}{3} + k \cdot (k+1) = \frac{k(k+1)(k+2)}{3}.$$

12. The base is $n = 2$. By the assumption

$$\frac{1}{1 \cdot 2} + \frac{1}{2 \cdot 3} + \ldots + \frac{1}{(k-1)k} = \frac{k-1}{k}.$$

Thus,

$$\frac{1}{1 \cdot 2} + \frac{1}{2 \cdot 3} + \ldots + \frac{1}{(k-1)k} + \frac{1}{k(k+1)} = \frac{k-1}{k} + \frac{1}{k(k+1)} = \frac{k}{k+1}.$$

13. Use induction on n. The base is $n = 1$. Let us prove the inductive step.

$$1 + x^2 + \ldots + x^k = \frac{x^{k+1} - 1}{x - 1}.$$

Thus,

$$1 + x^2 + \ldots + x^k + x^{k+1} = \frac{x^{k+1} - 1}{x - 1} + x^{k+1} = \frac{x^{k+2} - 1}{x - 1}.$$

14. We use induction on n. The base ($n = 1$) is quite clear. To prove the inductive step, we start as follows:

$$\frac{1}{a(a+b)} + \frac{1}{(a+b)(a+2b)} + \ldots + \frac{1}{(a+(k-1)b)(a+kb)} = \frac{k}{a(a+kb)}.$$

Thus,

$$\frac{1}{a(a+b)} + \frac{1}{(a+b)(a+2b)} + \ldots + \frac{1}{(a+(k-1)b)(a+kb)} + \frac{1}{(a+kb)(a+(k+1)b)}$$
$$= \frac{k}{a(a+kb)} + \frac{1}{(a+kb)(a+(k+1)b)} = \frac{k+1}{a(a+(k+1)b)}.$$

15. Let us use induction on n. The base is $n = 0$:

$$\frac{m!}{0!} = \frac{(m+1)!}{0!(m+1)}.$$

For the inductive step we have (by assumption):

$$\frac{m!}{0!} + \frac{(m+1)!}{1!} + \ldots + \frac{(m+k)!}{k!} = \frac{(m+k+1)!}{k!(m+1)}.$$

Thus,

$$\frac{m!}{0!} + \frac{(m+1)!}{1!} + \ldots + \frac{(m+k)!}{k!} + \frac{(m+k+1)!}{(k+1)!} = \frac{(m+k+1)!}{k!(m+1)!} + \frac{(m+k+1)!}{(k+1)!}$$
$$= \frac{(m+k+1)!}{k!}\left(\frac{1}{m+1} + \frac{1}{k+1}\right) = \frac{(m+k+2)!}{(k+1)!(m+1)}.$$

16. The base is $n = 2$. By the assumption

$$\left(1 - \frac{1}{4}\right)\left(1 - \frac{1}{9}\right) \cdots \left(1 - \frac{1}{k^2}\right) = \frac{k+1}{2k}.$$

Thus,

$$\left(1 - \frac{1}{4}\right)\left(1 - \frac{1}{9}\right) \cdots \left(1 - \frac{1}{k^2}\right)\left(1 - \frac{1}{(k+1)^2}\right)$$
$$= \frac{k+1}{2k}\left(1 - \frac{1}{(k+1)^2}\right) = \frac{k+2}{2(k+1)}.$$

17. The base is $n = 1$: $1^3 + 2^3 + 3^3 = 36$, and 36 is divisible by 9.

Now the inductive step. By the inductive assumption $k^3 + (k+1)^3 + (k+2)^3$ is divisible by 9. Thus,

$$(k+1)^3 + (k+2)^3 + (k+3)^3 = (k^3 + (k+1)^3 + (k+2)^3) + (k+3)^3 - k^3$$
$$= (k^3 + (k+1)^3 + (k+2)^3) + 9(k^2 + 3k + 3)$$

is also divisible by 9.

18. The base is $n = 1$: $3^4 + 8 - 9 = 80$, and 80 is divisible by 16.

Let us prove the inductive step. By the assumption $3^{2k+2} + 8k - 9$ is divisible by 16. We have

$$\left(3^{2k+4} + 8(k+1) - 9\right) - \left(3^{2k+2} + 8k - 9\right) = 3^{2k+2} \cdot 8 + 8 = 8(3^{2k+2} + 1).$$

The number $3^{2k+2} + 1$ is even, so $8(3^{2k+2} + 1)$ (and, hence, $3^{2k+4} + 8(k+1) - 9$) is divisible by 16.

19. The base is $n = 1$: $4^1 + 15 - 1 = 18$.

Let us prove the inductive step. We know that $4^k + 15k - 1$ is divisible by 9. Thus, we have

$$\left(4^{k+1} + 15(k+1) - 1\right) - \left(4^k + 15k - 1\right) = 4^k \cdot 3 + 15 = 3(4^k + 5).$$

The number 4^k has remainder 1 when divided by 3. So $4^k + 5$ is divisible by 3, and, thus, $3(4^k + 5)$ is divisible by 9.

20. The base is $n = 1$: $11^3 + 12^3 = 23 \cdot 133$.

Further, by the inductive assumption we know that $11^{k+2} + 12^{2k+1}$ is divisible by 133. Therefore,

$$11^{k+3} + 12^{2k+3} = 11(11^{k+2} + 12^{2k+1}) + 133 \cdot 12^{2k+1}.$$

Thus, $11^{k+3} + 12^{2k+3}$ is divisible by 133.

21. The base is $n = 1$: the number $2^3 + 1$ is divisible by 3^2.

Then we know that $2^{3^k} + 1$ is divisible by 3^{k+1}. Further

$$2^{3^{k+1}} + 1 = \left(2^{3^k}\right)^3 + 1 = \left(2^{3^k} + 1\right)\left(\left(2^{3^k}\right)^2 - 2^{3^k} + 1\right).$$

Thus, it remains to prove that $\left(2^{3^k}\right)^2 - 2^{3^k} + 1$ is divisible by 3. The number 2^{3^k} has remainder 2 when divided by 3. Hence, the remainder of $\left(2^{3^k}\right)^2 - 2^{3^k} + 1$ when divided by 3 is zero.

22. The base is $n = 0$, $n = 1$, and is obvious. To prove the inductive step from n to $n + 1$, we must show that $ab^{n+1} + c(n+1) + d$ is divisible by m. Let us use the fact that the previous member of the sequence $ab^n + cn + d$ is divisible by m, and multiply it by b. We have that $ab^{n+1} + cbn + bd$ is divisible by m, so it is left to prove that $c(n + 1 - bn) + d(1 - b)$ is divisible by m as well. Adding $(b-1)cn$, which is divisible by m, we get $c + d(1 - b)$, which is also divisible by m, since it can be represented as $(ab - a + c) - (a + d)(b - 1)$. This completes the proof of the inductive step.

23. The base is $n = 1$: the inequality $2^1 > 1$ is certainly true.

Now the inductive step. By the assumption $2^k > k$. Thus,

$$2^{k+1} = 2 \cdot 2^k > 2k \geq k + 1.$$

24. a) Answer: $n \geq 3$.

For $n = 1, 2$, we have $2^n < 2n + 1$. Let us prove by induction on n that for $n \geq 3$ the inequality $2^n > 2n + 1$ holds true.

The base is $n = 3$: $2^3 > 2 \cdot 3 + 1$. To prove the inductive step we start with $2^k > 2k + 1$. Then,

$$2^{k+1} = 2 \cdot 2^k > 4k + 2 > 2(k+1) + 1.$$

b) Answer: $n = 1$, $n \geq 5$.

For $n = 1$ we have $2^1 > 1^2$, and for $n = 2$, 3, and 4 we have $2^n \leq n^2$. Let us prove by induction on n that for $n \geq 5$ the inequality $2^n > n^2$ holds true.

The base is $n = 5$: $2^5 > 5^2$. Now the inductive step: we know that $2^k > k^2$. Thus,

$$2^{k+1} = 2^k + 2^k > k^2 + 2k + 1 = (k+1)^2$$

(we use the inequality $2^k > 2k + 1$ proved in Problem 24a).

25. The base is $n = 2$: $\frac{1}{3} + \frac{1}{4} = \frac{7}{12} > \frac{13}{24}$. Further, by the inductive assumption $\frac{1}{k+1} + \frac{1}{k+2} + \ldots + \frac{1}{2k} > \frac{13}{24}$. Thus,

$$\begin{aligned}&\frac{1}{k+2} + \frac{1}{k+3} + \ldots + \frac{1}{2k} + \frac{1}{2k+1} + \frac{1}{2k+2}\\ &\quad= \left(\frac{1}{k+1} + \frac{1}{k+2} + \ldots + \frac{1}{2k}\right) + \frac{1}{2k+1} + \frac{1}{2k+2} - \frac{1}{k+1}\\ &\quad> \frac{1}{k+1} + \frac{1}{k+2} + \ldots + \frac{1}{2k} > \frac{13}{24},\end{aligned}$$

since $\frac{1}{2k+1} + \frac{1}{2k+2} > \frac{2}{2k+2} = \frac{1}{k+1}$.

26. The base is $n = 2$, and $n = 3$: $4 > 1 + 2\sqrt{2}$, $8 > 1 + 3 \cdot 2$. Since

$$2^k > 1 + k\sqrt{2^{k-1}},$$

we obtain

$$2^{k+1} > 2 + 2k\sqrt{2^{k-1}} > 1 + \sqrt{2} \cdot k\sqrt{2^k}.$$

It remains to note that $\sqrt{2} \cdot k > k + 1$ for $k \geq 3$.

27. We need to prove that $|a_1 + a_2 + \ldots + a_n| \leq |a_1| + |a_2| + \ldots + |a_n|$ for any positive integer n and for any real numbers $a_1, a_2, \ldots, a_n$. We will prove the statement using the induction on n.

The base is $n = 1$, $n = 2$. For $n = 1$ the statement is evident. For $n = 2$ we have $|a_1 + a_2| \leq |a_1| + |a_2|$, which can be proved by a simple case-by-case analysis considering all four possible combinations of signs. Now, using the base, we obtain

$$|a_1 + a_2 + \ldots + a_k + a_{k+1}| \leq |a_1 + a_2 + \ldots + a_k| + |a_{k+1}|.$$

Then

$$|a_1 + a_2 + \ldots + a_k| + |a_{k+1}| \leq |a_1| + |a_2| + \ldots + |a_k| + |a_{k+1}|.$$

28. Use induction on n. The base is $n = 2$: $(1+x)^2 = 1 + 2x + x^2 > 1 + 2x$ for $x \neq 0$.

Now, by the inductive assumption $(1+x)^k > 1 + kx$, and we have

$$(1+x)^{k+1} > (1+x)(1+kx) = 1 + (k+1)x + kx^2 > 1 + (k+1)x$$

(remember that $1 + x > 0$).

29. The base is $n = 1$:

$$\frac{1}{2} \leq \frac{1}{\sqrt{3}}.$$

Now, by the assumption

$$\frac{1\cdot 3\cdot 5\cdot \ldots \cdot (2k-1)}{2\cdot 4\cdot 6\cdot \ldots \cdot 2k} \le \frac{1}{\sqrt{2k+1}}.$$

Therefore,

$$\frac{1\cdot 3\cdot 5\cdot \ldots \cdot (2k-1)}{2\cdot 4\cdot 6\cdot \ldots \cdot 2k}\cdot\frac{2k+1}{2k+2} \le \frac{1}{\sqrt{2k+1}}\cdot\frac{2k+1}{2k+2} = \frac{\sqrt{2k+1}}{2k+2}.$$

It remains to prove the inequality

$$\frac{\sqrt{2k+1}}{2k+2} \le \frac{1}{\sqrt{2k+3}}.$$

This inequality is equivalent to the following: $(2k+1)(2k+3) \le (2k+2)^2$, which is evident after expanding both sides.

33. The base is $n = 1$ and $n = 2$. The proof of the inductive step is also quite simple:

$$a_{k+1} = 3a_k - 2a_{k-1} = 3(2^k+1) - 2(2^{k-1}+1) = 2^{k+1}+1.$$

34. Indeed, $a_3 = 1$, $a_4 = -1$, $a_5 = -2$, $a_6 = -1$, $a_7 = 1$, and $a_8 = 2$. Hence for $n = 1$ and $n = 2$ we have $a_{n+6} = a_n$. Then a "strong induction" (see §4 of the chapter "Induction") gives us the proof.

35. We can check directly for the natural numbers 1 through 5. If x is the given natural number, then let F_n be the maximum Fibonacci number not greater than x. Then we have $0 \le x - F_n < F_{n-1}$ (since $x < F_{n+1} = F_n + F_{n-1}$), and therefore $x - F_n$ can be represented as the sum of several different Fibonacci numbers less than F_{n-1}.

36. We will prove the following statement by induction on n:

The remainders of the Fibonacci numbers F_n and F_{n+8} when divided by 3 are equal for all natural n.

The base is $n = 1$ and $n = 2$: $F_1 = 1$, $F_2 = 1$, $F_9 = 34$, $F_{10} = 55$, and we see that the remainders of F_1 and F_9 (and of F_2 and F_{10}) are equal.

The proof of the inductive step is very similar to the solution to Problem 34. By the inductive assumption, F_{k+8} and F_k (F_{k+7} and F_{k-1}, respectively) have the same remainders when divided by 3. Thus, the remainders of $F_{k+9} = F_{k+8} + F_{k+7}$ and of $F_{k+1} = F_k + F_{k-1}$ are equal.

It remains to calculate the remainders of the first eight Fibonacci numbers. They are

$$1, 1, 2, 0, 2, 2, 1, 0.$$

Thus, the nth Fibonacci number is divisible by 3 if and only if n is divisible by 4.

42. We will prove the required statement by induction on n. The base $n = 1$ is trivial.

To prove the inductive step we will first prove that using a calculator we can obtain a natural number less than n. To show this, we choose among n, m, and 0 two numbers of the same parity, and calculate their arithmetic mean x. At least one of the two chosen numbers is different from 0. Replace this non-zero number by x and repeat the operation for the new trio of numbers. We repeat this procedure until one of the positive numbers in the trio becomes less than n (this will eventually happen, because the sum of two positive numbers in the trio decreases after each operation).

Now, let l be the largest natural number less than n which can be obtained using a calculator. Suppose that $l \neq n-1$. By the assumption, all natural numbers 1 through l can be obtained using a calculator. If l and n are of the same parity, then we can calculate the arithmetic mean y of l and n, which contradicts the definition of l since $l < y < n$. If l and n are not of the same parity, we can calculate the arithmetic mean of $l-1$ and n and come to the same contradiction.

43. Hint: the inductive step follows from the formula

$$3(2^n + 1) - 2(2^{n-1} + 1) = 2^{n+1} + 1\ ,$$

which is true for any integer n.

44. Let us use induction on m. The base is $m = 1$: $2^{n-1} \geq n$. This inequality follows from the result of Problem 23, and can be proved by induction on n.

By the inductive assumption $2^{m+n-2} \geq mn$, and we have $2^{(m+1)+n-2} \geq 2mn \geq (m+1)n$.

45. Hint: first, prove that any square can be cut into several parts which can be arranged to form a rectangle with one side of unit length.

46. The inductive step can be proved as follows: split the given 2^{n+1} numbers into two halves each containing 2^n numbers. In each of these halves we can find 2^{n-1} numbers with the sum divisible by 2^{n-1}. Then, out of the remaining 2^n numbers, we can choose the third set of 2^{n-1} numbers whose sum is divisible by 2^{n-1}. Let the sums of the numbers in the three chosen sets be $2^{n-1}a$, $2^{n-1}b$, and $2^{n-1}c$. Among the numbers a, b, and c we can find two numbers of the same parity. The union of corresponding sets is a set of 2^n numbers whose sum is divisible by 2^n.

47. For n circles the answer is $n(n-1)+2$. To prove it, we can use induction on n. The base $(n = 1)$ is clear.

To prove the inductive step we temporarily remove the $(k+1)$st circle. By the inductive assumption the number of parts into which k circles dissect the plane is not greater than $k(k-1)+2$. Now we "restore" the removed circle. It intersects each of the k circles at no more than two points, and, thus, the number of parts of the plane increases by at most $2k$. The formula $k(k-1)+2+2k = k(k+1)+2$ completes this part of the proof.

An example of the required dissection can also be obtained by induction. We will construct examples (one for each natural number n) with the following properties:

a) no three circles meet at the same point;

b) the interiors of all n circles have a common point;

c) the number of parts into which our n circles dissect the plane equals $n(n-1)+2$.

The base $n = 1$ is again trivial.

To prove the inductive step we start with the configuration (which is assumed to exist) of k circles that dissect the plane into $k(k-1)+2$ parts. Let us add another circle which passes through a point lying inside all k circles of the configuration. We can choose this $(k+1)$st circle in such a way that it does not pass through the points of intersection of the other circles. This new configuration of $k+1$ circles satisfies all the required conditions.

For n triangles the answer is $3n(n-1)+2$. The proof is similar to the proof above, except that two triangles can intersect in at most 6 points.

To construct a configuration of n triangles dissecting the plane into $3n(n-1)+2$ parts, one can draw n congruent equilateral triangles in such a way that they have the same center and no three of them meet in a point.

48. We use induction on the number n of circles. The base is $n = 1$.

To prove the inductive step, temporarily remove the $(k+1)$st circle and its chord. By the inductive assumption the parts of the plane created by k circles with their chords can be colored using 3 colors (say, red, blue, and green) satisfying the given condition.

Now, let us replace the missing circle and its chord. The chord divides the interior of the circle into two parts. Let us change the colors in one part using the scheme: red $\to$ blue, blue $\to$ green, green $\to$ red, and change the colors in the other part using the scheme: red $\to$ green, green $\to$ blue, blue $\to$ red. The colors of the parts lying outside the $(k+1)$st circle remain unchanged. In the resulting coloring, the colors of adjacent regions of the plane are different.

49. The principle of mathematical induction implies the "well order principle". Indeed, suppose that the principle of mathematical induction holds true and the "well order principle" does not. Take a non-empty set S of natural numbers which does not contain a least element. Let us prove by induction that any natural number n does not belong to S (which contradicts the fact that S is non-empty).

The base is $n = 1$. If $1 \in S$, then 1 is the least element of S.

Now, the inductive step. By the assumption, the numbers $1, 2, \ldots, k$ do not belong to S. Then, if $k+1 \in S$, we have that $k+1$ is the least element of S.

Now let us prove that the well order principle implies the principle of mathematical induction.

Suppose that the well order principle is true while the principle of mathematical induction is not. Consider a series of propositions such that the first proposition is true and for any natural k the truth of the kth proposition in the series implies the truth of the $(k+1)$st proposition. Form the set of natural numbers n such that the nth proposition in the series is not true. Assume this set is non-empty, and let n_0 be its least element. Then the (n_0-1)th proposition is true, but the n_0th proposition is not. This contradiction to our assumption completes the proof.

10. DIVISIBILITY–2

3. Change the solution to Problem 2 by subtracting the equalities instead of adding them.

5. This is an immediate corollary of Problem 4.

8. $30^{99} \equiv (-1)^{99} \equiv -1 \pmod{31}$, $61^{100} \equiv (-1)^{100} \equiv 1 \pmod{31}$.

9. b)By direct multiplication, we can check that

$$a^n + b^n = (a+b)(a^{n-1} - a^{n-2}b + \ldots + (-1)^{n-1}b^{n-1}).$$

10. Consider the summands with the numbers k and $n-k$; that is, k^n and $(n-k)^n$. Since n is odd, the remainders of k^n and $(-k)^n$ have opposite signs. Thus, the sum of these two summands is divisible by n. Since the sum can be split into $(n-1)/2$ such pairs we obtain the required result.

11. No number of the form $8k+7$ can be represented as the sum of three squares. Indeed, a perfect square has a remainder of 0, 1, or 4 when divided by 8. It is easy to see that the sum of three such remainders cannot give us the remainder 7.

13. Use the identity $x^2 - y^2 = (x-y)(x+y)$.

14. Hint: show that if x is a convenient number, then $1000001 - x$ is also convenient.

15. The answers are a) no; b) no. Indeed, the last digit of a number determines the last digit of its square. After analyzing all possible last digits and their squares, we can see that the last digits of the squares can be equal only to 0, 1, 4, 5, 6, and 9. This can be expressed, of course, using the language of "mod 10".

16. There are several answers. For example, -1 or $n-1$. The remainder of the given number when divided by n is 1.

17. The answer is 5.

18. The answer is 2858.

19. a) Since k is divisible by 3 we conclude that $k - 1 \equiv 2 \pmod 3$. To complete the proof it suffices to remember that squares cannot have a remainder of 2 when divided by 3.

b) Since k is even but not divisible by 4 we have $k+1 \equiv 3 \pmod 4$. Now we use the fact that squares can be congruent only to 0 or 1 (mod 4).

20. No, since n^2+n+1 cannot be divisible by 5. Indeed, there are five congruence classes modulo 5. If $n \equiv 0 \pmod 5$, then $n^2+n+1 \equiv 1$, if $n \equiv 1 \pmod 5$, then $n^2+n+1 \equiv 3$, et cetera. So n^2+n+1 is never congruent to zero, i.e. n^2+n+1 cannot be divisible by 5, and therefore, by 1955.

22. We prove that the sum of the divisors is divisible by 3 and by 8. To prove that this sum is divisible by 3 we split it into pairs of divisors $(k, n/k)$ (notice that $k \neq n/k$ since n cannot be a perfect square. Why?) and prove that for each of these pairs the sum of the numbers in it—that is, $k + n/k$—is divisible by 3. Indeed, k cannot be divisible by 3 (otherwise, n would be divisible by 3, which is obviously impossible). Therefore, either $k \equiv 1 \pmod 3$ or $k \equiv 2 \pmod 3$. In the former case $n/k \equiv 2 \pmod 3$ and in the latter case $n/k \equiv 1 \pmod 3$ (we recall that $n+1$ is divisible by 3). Thus, in any case, $k + n/k$ is a multiple of 3. The second part of the proof (regarding divisibility by 8) is similar.

23. a) Let us consider the remainders of the members of the sequence when divided by 4. We have $a_1 \equiv a_2 \equiv 1 \pmod 4$. Consequently, $a_3 \equiv 1 \cdot 1 + 1 = 2$, $a_4 \equiv 2 \cdot 1 + 1 = 3$, $a_5 \equiv 3 \cdot 2 + 1 \equiv 3$, $a_6 \equiv 3 \cdot 3 + 1 = 2$, $a_7 \equiv 2 \cdot 3 + 1 \equiv 3$, and we have a cycle which does not contain zero remainders.

b) Consider the remainders of the members of the sequence when divided by a_n. Simple calculation shows that $a_{n+1} \equiv 1$, $a_{n+2} \equiv 1$, $a_{n+3} \equiv 2$, ... , $a_{n+6} \equiv 22$. Thus, $a_{n+6} - 22$ is divisible by a_n, and if $n+6 > 10$, then, obviously, $a_{n+6} > a_n > 1$.

25. Hint: all the powers of ten, starting from 100, are divisible by 4.

26. A number is divisible by 2^n (or by 5^n) if and only if the number formed by its last n digits is divisible by 2^n (or by 5^n).

28. The last two digits of the square of n depend only on the last two digits of n itself. Suppose $n = \overline{\ldots ab}$, and we have $\overline{ab}^2 = (10a+b)^2 = 100a^2 + 20ab + b^2$. It is clear that the tens digit of the number b^2 must be odd. A case-by-case analysis shows that the units digit must then be equal to 6.

29. Hint: consider remainders modulo 16.

32. The answers are a) no; b) no. Use remainders modulo 9.

33. The answer is 7. Use remainders modulo 9.

34. The original number and the reversed number have equal sums of digits and, therefore, equal remainders when divided by 9.

35. This can be done in six different ways: 1155, 4155, 7155, 3150, 6150, 9150. Indeed, the last digit must be 0 or 5. Then our number is divisible by 3; that is, the sum of its digits must be divisible by 3.

36. There are two such numbers: 6975 and 2970. See the solution to the previous problem.

37. This is the number 1023457896. Hint: first, any number which has all 10 digits in its decimal representation is divisible by 9. Second, divisibility by 4 depends only on the last two digits of a number. Therefore, the required number starts with $10\ldots$ and must end with an even digit. A simple case-by-case analysis leads us to the answer.

38. Hint: find a cycle in the remainders of the numbers 2^n when divided by 3, and also in the units digits of these numbers.

39. The answer is no. By the usual test (see Problem 31) we have $1970 \equiv 8 \pmod 9$. But no perfect square is congruent to 8 modulo 9.

40. Since after the first subtraction the result is divisible by 9, all the numbers we obtain in the process have the sum of their digits no less than 9. Therefore, if the original number was not greater than $891 = 9 \cdot 99$ then the proof is now obvious. The investigation of the rest of the set of three-digit numbers is left to the reader.

41. Since $4444^{4444} < 10^{4444}$ it is easy to see that $A < 44440$. Therefore, $B < 5 \cdot 9 = 45$, which implies that the sum of its digits is a one-digit number. We also know that A and B (and the sum of its digits as well) are congruent to 4444^{4444} modulo 9. That is, they all have remainder 7 when divided by 9. Thus, the sum of the digits of B must be 7.

43, 44. These numbers are divisible by 11.

45. The number $\overline{aabb}$ is divisible by 11 while $\overline{cdcdcdcd}$ is not.

46. The set $\{1, 2, 3, 4, 5, 6\}$ cannot be split into two triples in such a way that the difference of the sums of the numbers in these triples is divisible by 11.

47. These two numbers have equal remainders when divided by 9, and also when divided by 11.

48. The answer is no. Indeed, if you multiply any of the given four digits by 9, then you cannot obtain one of these digits again in the units place of the answer.

49. Indeed, $\overline{aba} = 101a + 10b = 7(14a + b) + 3(a + b)$.

50. Since $2(a+b+c) \equiv 0 \pmod 7$, we get $\overline{abc} = 100a+10b+c \equiv 2a+3b+c \pmod 7 \equiv b - c \pmod 7$. Therefore, $\overline{abc}$ is divisible by 7 if and only if $b - c$ is divisible by 7. Taking into account that both digits b and c are less than 7, the only way this difference can be divisible by 7 is if the two digits are equal.

51. a) Since b) we have $\overline{abcdef} = 1000\overline{abc} + \overline{def} \equiv \overline{def} - \overline{abc} \pmod 7$.

b), c) A number is divisible by 7 (or by 13) if and only if the following operation gives us a number divisible by 7 (or by 13): starting at the right of the number, group the digits in threes and alternately add and subtract the resulting numbers. Example: 10345678. The operation described gives us $678 - 345 + 10 = 343$ which is divisible by 7. Thus, the original number is divisible by 7 as well. Actually, it equals $7 \cdot 1477954$. The proof of these divisibility tests is left to the reader; they use the fact that 1001 is divisible by 7 and by 13.

52. A number is divisible by 37 if and only if the sum of the numbers formed by the triples of its consecutive digits is divisible by 37. Example: 830946. The described operation gives us $830 + 946 = 1776$ which is divisible by 37. Therefore we can conclude that 830946 is also divisible by 37. Indeed, it equals $37 \cdot 22458$. The proof is similar to that of Problem 51. It uses the fact that $1000 \equiv 1 \pmod{37}$.

53. The answer is no. Since $\overline{abc} - \overline{cba} = 99(a - c)$, where a and c are different digits, we have a number which is divisible by 11, but not by 11^2.

54. Answer: this number is written with three hundred 1's. Indeed, it is divisible by 3 and it is divisible by $333\ldots33$ (one hundred 3's), which are co-prime. To prove that this is the minimum number required we notice first that the required number must have a number of digits divisible by 100—otherwise it would not be divisible by $111\ldots11$ (one hundred 1's). Secondly, the numbers $111\ldots11$ (one hundred 1's) and $111\ldots11$ (two hundred 1's) are not divisible by 3.

55. The answer is no. Assuming the opposite, let us consider remainders modulo 5. Since the sum of the first n natural numbers is $n(n+1)/2$, we have that $2\big(n(n+1)/2+1\big) = n(n+1)+2$ must be divisible by 5 (indeed, the last digits of $n(n+1)/2+1$ would be ... 1990). Substituting all five possible remainders modulo 5, that is, 0, 1, 2, 3, 4, and 5, we observe this is not true, which proves that the answer is no.

57. The answer is 69. Write $102a + b = 90a + 9b$. Simplify to find $3a = 2b$, and remember that a and b are digits.

58. Since any number of the form $\overline{aabb}$ is divisible by 11, we know that the square root of our number must be a two-digit number divisible by 11 (that is, with equal digits). Calculating squares for these 9 numbers 11 through 99, we see that the only answer is $7744 = 88^2$.

59. The answers are 625 and 376. Hint: the units digit must be 0, 1, 5, or 6. Then analyze the tens digit similarly, then the hundreds digit. Remember that neither 000 nor 001 is a valid three-digit number.

60. We can prove that sooner or later this number will be divisible by 11. Indeed, if we denote some number in this series by x, then the next number will be $100x+43 \equiv x - 1 \pmod{11}$. This means that after no more than 10 operations the current number will be congruent to zero modulo 11.

61. First, $10001 = 73 \cdot 137$, which is not obvious but nevertheless true. Second, to prove that any other number $10001\ldots10001$ of the series is composite, we multiply it by 1111. The result has $4k$ digits ($k > 2$) and, therefore, is divisible by $x = 1000\ldots001 = 10^{2k} + 1$ (indeed,

$$\underbrace{111\ldots11}_{4k \text{ digits}} = \underbrace{1000\ldots001}_{2k+1 \text{ digits}} \cdot \underbrace{111\ldots11}_{2k \text{ digits}}) .$$

Finally, we use the fact that x is greater than 1111 and less than the original number. Therefore, the original number must be divisible by $x/\gcd(x, 1111) > 1$.

63. This equation does not have integers roots. Its left side is always divisible by 3, but its right side never is.

65. The solutions are of the form $\{x = 16k - 2,\ y = -7k + 1\}$, where k takes on all integer values.

66. Our problem can be reduced to the solution of an ordinary Diophantine equation. Set $p = z$, and we have

$$2x + 3y = 11 - 5p\ ,$$

where we consider p as an unknown parameter, not as a variable. Using the same technique as before, we have the following answer: $x = 5p+3q-11$, $y = 11-5p-2q$, and, of course $z = p$, where p and q are any integers.

There are no solutions in natural numbers, since if any of the numbers x, y, and z is greater than 1, then the sum $2x + 3y + 5z$ is greater than 11.

67. It is possible to move the pawn to the neighboring box if and only if the numbers m and n are relatively prime. Indeed, if we make k moves to the right (shifting the pawn m boxes to the right with each of these moves) and l moves to the left, then the resulting shift equals $km - ln$ boxes to the right (a negative result means a shift to the left). The number 1 can be represented by such an expression if and only if the numbers m and n are relatively prime. The question about the minimum number of moves to get to the neighboring box is far more complicated. Hint: it is convenient to begin with the following reformulation. Given two relatively prime natural numbers m and n, find natural numbers k and l such that $mk - nl = 1$ and the absolute value of the sum $|k + l|$ is as small as possible.

68. The answers are $(-4, 9)$, $(14, -21)$, $(4, -9)$, and $(-14, 21)$. Analyze all possible representations of the prime number 7 as the product of two integers.

70. There are no integer solutions. Indeed, in the equation $(x-y)(x+y) = 14$ both factors in the left side are of the same parity, and therefore their product must be either odd (if both expressions $x + y$ and $x - y$ are odd) or divisible by 4 (if both $x+y$ and $x-y$ are even). But the number 14 belongs to neither of these two types.

71. The answers are $(2, 0)$, $(2, 1)$, $(-1, 0)$, $(-1, 1)$, $(0, 2)$, $(1, 2)$, $(0, -1)$, $(1, -1)$. We transform the original equation into $x(x - 1) + y(y - 1) = 2$. Since the product $t(t - 1)$ is never negative and is greater than 2 if $t > 2$ or $t < -1$, we have only a few pairs (x, y) to consider.

73. There are no integer solutions. Hint: use remainders modulo 7.

75. There are no integer solutions. Hint: use remainders modulo 5.

76. There are no integer solutions. Hint: use remainders modulo 8.

79. Answer: there are three families of solutions $(1, a, -a)$, $(b, 1, -b)$, $(c, -c, 1)$, where a, b, and c are arbitrary integers. And there are three more solutions: $(1, 2, 3)$, $(2, 4, 4)$, $(3, 3, 3)$. Hint: if all numbers are positive, at least one of them is not greater than 2 or they all are equal to 3. If one of the numbers—say, a—is negative, then $1/b + 1/c > 1$, and this means that either b or c is 1.

80. The answers are $x = \pm 498$, $y = \pm 496$ and $x = \pm 78$, $y = \pm 64$ (the signs can be chosen independently). To prove this, we rewrite the equation as follows: $(x - y)(x + y) = 2 \cdot 2 \cdot 7 \cdot 71$ (the number 71 is prime). We can temporarily assume that x and y are positive (later we can supply these numbers with arbitrary signs). We have only two representations of the number 1988 as the product of two positive integers of equal parity (see the solution to Problem 70): $1988 = 2 \cdot 994$ and $1988 = 14 \cdot 142$. Setting the factors $x - y$, $x + y$ equal to these completes the solution.

81. If $n = pq$ (where p, $q > 1$), then $1/n = 1/(n-1) - 1/n(n-1)$ and $1/n = 1/p(q-1) - 1/pq(q-1)$. If n is prime, then from the original equation $n(y-x) = xy$ and, therefore, xy is divisible by n. So, either x or y is divisible by n. It is clear that y is divisible by n since otherwise $x \geq n$ and $1/x - 1/y$ cannot be equal to $1/n$. Thus $y = kn$ and $x = kn/(k+1)$, and $k = n - 1$. Thus, there is only one representation of $1/n$: $1/n = 1/(n-1) - 1/n(n-1)$.

82. Answer: there are no integer solutions. Hint: rewrite the equation as $x^3 = (2y-1)(2y+3)$ and make use of the fact that the factors on the right-hand side are relatively prime.

83. This is the famous Pythagorean equation. Answer: all the solutions can be described as follows:

$$x = (a^2 - b^2)c\,,$$
$$y = 2abc\,,$$
$$z = (a^2 + b^2)c\,,$$

where a, b, and c are arbitrary integers. Hint: first, make x, y, and z pairwise relatively prime by dividing them by their common G.C.D.

For a complete solution, see [**90**], Chapter 17.

84. This is another famous equation, called Pell's equation after the XVII century English mathematician. Hint: first, we will look for non-negative solutions only. One of these is easy to find: it is the pair (1, 0), and we can generate all other solutions starting from this one. More precisely, if the pair (a, b) is a solution to our equation, then the pair $(3a + 4b, 2a + 3b)$ is the next solution.

For a complete solution of the problem, see [**90**], Chapter 17.

86. We know that $ka - kb$ is divisible by kn. Thus, $k(a - b) = mkn$, and we have $a - b = mn$, which proves the result.

87. Fermat's "little" theorem implies that this remainder is 1.

89. We have $300^{3000} = (300^{500})^6 \equiv 1 \pmod{7}$. Similarly, $300^{3000} \equiv 1 \pmod{11}$ and also $\pmod{13}$. Therefore, $300^{3000} - 1$ is divisible by 7, by 11, and by 13; that is, by 1001.

90. The answer is 7. Hint: use Fermat's "little" theorem.

92. Hint: prove that the given number is divisible by 31.

93. It is sufficient to write the following short chain of equalities and congruences: $(a+b)^p \equiv (a+b) = a + b \equiv a^p + b^p \pmod{p}$.

94. Hint: prove that for any integer x the congruence $x^5 \equiv x \pmod{30}$ holds true by showing that $x^5 - x$ is divisible by 2, 3, and 5.

95. a) Hint: prove that $p^q + q^p - p - q$ is divisible by p and by q.

96. Let us set $b = a^{p-2}$. Then, $ab = a^{p-1} \equiv 1 \pmod{p}$.

97. Let us split all the numbers 2 through $p - 2$ into pairs such that the product of two members of any pair is congruent to 1 modulo p (we leave to the reader the proof of the possibility of such a splitting). Thus, the product of all the numbers 2 through $p - 2$ is congruent to 1. Hence $p! \equiv 1 \cdot (p-1) = p - 1 \equiv -1 \pmod{p}$.

98. Since $(n^8+1)(n^8-1) = n^{16}-1 \equiv 0 \pmod{17}$, one of the factors must be divisible by 17.

99. a) The number $111\ldots11$ (p ones) is equal to $(10^p-1)/9$. But 10^p-1 is not divisible by p, since $10^p-1 \equiv 10-1 = 9 \pmod{p}$.

b) The number $111\ldots11$ ($p-1$ ones) is equal to $(10^{p-1}-1)/9$, and $10^{p-1}-1$ is divisible by p, since p is relatively prime to both 10 and 9.

100. Hint: use the following congruences: $10^p \equiv 10 \pmod{p}$, $10^{2p} \equiv 100 \pmod{p}$, $\ldots$, $10^{8p} \equiv 10^8 \pmod{p}$.

11. COMBINATORICS–2

7. The answer is $\binom{10}{3} = 120$. This is a straightforward corollary of the definition of the number of combinations.

8. The officer can be chosen in three ways, the 2 sergeants in $\binom{6}{2}$ ways, and the 20 privates in $\binom{60}{20}$ ways. Thus a group for the assignment can be chosen in $3\cdot\binom{6}{2}\cdot\binom{60}{20}$ ways.

9. a) Each triangle with vertices at the marked points has either one vertex on the first line and two vertices on the second, or two vertices on the first line and one on the second. There are $10\cdot\binom{11}{2}$ triangles of the first kind and $\binom{10}{2}\cdot 11$ of the second kind. Therefore the answer is $10\cdot\binom{11}{2}+11\cdot\binom{10}{2}$.

b) Answer: $\binom{10}{2}\cdot\binom{11}{2} = 2475$.

10. Add up the numbers of ways to choose exactly 0, 1, ... , and 5 words from the given set. The answer is $\binom{15}{0}+\binom{15}{1}+\binom{15}{2}+\binom{15}{3}+\binom{15}{4}+\binom{15}{5} = 4944$.

11. Let us choose three couples first. This can be done in $\binom{4}{3}$ ways. Three representatives from these couples can be chosen in 2^3 ways (either husband or wife from each of the three couples). Thus there are $\binom{4}{3}\cdot 2^3 = 32$ ways to choose a committee.

12. There are three possible cases: only Pete is on the team, only John is on the team, or neither one of them is on the team. There are $\binom{29}{10}$ different teams with Pete but without John (because ten of Pete's teammates can be chosen only from the other 29 students of the class). Similarly, there are $\binom{29}{10}$ teams with John but without Pete. And finally there are $\binom{29}{11}$ teams without either. Therefore the answer is $\binom{29}{10}+\binom{29}{10}+\binom{29}{11}$.

13. Since the order of the vowels, as well as of the consonants is known, everything is defined by the places occupied by the vowels. There are $\binom{7}{3} = 35$ ways to choose 3 places for the vowels in a word consisting of 7 letters.

14. There are $\binom{12}{5}$ teams with no boys at all, $10\cdot\binom{12}{4}$ teams with 1 boy and 4 girls, $\binom{10}{2}\cdot\binom{12}{3}$ teams with 2 boys and 3 girls, and $\binom{10}{3}\cdot\binom{12}{2}$ teams with 3 boys and 2 girls. Thus there are $\binom{12}{5}+10\cdot\binom{12}{4}+\binom{10}{2}\cdot\binom{12}{3}+\binom{10}{3}\cdot\binom{12}{2} = 23562$ different teams which satisfy the conditions of the problem.

15. First, let us choose the places for 12 white checkers on the 32 black squares of the chessboard. This can be done in $\binom{32}{12}$ ways. After the white checkers are placed, there are $\binom{20}{12}$ ways to put 12 black checkers on the 20 free black squares. Answer: $\binom{32}{12}\cdot\binom{20}{12} = 32!/12!12!8!$.

16. The solution is analogous to that of Problem 6. Answers are a) $\binom{15}{5}\cdot\binom{10}{5}\cdot\binom{5}{5}/3!$; b) $\binom{15}{5}\cdot\binom{10}{5}/2$.

17. a) Let us choose one of the four aces, then choose separately nine other cards from the 48 cards which are not aces. Since the first choice can be made in 4 ways and the second in $\binom{48}{9}$ ways, the result is $4 \cdot \binom{48}{9}$

b) There are $\binom{52}{10}$ ways to choose any 10 cards from the deck, and $\binom{48}{10}$ ways to choose 10 cards, none of which are aces. Thus there are $\binom{52}{10} - \binom{48}{10}$ ways to choose 10 cards so that there is at least one ace.

18. There are two cases, depending on the parity of the first digit of the number. In each of the cases you can calculate the number of ways by choosing the places for the odd digits. The answer is: $\binom{5}{2} \cdot 5^6 + \binom{5}{3} \cdot 4 \cdot 5^5$.

19. Hint: find all the representations of the numbers 2, 3, and 4 as the sums of several natural numbers. Do not forget that the first digit cannot be zero. The answers are: a) 10; b) $1+\binom{9}{2}+9+9+1 = 55$; c) $1+2\binom{9}{1}+\binom{9}{1}+\binom{9}{2}\cdot 3!/2!+\binom{9}{3} = 220$.

21. a) Answer: $\binom{45}{6}$.

b) Let us suppose that the results are already known. Now we have to choose exactly three numbers from the six "lucky" numbers and three numbers from 39 "unlucky" ones. Thus the result is $\binom{6}{3} \cdot \binom{39}{3} = 182780$.

22. The answer is the number of all subsets of a 10-element set; that is, $2^{10} = 1024$.

23. A way to go down the flight of stairs is simply the choice of several steps you are going to step on. Therefore, the question is equivalent to the calculation of the number of the subsets of a 7-element set, and the answer, of course, is $2^7 = 128$.

25. Let us prove that the number of subsets of the given set of objects with evenly many elements is the same as the number of subsets with oddly many objects. We begin by choosing one of the objects—say, A—which will play a special part further in the solution. Now, we will split all the subsets in pairs in such a way that each pair consists of two subsets, one of which always has evenly many elements and another which contains oddly many elements. To do this, we consider one arbitrary subset of the given set of objects and, if it contains A as its element, we remove A from it; if it does not, we add A to it. The resulting subset will be in the same pair with the original one, and the number of elements in these two subsets have different parity. It is easy to see that if the subset S generates the subset S', then this construction applied to S' gives us S. Therefore, we have the required splitting, and the proof is complete.

26. Use the result of Problem 25.

27–28. Hint: use induction on the number of Pascal's triangle numbers on the diagonal in question.

29. Apply the results of Problems 27 and 28.

30. The number of ways of going downward from the "summit" of Pascal's triangle to the number occupying the nth place in the $2n$th row equals $\binom{2n}{n}$. Each of these ways passes through exactly one of the numbers in the nth row. Since the number of such ways passing through the kth number is equal to $\binom{n}{k}\binom{n}{n-k} = \left(\binom{n}{k}\right)^2$, we can add up these numbers of ways to obtain the required total. This geometric interpretation of the algebraic equality to be proved demonstrates one of the most beautiful tricks in elementary combinatorics.

34. Let us call the 12 pennies "balls" and the five purses "boxes". Now the problem is almost the same as Problem 31. Answer: $\binom{12-1}{5-1} = \binom{11}{4} = 330$.

35. This time the 12 books are the "balls", and the three colors are the "boxes". The problem turns into a question similar to Problem 32. Answer: $\binom{12+3-1}{3-1} = \binom{14}{2} = 91$.

36. To cut the necklace into 8 parts it is necessary to choose 8 places from the 30 where the cuts are to be done. Therefore the result is $\binom{30}{8}$.

37. If candidates are "boxes" and voters are "balls", the problem is similar to Problem 32. Answer: $\binom{34}{4}$.

38. Hint: regard the postcards as "balls" and the types as "boxes". Answers: a) $\binom{21}{9}$; b) $\binom{17}{9}$.

39. a) The first passenger can get off the train at any one of n stops. The second passenger can also get off the train at any one of n stops. Thus there are $n \cdot n = n^2$ different ways to get off the train for these two passengers. Since the third passenger can choose any of n stops, there are $n \cdot n^2 = n^3$ different ways for three passengers to leave the train. It is clear now that the same argument for the remaining passengers leads to the answer n^m.

b) This is again a problem about "balls" (passengers) and "boxes" (stops). Thus the answer is $\binom{n+m-1}{n-1}$.

40. This problem is identical to the problem of representing the number 20 as the sum of three non-negative integers. Answer: $\binom{22}{2} = 231$.

41. Hint: find the answers for the black and for the white balls separately, then multiply the results. Answer: $\binom{16}{8} \cdot \binom{10}{8}$.

42. Hint: divide the process of the distribution of the fruits into four steps: apples, orange, plum, and tangerine. Answer: $\binom{8}{2} \cdot 3 \cdot 3 \cdot 3 = 756$.

43. Since there are $\binom{9}{5}$ different ways to put balls of each of three colors into six different boxes, the result is $\binom{9}{5}^3$.

44. a) Each of n voters can choose any of n candidates. Thus the result is n^n.

b) Consider the members of a community as "boxes", and the votes as the "balls". Answer: $\binom{2n-1}{n-1}$.

45. Let us temporarily repaint the red and the green balls black, and line the black balls up in a row. Arranging 10 black balls and 5 blue balls so that no two blue balls are next to each other is the same as placing 5 blue balls into 11 "buckets" between the black balls, and at the ends of the row of the black balls, so that no two blue balls are in the same buckets. That is, we just choose 5 buckets out of 11—this can be done in $\binom{11}{5}$ ways. Finally, there are $\binom{10}{5}$ ways to repaint the black balls red and green. Thus, the answer is $\binom{11}{5}\binom{10}{5} = 116424$.

46. Hint: we know that $1000000 = 2^6 \cdot 5^6$. Each factor can be completely determined by the number of 2's and 5's in its decomposition. The total number of 2's is 6, and the total number of 5's is the same. Answer: $\left[\binom{8}{2}\right]^2 = 784$.

47. Let us remove the chosen books and consider the seven remaining books. Between any two of them and at the ends of the row, we either have a gap (caused by a missing book) or we don't. The set of the gaps uniquely specifies the set of the chosen books. The answer therefore is $\binom{8}{5}$.

48. Since there are only two blue beans, the type of necklace is fully determined by the distance between these two beans as measured by the minimum number of the beans (regardless of their color) between them. This quantity can take only three values 0, 1, and 2. Hence there are only 3 different types of necklaces.

49. These are applications of the basic results of the chapter. The answers are a) $\binom{30}{4} = 27405$; b) $30 \cdot 29 \cdot 28 \cdot 27 = 657720$.

50. Let us calculate the number of all six-letter words first. Any of 26 letters can be in each of 6 places. Thus it is possible to write 26^6 different six-letter words using the 26 letters of the English alphabet. And there are 25^6 words without the letter A (in this case only 25 letters can be used). Therefore, there are $26^6 - 25^6 = 64775151$ words containing at least one letter A.

51. It is possible to start drawing the path at any of the six vertices of the hexagon. The second point can be chosen in 5 ways, and so on. Thus, there are 6! ways to draw a path. But each path was counted exactly six times during this calculation, since each of its vertices could be chosen as the first one. Therefore, there are $6!/6 = 5!$ paths.

52. a) A number divisible by 4 and written with the given digits must end with 12, 24, or 32. In each of these cases there are two different ways to use two other digits at the beginning of the number. The answer is $3 \cdot 2 = 6$.

b) In this case the number must end with 12, 24, 32, or 44. Each of 4 digits can be used in any of the two remaining places. Therefore the result is $4 \cdot 4^2 = 64$.

53. Everything is defined by specifying the three days when the father gives pears to his daughter. Since there are $\binom{5}{3}$ ways to choose three out of the five days, the answer is $\binom{5}{3}$.

54. There are $\binom{20}{6}$ ways to choose 6 actors for the first performance. In each of these cases 14 actors did not take part in the first performance. Thus there are $\binom{14}{6}$ ways to choose 6 actors for the second performance. The answer is $\binom{20}{6}\binom{14}{6}$.

55. In every decimal place each of the digits is used exactly $4^2 = 16$ times. Answer: $16 \cdot 1111 \cdot (1+2+3+4) = 17760$.

56. The 6 cards can be distributed among the four suits in two ways: $1+1+1+3$ or $1+1+2+2$. Let us calculate the number of different choices for the first variant. The suit containing 3 cards can be chosen in 4 ways. There are $\binom{13}{3}$ ways to choose 3 cards from this suit. One card from each of the remaining suits can be chosen in 13 ways. Thus, the result is $4 \cdot \binom{13}{3} \cdot 13^3$. A similar calculation leads to the result $\binom{4}{2} \cdot \binom{13}{2}^2 \cdot 13^2$ for the second variant of distribution. Therefore, the answer is $4 \cdot \binom{13}{3} \cdot 13^3 + \binom{4}{2} \cdot \binom{13}{2}^2 \cdot 13^2$.

57. Answer: $\binom{6}{3} \cdot \binom{13}{3} = 5720$. See the solution to Problem 41.

58. It is obvious that there are 10 one-digit integers satisfying the conditions of the problem. Let us calculate how many two-digit integers there are. The first digit of a two-digit integer can be any digit but 0. The second digit can be any of 9 digits which differ from the first one. Therefore there are 9^2 two-digit integers with two different digits. There are 9^3 three-digits integers which satisfy the conditions of the problem, because there are 9 digits (any digit but the one used as the second one) to choose from for the third place. Continuing in the same way, we get the final result: $10 + 9^2 + 9^3 + 9^4 + 9^5 + 9^6$.

59. Let us start with another problem:

How many ways are there to take 18 cards from a deck of 36 cards (including 4 aces) so that these 18 cards contain exactly 2 aces?

First, we choose two of the four aces. Second, we choose sixteen other cards from the 32 non-aces. Therefore, the answer to the new problem is $\binom{4}{2} \cdot \binom{32}{16}$. Let us now notice that by choosing 18 cards from the deck we have divided the deck into halves. But each possible division has been counted twice. Thus, the result is $\binom{4}{2} \cdot \binom{32}{16}/2$.

60. a) The rook can either visit or not visit each of the 28 non-border boxes. The answer, therefore, is 2^{28}.

b) The answer is the number of representations of the number 29 as the sum of seven natural numbers whose order is significant. Answer: $\binom{28}{6}$.

61. Let us suppose that none of the rowers has been chosen from those ten who wanted to be on the left side. Then four rowers for the left side have been chosen from the nine who can sit on either side. And the four rowers for the right side have been chosen from seventeen (the twelve who want to sit on the right side, and those five from the nine without preferences who were not chosen for the left side). Thus in this case there are $\binom{10}{0} \cdot \binom{9}{4} \cdot \binom{17}{4}$ ways to make a choice. Now let us suppose that exactly one rower has been chosen from those who want to sit on the left side. Then another three left-side rowers have been chosen from the nine. And four right-side rowers have been chosen from eighteen. This gives us $\binom{10}{1} \cdot \binom{9}{3} \cdot \binom{18}{4}$ choices for this case. Taking into account three final cases (two, three, or four rowers to choose from those ten who wanted to be on the left side) we get the final result: $\binom{10}{0} \cdot \binom{9}{4} \cdot \binom{17}{4} + \binom{10}{1} \cdot \binom{9}{3} \cdot \binom{18}{4} + \binom{10}{2} \cdot \binom{9}{2} \cdot \binom{19}{4} + \binom{10}{3} \cdot \binom{9}{1} \cdot \binom{20}{4} + \binom{10}{4} \cdot \binom{9}{0} \cdot \binom{21}{4}$.

62. The rectangle can be defined without ambiguity by its upper left and lower right vertices. To contain the marked box, the upper left vertex must be in a row with a number less than or equal to p and in a column with a number less than or equal to q. The lower right vertex must be in a row with a number greater than or equal to p and in a column with a number greater than or equal to q. Thus there are $p \cdot q$ different positions for the upper left vertex and there are $(m - p + 1) \cdot (n - q + 1)$ different positions for the lower right vertex. Therefore there are $p \cdot q \cdot (m - p + 1) \cdot (n - q + 1)$ rectangles containing the marked box.

63. The grasshopper has to make 27 jumps; 9 jumps in each direction. Let us denote jumps in the first direction by the letter A, jumps in the second direction by the letter B, and jumps in the third direction by the letter C. Now each route of the grasshopper can be defined without ambiguity by the 27-letter word in which each of the letters A, B, C is used exactly 9 times, and the problem is reduced to these words. Doing this in the same way as in Problems 17–21 from the chapter "Combinatorics–1" we obtain the answer: $27!/(9!)^3$.

12. INVARIANTS

4. Answer: $21! - 1$.

5. We can use the following quantity as an invariant: each sparrow is supplied with a special index equal to the number of the tree it is currently sitting on (counting from left to right). Then the sum S of these indices is the required quantity. Indeed, after the flights of any two birds only their indices change—one increases by some number x and another decreases by the same number. Thus, the sum S is invariant. Initially, the value of S is $1+2+3+4+5+6=21$, and if all the sparrows are on the same tree with number k, then the value of S is $6k$. Since 21 is not divisible by 6 we can conclude that the sparrows cannot all gather on one tree.

On the other hand, if there are seven sparrows and seven trees, then initially $S = 28$ which is divisible by 7, and we cannot exclude the possibility of all the sparrows gathering on one tree. In fact, the reader can easily construct a sequence of flights which results in the required situation: all the sparrows are together on the middle tree.

6. Hint: prove that the parity of the number of black boxes in the table is invariant.

7. Hint: prove that the parity of the number of black boxes among the four corner boxes is invariant under recolorings.

8. This problem can be solved in just the same way as Problem 7, if we consider a set of four boxes with the same property. One such set is the four boxes forming a 2×2 square in the upper left corner of the table.

9. Hint: use as the invariant the parity of the sum of all the numbers on the blackboard.

13. a) Use the following coloring of the board: the rows with the odd numbers are colored black, and the rows with the even numbers are colored white. Then the 1×4 polyminos always cover an even number of white boxes regardless of their position on the board. Furthermore, the one special polymino always covers an odd number of white boxes. These two facts together imply that the entire number of white boxes covered is odd, but this number must be 32. This contradiction completes the proof.

c) Apply the analogous coloring using four colors. Since each polymino covers either four boxes of the same color or four boxes colored with four different colors, we can conclude that the difference between the numbers of the boxes of color A and of color B is divisible by 4 (regardless of which colors A and B we choose). An easy calculation shows that there are 2652 boxes of the 1st color, 2652 boxes of the 2nd color, 2550 boxes of the 3rd color, and 2550 boxes of the 4th color. The difference between the number of the boxes of the 1st and the 3rd color is 102 which is not divisible by 4. This completes the proof.

14. Let us consider the coloring using 4 colors shown in Figure 154. Then each 2×2 tile contains exactly one box of color 1, and each 1×4 tile contains none or two boxes of color 1. Therefore, the parity of the number of the 2×2 polyminos coincides with the parity of the number of boxes of color 1. This proves the statement: after the parity of the number of 2×2 polyminoes changed (when one was lost) we cannot cover the same board without overlapping.

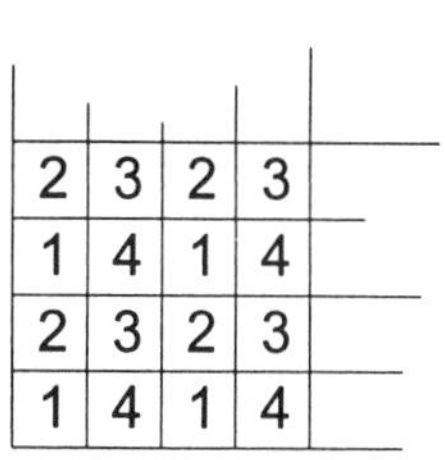

2	3	2	3
1	4	1	4
2	3	2	3
1	4	1	4

FIGURE 154

16. Let us use as the invariant the remainder of the number of heads of the Dragon when divided by 7. Using either of the swords does not change this remainder, and since 100 and 0 are not congruent modulo 7 the answer is, alas, negative.

17. The remainder when divided by 11 of the difference between the number of dallers and the number of dillers belonging to the businessman can be used as the invariant. Since initially this difference is 1, the businessman can never make this difference equal to zero.

18. Since after every operation of Dr. Gizmo's machine the number of coins increases by 4, then the remainder of the number of coins when divided by 4 is invariant. But 26 and 1 have different remainders when divided by 4. Therefore we cannot end up with 26 coins.

20. The answer is 8. Since the remainders of a natural number and of the sum of its digits when divided by 9 are the same, the remainder of 8^{1989} coincides with the remainder of the final result x. Hence, x has remainder 8 modulo 9, and we know that x is a digit. We conclude that $x = 8$.

21. Its type is B. Consider the parities of the differences $N(A) - N(B)$, $N(B) - N(C)$, and $N(C) - N(A)$, where $N(X)$ is the number of type X amoebae. These parities do not change in the course of the merging process. This means, in particular, that in the end (when there is only one amoeba in the tube) the numbers of A-amoebae and C-amoebae have the same parity, which is possible only if the only amoeba left belongs to type B.

22. After each move the sum of the numbers of the row and the column of the square the pawn is on either decreases by 2 or increases by 1. Thus, the remainder of this sum when divided by 3 increases by 1 each time. Since there are n^2-1 moves in all, and the final sum must be 1 more than the original, we get that $n^2 - 2$ must be divisible by 3. This is impossible (a perfect square cannot have a remainder 2 when divided by 3) and therefore such a route for the pawn is impossible.

Remark. We would like to draw your attention to the fact that we did not use the word "invariant" in this solution. However, some quantity is invariant. Can you find it?

23. Since the sum in each row is 1, and we have m rows, the sum of all the numbers in the table is m. On the other hand, the sum in each column is 1, and the table has n columns. Hence, the sum of the numbers is n. But the sum of the numbers in the table does not depend on the way it is calculated (in this sense, this problem concerns the idea of invariant). Therefore, $m = n$.

24. The answer is no. Hint: use the parity of the number of glasses standing upside down as the invariant.

25. Let us mark four vertices of the cube such that no two of them are connected by an edge (this is not difficult). Then consider the difference between the sum of the numbers on the marked vertices and the sum of the numbers on the other vertices. This difference is invariant under the operations described. Using this invariant, we can easily prove that the answer to both questions is negative.

26. Let us number the sectors with the numbers 1 through 6 in the clockwise direction starting at some sector. Then consider the difference between the sum of the numbers in sectors 1, 3, and 5, and the sum of the numbers in the sectors 2, 4, and 6. This quantity is invariant, and its initial value is ± 1. Thus, it cannot be equal to 0, and the answer is negative.

27. We can only obtain cards (a, b) such that $a < b$ and $b - a$ is divisible by 7.

28. The answer is no. Let us introduce the quantity S equal to the sum of the number of stones and the number of heaps. It is not hard to show that S is invariant, and its initial value is 1002. If there are k heaps with exactly 3 stones in each of them, then the value of S is $k + 3k = 4k$, which cannot be equal to 1002, since 1002 is not divisible by 4.

29. The answer is no. Hint: use the following quantity as an invariant: the parity of the number of pairs (a, b) where the number a occupies the place to the right of the number b and $a > b$.

30. The sum of the squares of the numbers in a trio does not change after any of the operations described. Using this quantity as the invariant, we can easily see that the answer is no (the values of the invariant for the given trios are different: $6 + 2\sqrt{2} \neq 13/2$).

13. GRAPHS–2

2. Since there are 4 edges leaving four of the vertices we conclude that each of these is connected with every other. But that would mean that the fifth vertex is also connected with all the other vertices; that is, its degree is also 4. This contradiction completes the proof.

3. Hint: use induction on n. The base $(n = 1)$ is easy. To prove the inductive step, consider graph G with $2n$ vertices which satisfies the condition of the problem, and add to it two other vertices A and B which, so far, are not connected to any of G's vertices. Graph G has two families V_1 and V_2 of vertices each containing n vertices with degrees equal to 1, 2, ... , n. Connect one of the new vertices—say, A—to all vertices of family V_1 and to the vertex B. The resulting graph is a graph with $2n + 2$ vertices satisfying the required condition.

4. a) Yes, since in such a graph each vertex is connected with every other; that is, the graph is isomorphic to the complete graph with 10 vertices.

b) No—see Figure 155.

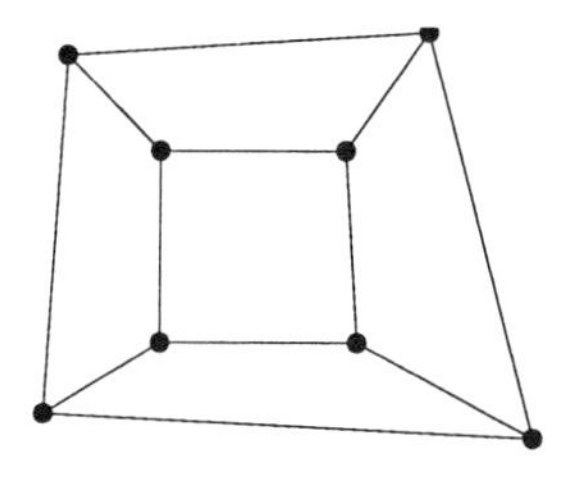

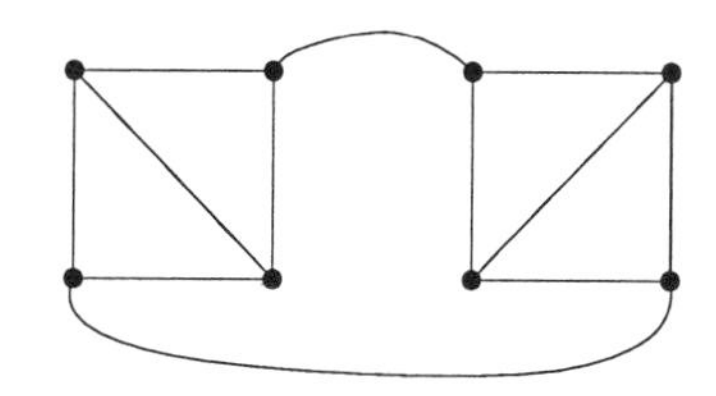

FIGURE 155

c) No—see Figure 156.

5. Assume that such a deletion is possible. If this edge connects two vertices with equal degrees, then each of the connected components would have an odd number of odd vertices, which is impossible. Otherwise, only one component will have a vertex with degree 2, and the components cannot be isomorphic.

9. Consider any of the connected components of the given graph. It is not a tree since it does not contain a pendant vertex. Therefore, it contains a cycle.

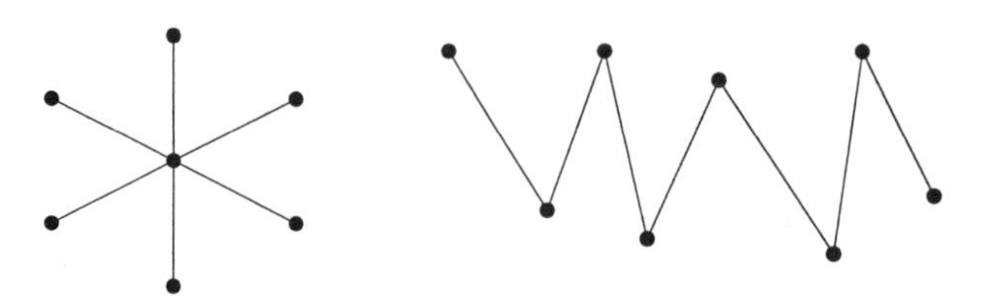

FIGURE 156

10. Suppose the ends of the deleted edge are connected by a simple path in the new graph. Then this path together with the deleted edge gives us a cycle in the old graph—a contradiction.

12. Hint: if it were not a tree, then we could obtain a tree from it by deleting some edges.

14. The total number of roads in the country is $\binom{30}{2}$. Since the number of edges in a tree with 30 vertices is 29 we get the answer: $30 \cdot 29/2 - 29 = 406$.

15. Hint: take a maximal tree of the given graph and delete any of its pendant vertices.

16. Hint: let us consider any maximal tree of the graph. Then double each edge of this tree (there are 99 of them). Though the result is not exactly a graph, Euler's theorem about drawing a graph with a pencil without lifting it from the paper applies to this "multi-graph" as well.

17. Consider the planar graph whose vertices are the lakes, whose edges are the canals, and whose faces are the islands. Since $V - E + F = 2$, $V = 7$, and $E = 10$, we have $F = 5$. However, one of the faces is the outer face, which is not an island. Answer: 4.

19. Hint: every face is bounded by at least three edges.

24. The inequality $3V - 6 \geq E$ does not hold true for this graph, and therefore this graph is not planar.

25. Assume the opposite. Then $2E \geq 6V$; that is, $E \geq 3V$, which contradicts the proved inequality.

26. Suppose both graphs are planar. Then they have no more than $(3 \cdot 11 - 6) + (3 \cdot 11 - 6) = 54$ edges together. However, the complete graph with 11 edges must have 55 edges, and we have a contradiction.

27. Hint: first, prove the inequality $E \leq 3V - 6$ using the fact that the degree of every vertex is at least 3. Denoting the number of pentagons by a and the number of hexagons by b, we have $5a + 6b + 7 = 2E \leq 6F - 12 = 6(a + b + 1) - 12$. Hence, $a \geq 13$.

29. a) No, since the graph has 12 odd vertices, which means that we need at least 6 paths to form the grid.

b) Yes, it is possible. We leave it to the reader to construct an example.

30. The graph formed by the circles (the points of intersection are its vertices, and the arcs of the circles are its edges) is connected, and all the degrees of its vertices are even.

31. The proof can be carried out using induction on n. The base $n = 0$ is obvious. To prove the inductive step we choose two odd vertices A and B and connect them mentally with a new edge. After that the new graph has only $2n-2$ odd vertices and can be drawn in such a way that the pencil will be lifted exactly $n-1$ times. When, in the process of this drawing, we must go along the mentally added edge AB (which does not really exist), we simply lift the pencil from paper and put it down at the other end of this edge.

32. Hint: find two scientists who are not acquainted, and consider all the people they are acquainted with.

33. Assuming the opposite, we have that for any number 68 through 101 there are exactly three students who have exactly so many acquaintances. Then the number of students having an odd number of acquaintances is odd, which is impossible.

34. Hint: let us connect these two vertices by a path. Then, if its length is a, the distances from any vertex to these two must differ by a number of the same parity as a.

35. Hint: prove that any tree with 6 vertices is isomorphic to one of the graphs shown in Figure 157.

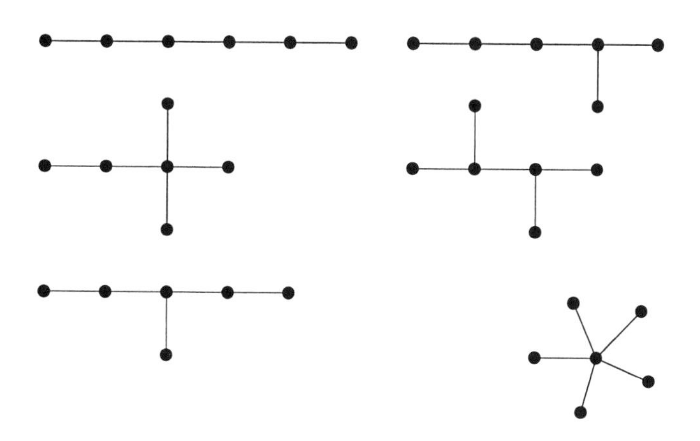

FIGURE 157

36. a), b) These items are corollaries to items c) and d).

c) Let us assume the opposite. Consider an arbitrary town X and a town A which cannot be reached from X by airplane with no more than one transfer, and a town B which cannot be reached by train with no more than one transfer. Now, notice that the towns A and B are connected by some kind of transportation. We can assume without loss of generality that A and B are connected by train. By assumption X and A are connected directly by train, and, therefore, we can reach B from X by train by transferring at A, which is a contradiction.

d) Hint: suppose we cannot fly from A to B with no more than two transfers, and we cannot go by train from C to D with no more than two transfers. Consider the graph formed by these four towns.

37. See Problem 28 from the chapter "The Pigeon Hole Principle".

38. Take an arbitrary vertex and notice that there are at least 6 edges of the same color leaving it. Now use the result of Problem 37.

39. Suppose there is a vertex with 6 blue edges leaving it. Then we can use the result of Problem 37. If there is a vertex with no more than 4 blue edges leaving it (it is impossible for all nine vertices to have five blue edges leaving them), then there are at least 4 red edges leaving it.

40. There are no less than 9 edges of the same color leaving any given vertex. Now use the result of Problem 39.

42. Denote the number of the roads entering the capital by a. Then the total number of all "incoming" roads is equal to $21 \cdot 100 + a$, and the total number of "outgoing" roads is no more than $20 \cdot 100 + (100 - a)$. Therefore $21 \cdot 100 + a \leq 20 \cdot 100 + (100 - a)$; that is, $2a \leq 0$. Finally, $a = 0$.

43. Hint: number the towns and mark each road as one-way in the direction leading from the town with the smaller number to the town with the greater number.

44. Hint: first, consider the vertices connected with the chosen vertex A, then the new vertices connected with these, et cetera. In the process of extending this "web" (as long as it is possible) we orient the edges which connect the newly added vertices with the older ones in the direction from the older to the newer endpoints.

45. Hint: consider an Euler path passing through all the edges of the graph and orient the edges according to their order in the cycle.

46. Hint: prove that there exists a closed path along the arrows which passes along each edge exactly once. This can be done by considering the closed path with the largest possible number of edges.

48. Assume that team A won the tournament. If there exists a team which defeated A as well as all the teams who lost to A, then this team would have scored more points than A, which is impossible. This proves both parts of the problem.

49. Hint: proceed by induction on the number of towns. The base (for three towns) can be proved using a simple case-by-case analysis. To prove the inductive step, temporarily remove the town which has roads both entering it and leaving it.

50. We prove this by induction on the number of teams n. The base $n = 2$ is easy. To prove the inductive step let us temporarily remove one of the teams X and number the other $n - 1$ teams as required. If X has defeated team 1 or has lost to team $n - 1$, then we can easily add X to the chain. Assuming the opposite we have that X lost to team 1 and defeated team $n - 1$. Therefore there must exist such an integer k that X lost to team k and defeated team $k + 1$—otherwise X must have lost to team 2, therefore to team 3, therefore to team 4, et cetera. Having found such a k we can insert X into the existing chain by "cutting in" between teams k and $k + 1$ and create the chain of teams we need.

51. Suppose A and B won an equal number of games, and B defeated A. Then if every team C which lost to A also lost to B, then B must have more points than A. Thus, there exists a team C such that A defeated C but C defeated B.

52. Hints: a) If we cannot reach town B from town A consider the towns where the roads leaving A lead, together with the towns where the roads entering B leave from.

b) Using the same notation as before we can assume that there is no road $A \to B$, and that there is no town C such that there are roads $A \to C$ and $C \to B$. Let us find the forty towns $A_1, A_2, \ldots, A_{40}$ that the roads leaving A lead to, and forty other (!) towns $B_1, B_2, \ldots, B_{40}$ that the roads coming to B come from.

There are 1600 roads which leave the towns A_i. At the same time the total number of roads connecting A_i with each other does not exceed $40 \cdot 39/2 = 780$, and the number of the roads leaving from them to the remaining 19 towns is no more than $40 \cdot 19 = 760$. Since $1600 > 1540 = 780 + 760$ it follows that there must be a road from A_i to B_j.

53. Hint: if we have removed the edge between vertices A and B, then let us choose two arbitrary vertices and consider three cases: none of these vertices coincides with A or B; one of them is either A or B; or in fact they are A and B.

14. GEOMETRY

1. Hint: try to find the possible length of the third side in a triangle whose two sides have lengths a and b.

2. Hint: prove the inequalities $AM > AB - BC/2$ and $AM > AC - BC/2$.

3. Consider the circle inscribed in the triangle and the lengths of the resulting segments when the meeting points divide the sides of the triangle. We have three pairs of equal segments, whose lengths are the required x, y, and z.

4. Let us assume the opposite. Then one of the angles is larger than the other one, and the corresponding side must be longer than the other. This contradiction completes the proof.

5. Hint: use the inequalities $\angle BAM < \angle ABM$ and $\angle CAM < \angle ACM$.

6. This is a simple exercise in the use of Inequality № 1. If $a + b > c$, then $a + 2\sqrt{ab} + b > c$, or $(\sqrt{a} + \sqrt{b})^2 > (\sqrt{c})^2$, and $\sqrt{a} + \sqrt{b} > \sqrt{c}$.

7. Since $AB + CD < AC + BD$ (by the way, why?) we can obtain the required result by adding this inequality to the inequality given in the statement of the problem.

9. Since $\angle A > \angle A_1$ we have $BD > B_1D_1$, and hence $\angle C > \angle C_1$. If $\angle B > \angle B_1$, then similarly $\angle D > \angle D_1$ which is impossible since the sum of the angles' measures in the quadrilaterals must be the same.

10. Hint: construct parallelogram $ABCD$, three vertices of which coincide with the vertices of triangle ABC, by extending the median its own length to D. Then apply Inequality № 2.

11. Answer: No. It would follow from Inequality № 2 that $\angle BAC > \angle BCA = \angle DCE > \angle DEC = \ldots > \angle KAI = \angle BAC$—which is a contradiction!

12. a) Hint: place three copies of the triangle on the plane so that their legs coincide. Then look for an equilateral triangle.

b) Find point E on side AB such that $AE = AC$. Then prove that $EB > CE > AC$.

13. Hint: if we denote the outer perimeter by a, the perimeter of the star by b, and the inner perimeter by c, then $a > c$, $a + c < b$, and $2a > b$. Now a case-by-case analysis completes the proof.

14. "Fold out" the perimeter of the quadrilateral as shown in Figure 158.

15. In three steps move the second triangle so that all three of its vertices coincide with the vertices of the first triangle (one vertex at a time).

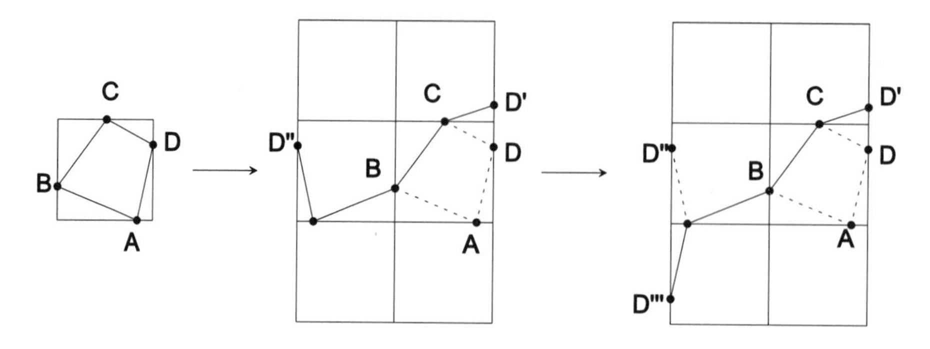

FIGURE 158

16. a) Prove that any point D will be fixed.

b) Apply the result of a).

17. a) Answer: this is a translation again.

b) Take two parallel lines which are perpendicular to the direction of the translation and such that the distance between them equals half the length of the translation.

c) This motion cannot be a rotation since there is no point which is left in place. It cannot be a translation since the distance between a point and its image is not constant. And, finally, it is not a line reflection: if it were, then for any point A and its image A' the perpendicular bisector of segment AA' would be some constant line not dependent on the choice of point A.

18. The answer is yes. It suffices to map one of the centers onto the other.

19. Only the identity rotation can map a half-plane onto itself: otherwise where would the boundary go? A line reflection can do it if the line is perpendicular to the boundary of the half-plane.

20. Yes, this is true. Indeed, the composition of eight such rotations is a rotation of 24 degrees about the same point. Further, the composition of three such superpositions will be a rotation of 72 degrees about point O.

21. Hint: reflect the triangle in the given point. Where the triangle and its image coincide are the endpoints of the required segment.

22. Hint: use a translation.

24. Hint: since the lines (AB), (CD), and (MN) meet at one point, the reflection in line (MN) maps lines (AB) and (CD) onto each other.

25. Hint: use a rotation of 90 degrees which maps the square onto itself.

26. Hint: use a rotation of 60 degrees about point P, and look at the point where the first line and the image of the second line meet.

27. a) Hint: if the two given points X and Y are on different sides of line L, then $M = (XY) \cap L$; else $M = (XY_1) \cap L$, where Y_1 is symmetric to Y with respect to L.

b) Hint: if X and Y lie on different sides of L, then $M = (XY) \cap L$; otherwise, $M = (XY_1) \cap L$, where Y_1 is symmetrical to Y with respect to L.

28. Hint: reflect the first axis in the second. Prove that the image must be an axis of symmetry for the triangle. Also, show that the two original lines cannot be perpendicular.

29. a) A, B, C, D, E, H, I, K, M, O, T, U, V, W, X, Y.

b) H, I, N, O, S, X, Z.

The answers depend on how you write the letters. Here we use the standard roman typeface.

30. The answer is no. See the hint to Problem 28.

31. Hint: these are the points from which segment OS (O is the center of the rotation) subtends an angle of $90° + \alpha/2$ or $90° - \alpha/2$ (α is the angle of rotation). The reader can try to describe this locus more precisely.

34. No. Hint: if two angle bisectors were perpendicular, then the sum of these two angles of the triangle would be $180°$.

35. Let line L be the common perpendicular to lines AB and CD, which passes through the center of the circle. Then segments AC and BD are symmetric with respect to line L.

36. Hint: the sum of two opposite angles in an inscribed quadrilateral must be 180 degrees. Answer: 60, 90, and 120 degrees.

38. It is not difficult to see that the measure of angle AOD is $60°$. Further, triangle DOC is isosceles, so $\angle DOC = 75°$. Therefore, $\angle AOC = 135°$.

39. Angles ABC and ABD are equal to $90°$. Thus $\angle CBD = 180°$.

40. a) Let $a = |AB|$, $b = |BC|$, $c = |CD|$, and $d = |DA|$. Then b is not greater than the altitude to AB in triangle ABC. Hence $S(ABC) \leq ab/2$, and $S(CDA) \leq cd/2$. Adding these inequalities, we are done.

b) Use a) and the fact that quadrilateral $ABCD$ can be turned into a quadrilateral with the same area but with the sides in a different order: a, b, d, and c (just cut it along diagonal AC and "turn over" one of the halves).

42. Since $bc/2 \geq 1$ we get $b^2 \geq 2$.

43. Yes, this is possible. Consider triangle ABC, where $AC = 2002 + \varepsilon$, and $AB = BC = 1001$ (where ε is some sufficiently small positive number; for example, $\varepsilon = 0.1$).

44. Cut $ABCD$ along diagonal AC and prove this equality for each of the halves separately. Do not forget that the sides of $KLMN$ are parallel to the diagonals of $ABCD$.

45. It is not difficult to show that the area of a quadrilateral whose diagonals are perpendicular is half the product of the diagonals (the result for a rhombus, given in many regular texts, is a special case). Here, $|ABCD| = 12$.

46. Answer: 7. Hint: prove that the area of each of the three additional triangles is 2.

47. The equality of the areas is equivalent to the equality of the altitudes dropped to BM from A and C respectively. This, in turn, is equivalent to the assumption that BM bisects AC.

48. Use the result of Problem 44.

49. Hint: prove that triangles ABD and ACD have the same area.

50. If we are given point O inside equilateral triangle ABC, then we can calculate the area of ABC as the sum of the areas of triangles OAB, OBC, and OAC. These areas can be found using the perpendiculars dropped from O to the sides of the triangle.

55. No, he is wrong. For example, we can prove that there are points on the plane satisfying the given property and lying arbitrarily far from the given point (in fact, the given set is a parabola).

56. The reason why the proof is wrong is that the figure is wrong, and point M lies outside the triangle.

57. Since P is equidistant from A and B as well as from C and D we have that triangles PAD and PBC are congruent. Thus, medians PM and PN in these triangles are equal as well.

58. One way to do this is to prove that this perpendicular bisector divides the greater leg of the triangle into two segments such that the length of one of them equals the length of the specified part of the bisector, and the other one is twice as long. Or, the lengths can be calculated directly, in terms of one of the sides of the original triangle.

59. Hint: make use of the fact that each of the segments of this broken line is a median to the hypotenuse in some right triangle and therefore equals half of this hypotenuse.

15. NUMBER BASES

Answers to the exercises

1. a) 2; b) n.

2. $10101_2 = 21$, $10101_3 = 91$, $211_4 = 37$, $126_7 = 69$, $158_{11} = 184$.

3. $100_{10} = 1100100_2 = 10201_3 = 1210_4 = 400_5 = 244_6 = 202_7 = 144_8 = 121_9$.

4. $111_{10} = A1_{11}$.

5. Here is the multiplication table in the base 5 number system:

$$\begin{array}{c} \\ 0 \\ 1 \\ 2 \\ 3 \\ 4 \end{array}\begin{array}{c} \begin{array}{ccccc} 0 & 1 & 2 & 3 & 4 \end{array} \\ \begin{pmatrix} 0 & 0 & 0 & 0 & 0 \\ 0 & 1 & 2 & 3 & 4 \\ 0 & 2 & 4 & 1 & 3 \\ 0 & 3 & 1 & 4 & 2 \\ 0 & 4 & 3 & 2 & 1 \end{pmatrix} \end{array}$$

6. a) 11001_2; b) 21202_3.

7. a) 2626_7; b) 1003_7.

Problems

1. Answer: in the base 12 system (duodecimal system). Hint: digits 3 and 4 always represent numbers 3_{10} and 4_{10}, and their product equals 12_{10}.

2. a) Yes, such a system exists. This is the base 7 system. See the hint to the previous problem.

b) Answer: No. This equality could be true only in the base 5 number system, but there is no digit 5 in this system.

3. A number is even if and only if

a) there is an even number of 1's in its base 3 representation (that is, the sum of its digit is even). Indeed, a number equals the sum of powers of 3 multiplied by the digits, which can be 0, 1, or 2. The summands with digits 0 and 2 are even, and therefore the parity of the sum depends on the number of the summands with digits 1.

b) for even n its base n representation ends with an even digit; for odd n the sum of its digits is even. The proof for the latter case is similar to the proof for part a). In case of even n, when a number is represented as a sum of powers of n multiplied by its digits, all summands starting from the second are even since they are divisible by n. Therefore, the parity of the sum is determined by the parity of the units digit.

4. The answer is $23451 + 15642 = 42423$ (the base 7 number system).

5. Let n be the base of the system. Then $n^2 = (2n+4) + (3n+2)$; that is, $n^2 - 5n - 6 = 0$. Therefore $n = -1$ or $n = 6$. Answer: $n = 6$.

6. a) In the base n number system the representation of a number ends with k zeros if and only if this number is divisible by n^k.

b) Let m be some divisor of n. The last digit of the base n representation of a number is divisible by m if and only if the number itself is divisible by m.

7. a) Let m be a divisor of $n-1$. Then the sum of the digits in the base n representation of a number is divisible by m if and only if the number itself is divisible by m.

b) The "alternating" sum (with alternating signs) of the digits in the base n representation of a number is divisible by $n+1$ if and only if the number itself is divisible by $n+1$.

c) Let m be some divisor of $n+1$. The alternating sum (with alternating signs) of the digits in the base n representation of a number is divisible by m if and only if the number itself is divisible by m.

12. Hint: the subset is the same as in Problem 11.

13. a) This is the same game of Nim, with eight heaps instead of three. The strategy and the proof are exactly the same. However, there is another, much simpler proof which shows that the second player wins. Indeed, all the second player must do is to maintain line symmetry on the board (with respect to the line separating the fourth and the fifth columns).

b) The second player cannot lose. The reason is the same as before. Actually, this game has nothing to do with the game of Nim—it is just a joke. In fact, this game can last forever, but this is not important.

16. INEQUALITIES

2. a) We have $3^2 = 9 > 8 = 2^3$, and therefore $3^{200} > 2^{300}$.

b) We have $2^{10} = 1024 < 2187 = 3^7$, and therefore $2^{40} < 3^{28}$.

c) The number 4^{53} is greater.

4. Answer: $8^{91} > 7^{92}$.

6. Answer: $|2^3 - 3^2| = 1$, $|3^3 - 5^2| = 2$, $|6^2 - 2^5| = 4$, $|3^3 - 2^5| = 5$, $|2^7 - 5^3| = 3$, $|11^2 - 5^3| = 4$, and $|11^2 - 2^7| = 7$.

8. Let us denote the number in the numerator of either fraction by x. Then the fraction itself is $a = x/(10x - 9)$, so that $1/a = 10 - 9/x$. This implies that as x increases, a decreases. Thus the first fraction is greater.

9. Let us answer the more general question: when is x/y greater than $(x+1)/(y+1)$? If x and y are positive, then

$$\frac{x}{y} - \frac{x+1}{y+1} = \frac{x-y}{y(y+1)} .$$

Hence everything depends on whether or not x is greater than y. In our case $y > x$, which means that $1234568/7654322$ is the greater of the two given numbers.

10. The number 100^{100} is greater, since $100^2 > 150 \cdot 50$.

11. Answer: $(1.01)^{1000} > 1000$. Indeed, $(1.01)^8 > 1.08$; $(1.01)^{1000} = ((1.01)^8)^{125} > (1.08)^{125}$. Furthermore, $(1.08)^5 > 1.4$; $(1.01)^{1000} > (1.4)^{25} > (1.4)^{24} > (2.7)^8 > 7^4 = 2401 > 1000$.

13. Since $99! > 100$, it is clear that $A < B$.

17. We can write $1 + x - 2\sqrt{x} = (\sqrt{x} - 1)^2 \geq 0$.

19, 20. Carrying everything over to one side, we can reduce the given inequality to $(x-y)^2 \geq 0$.

21. Carrying everything over to one side and multiplying by the denominator we have $(x-y)^2 \geq 0$.

23. Multiply the following three inequalities: $a + b \geq 2\sqrt{ab}$, $b + c \geq 2\sqrt{bc}$, $c + a \geq 2\sqrt{ca}$.

24. Hint: use the fact that $(\sqrt{ab} - \sqrt{ac})^2 + (\sqrt{ac} - \sqrt{bc})^2 + (\sqrt{bc} - \sqrt{ab})^2 \geq 0$.

25. We have $x^2 + y^2 + 1 - xy - x - y = ((x-y)^2 + (x-1)^2 + (y-1)^2)/2 \geq 0$.

27. We have $x^4 + y^4 + 8 = x^4 + y^4 + 4 + 4 \geq 4\sqrt[4]{x^4y^4 \cdot 4 \cdot 4} = 8xy$.

28. We can write $a+b+c+d \geq 4\sqrt[4]{abcd}$; $1/a + 1/b + 1/c + 1/d \geq 4\sqrt[4]{1/abcd}$. Now we just multiply these inequalities.

29. $a/b + b/c + c/a \geq 3\sqrt[3]{(a/b) \cdot (b/c) \cdot (c/a)} = 3$.

42. Hint: the inequality can be proved by adding up two simpler inequalities:

$$(2^k - 1)(2^l - 1)(2^m - 1) > 0,$$
$$2^{k+l+m} > 2^k + 2^l + 2^m ,$$

since $2^{k+l+m} > 2^{k+2} = 4 \cdot 2^k > 2^k + 2^l + 2^m$ (if $k \geq l \geq m$).

43. We have $ab + bc + ca = ((a+b+c)^2 - a^2 - b^2 - c^2)/2 = -(a^2 + b^2 + c^2)/2 \leq 0$.

45. Carrying all the terms over to one side we get $(x-y)(\sqrt{x} - \sqrt{y})/\sqrt{xy} \geq 0$.

47. Here is the main idea: if the permutation (c_i) is not the identity, then there exist indices i and j such that $c_i > c_j$ and $i < j$. Then by switching c_i and c_j we can increase the sum of the products. Indeed,

$$c_i a_i + c_j a_j - c_j a_i - c_i a_j = (a_i - a_j)(c_i - c_j) < 0 .$$

Thus, using these transpositions, we can make the permutation (c_i) into the identity permutation without decreasing the sum of the products during this process.

53. The base is easy. The proof of the inductive step goes as follows: $1 + 1/\sqrt{2} + \ldots + 1/\sqrt{n-1} + 1/\sqrt{n} < 2\sqrt{n-1} + 1/\sqrt{n} < 2\sqrt{n}$ since $1/\sqrt{n} < 2(\sqrt{n} - \sqrt{n-1}) = 2/(\sqrt{n} + \sqrt{n-1})$.

54. The solution is similar to the previous one (just change the direction of the inequalities).

58. The base $n = 4$ can be checked "manually". The inductive step: $(n+1)! = (n+1)n! > 2^n(n+1) > 2 \cdot 2^n = 2^{n+1}$.

59. The base $n = 1$ is easy. The inductive step: $2^{n+1} = 2{\cdot}2^n > 2{\cdot}2n = 4n > 2(n+1)$ (if $n > 1$).

60. Answer: the inequality holds true for $n \geq 10$. Hint: you can check that it is true directly for $1 \leq n \leq 10$. To prove the inductive step, show that while 2^{n+1} is twice as large as 2^n, $(n+1)^3$ is less than $2n^3$.

APPENDIX C

References

1. General texts

1. S.Barr, *Rossypi golovolomok*, Mir, 1978. (Russian)
2. G.Bizám, Y. Herczeg, *Igra i logika*, Mir, 1975. (Russian)

*3. G.Bizám, Y.Herczeg, *Mnogotsvetnaya logika*, Mir, 1978. (Russian)

4. N.Ya.Vilenkin, *Rasskazy o mnozhestvakh*, Nauka, 1969. (Russian)

*5. M.Gardner, *Mathematical puzzles and diversions*, Bell and Sons, London.

*6. M.Gardner, *New mathematical diversions from Scientific American*, Simon & Shuster, New York, 1966.

*7. M.Gardner, *Matematicheskie novelly*, Mir, 1974. (Russian)

8. M.Gardner, *aha! Gotcha*, San Francisco, W.H.Freeman, 1982.
9. M.Gardner, *A nu-ka, dogadaisya!*, Mir, 1984. (Russian)
10. M.Gardner, *The unexpected hanging and other mathematical diversions*, Simon and Shuster, 1969.
11. M.Gardner, *Time travel and other mathematical bewilderments*, San Francisco, W.H.Freeman, 1988.
12. E.G.Dynkin, V.A.Uspenskii, *Matematicheskie besedy*, GITTL, 1952. (Russian)
13. B.A.Kordemskii, *Matematicheskaya smekalka*, GITTL, 1958. (Russian)
14. H.Lindgren, *Geometric dissections*, Princeton, Van Nostrand, 1964.
15. H.Rademacher, O.Toeplitz, *Von Zahlen und der Figuren*, Princeton University Press, 1957. (English)
16. R.Smullyan, *What is the name of this book?*, Prentice-Hall, 1978.
17. R.Smullyan, *The lady or the tiger*, New York, Knopf, 1982.
18. R.Smullyan, *Alice in the Puzzle-land*, New York, Morrow, 1982.
19. V.L.Ufnarovskii, *Matematicheskiy akvarium*, Shtiintsa, 1987. (Russian)

2. For teachers

*20. V.A.Gusev, A.I.Orlov, A.L.Rozental', *Vneklassnaya rabota po matematike v 6-8 klassakh*, Prosveshchenie, 1977, 1984. (Russian)

21. Various authors, *Matematicheskiy kruzhok. Pervyi god obucheniya, 5-6 klassy*, Izd-vo APN SSSR, 1990, 1991. (Russian)
22. G.Polya, *How to solve it; a new aspect of mathematical method*, Princeton University Press, 1945.
23. G.Polya, *Mathematical discovery: on understanding, learning, and teaching problem solving*, New York, Wiley, 1981.

*24. G.Polya, *Mathematics and plausible reasoning*, Princeton University Press, 1968.

25. K.P.Sikorskii, *Dopolnitel'nye glavy po kursu matematiki 7-8 klassov dlya fakul'tativnykh zanyatiy*, Prosveshchenie, 1969. (Russian)

3. For younger students

26. S.Bobrov, *Volshebnyi dvurog*, Detskaya literatura, 1967. (Russian)
27. V.Levshin, *Tri dnya v Karlikanii. Skazka da ne skazka*, Detskaya literatura, 1964. (Russian)
28. V.Levshin, E.Aleksandrova., *Chernaya maska iz Al'-Dzhebry. Puteshestvie v pis'makh s prologom*, Detskaya literatura, 1965. (Russian)
29. V.Levshin, *Fregat kapitana Edinitsy*, Detskaya literatura, 1968. (Russian)
30. V.Levshin, *Magistr Rasseyannykh Nauk: matematicheskaya trilogiya*, Detskaya literatura, 1987. (Russian)

4. Problem books

*31. I.L.Babinskaya, *Zadachi matematicheskikh olimpiad*, Nauka, 1975. (Russian)
*32. D.Yu.Burago, S.M.Finashin, D.V.Fomin, *Fakul'tativnyi kurs matematiki dlya 6-7 klassov v zadachakh*, Izd-vo LGU, 1985. (Russian)
*33. N.B.Vasil'ev, V.L.Gutenmakher, Zh.M.Rabbot, A.L.Toom, *Zaochnye matematicheskie olimpiady*, Nauka, 1986. (Russian)
34. N.B.Vasil'ev, S.A.Molchanov, A.L.Rozental', A.P.Savin, *Matematicheskie sorevnovaniya (geometriya)*, Nauka, 1974. (Russian)
35. G.A.Gal'perin, A.K.Tolpygo, *Moskovskie matematicheskie olimpiady*, Prosveshchenie, 1986. (Russian)
36. P.Yu.Germanovich, *Sbornik zadach po matematike na soobrazitel'nost'*, Uchpedgiz, 1960. (Russian)
37. E.B.Dynkin, S.A.Molchanov, A.L.Rozental', *Matematicheskie sorevnovaniya. Arifmetika i algebra*, Nauka, 1970. (Russian)
38. E.B.Dynkin, S.A.Molchanov, A.L.Rozental', A.K.Tolpygo, *Matematicheskie zadachi*, Nauka, 1971. (Russian)
39. G.I.Zubelevich, *Sbornik zadach Moskovskikh matematicheskikh olimpiad (V–VIII klassy)*, Prosveshchenie, 1971. (Russian)
*40. A.A.Leman, *Sbornik zadach Moskovskikh matematicheskikh olimpiad*, Prosveshchenie, 1965. (Russian)
41. A.I.Ostrovskii, *75 zadach po yelementarnoy matematike – prostykh, no ...*, Prosveshchenie, 1966. (Russian)
42. V.V.Prasolov, *Zadachi po planimetrii. Chasti 1, 2*, Nauka, 1986. (Russian)
43. I.S.Rubanov, V.Ya.Gershkovich, I.E.Molochnikov, *Metodicheskie materialy dlya vneklassnoy raboty so shkol'nikami po matematike*, Leningradskiy dvorets pionerov, 1973. (Russian)
44. I.N.Sergeev, S.N.Olekhnik, S.B.Gashkov, *Primeni matematiku*, Nauka, 1989. (Russian)
*45. D.O.Shklyarskii, N.N.Chentsov, I.M.Yaglom, *Izbrannye zadachi i teoremy yelementarnoy matematiki. Arifmetika i algebra*, Nauka, 1965. (Russian)
46. H.Steinhaus, *One hundred problems in elementary mathematics*, New York, Basic Books, 1964.

5. For "Combinatorics"

*47. N.Ya.Vilenkin, *Kombinatorika*, Nauka, 1964. (Russian)
48. N.Ya.Vilenkin, Kvant **1** (1971). (Russian)
*49. N.Ya.Vilenkin, *Populyarnaya kombinatorika*, Nauka, 1975. (Russian)
50. I.I.Ezhov et al., *Yelementy kombinatoriki*, Nauka, 1977. (Russian)

51. V.A.Uspenskii, *Treugol'nik Paskalya ("Populyarnye lektsii po matematike", 43)*, Nauka. (Russian)

6. For "Divisibility"

52. M.I.Bashmakov, *Nravitsya li vam vozit'sya s tselymi chislami?*, Kvant **3** (1971). (Russian)
53. V.N.Vaguten, *Algoritm Evklida i osnovnaya teorema arifmetiki*, Kvant **6** (1972). (Russian)
54. N.N.Vorob'ev, *Priznaki delimosti ("Populyarnye lektsii po matematike", 39)*, Nauka, 1963. (Russian)
55. A.O.Gel'fond, *Reshenie uravneniy v tselykh chislakh ("Populyarnye lektsii po matematike", 8)*, Nauka, 1983. (Russian)
*56. A.A.Egorov, *Sravneniya po modulyu i arifmetika ostatkov*, Kvant **5** (1970). (Russian)
*57. L.A.Kaluzhnin, *Osnovnaya teorema arifmetiki ("Populyarnye lektsii po matematike", 47)*, Nauka, 1969. (Russian)
*58. O.Ore, *Invitation to number theory*, New York, Random House, 1967.

See also [**25**].

7. For " The Pigeon Hole Principle"

59. V.G.Boltyanskii, *Shest' zaitsev v pyati kletkakh*, Kvant **2** (1977). (Russian)
60. A.I.Orlov, *Printsip Dirikhle*, Kvant **3** (1971). (Russian)

See also [**19, 20**]; [**42**], chapter 20; [**43**].

8. For "Graphs"

61. L.Yu.Berezina, *Grafy i ikh primenenie*, Prosveshchenie, 1979. (Russian)
*62. O.Ore, *Graphs and their use*, New York, Random House, 1963.
63. R.Wilson, *Introduction to graph theory*, New York, Academic Press, 1972.

See also [**12, 20**].

9. For "Geometry"

64. J.Hadamard, *Leçons de geometrie elementaire*, Paris, Armand Colin et Cie., 1898–1901. (French)
*65. V.A.Gusev et al., *Sbornik zadach po geometrii dlya 6–8 klassov*, Prosveshchenie, 1975. (Russian)
66. P.R.Kantor, Zh.M.Rabbot, *Ploshchadi mnogougol'nikov.*, Kvant **2** (1972). (Russian)
67. H.S.M.Coxeter, *Introduction to geometry*, New York, Wiley, 1961.
68. H.S.M.Coxeter, S.L.Greitzer, *Geometry revisited*, New York, Random House, 1967.
69. D.Pedoe, *Geometry and visual arts*, New York, Dover Publications, 1983.
70. D.O.Shklyarskii, N.N.Chentsov, I.M.Yaglom, *Izbrannye zadachi i teoremy planimetrii*, Nauka, 1967. (Russian)
71. I.M.Yaglom, *Geometricheskie preobrazovaniya, t.1.*, Gostekhizdat, 1955. (Russian)

See also [**34, 42, 44**].

10. For "Games"

72. E.Ya.Gik, *Zanimatel'nye matematicheskie igry*, Znanie, 1987. (Russian)
73. V.N.Kasatkin, L.I.Vladykina, *Algoritmy i igry*, Radyan'ska shkola, 1984. (Russian)
*74. A.I.Orlov, *Stav' na minus*, Kvant **3** (1977). (Russian)

See also [**20**]; [**5**], chapters 8, 14, 39; [**6**], chapters 5, 30, 33; [**7**], chapters 7, 18, 23, 26, 33, 36.

11. For "Induction"

75. N.N.Vorob'ev, *Chisla Fibonachchi ("Populyarnye lektsii po matematike", 6)*, Nauka, 1978. (Russian)
76. L.I.Golovina, I.M.Yaglom, *Induktsiya v geometrii ("Populyarnye lektsii po matematike", 21)*, GITTL, 1956. (Russian)
77. A.I.Markushevich, *Vozvratnye posledovatel'nosti ("Populyarnye lektsii po matematike", 1)*, Nauka, 1975. (Russian)
*78. I.S.Sominskii, *Metod matematicheskoy induktsii ("Populyarnye lektsii po matematike", 3)*, Nauka, 1965, 1974. (Russian)
*79. I.S.Sominskii, L.I.Golovina, I.M.Yaglom, *O matematicheskoy induktsii*, Nauka, 1967. (Russian)

See also [**19**]; [**42**], chapter 27.

12. For "Invariants"

*80. Yu.I.Ionin, L.D.Kurlyandchik, *Poisk invarianta*, Kvant **2** (1976). (Russian)
81. A.K.Tolpygo, *Invarianty*, Kvant **12** (1976). (Russian)

See also [**19**]; [**42**], chapter 22.

13. For "Number bases"

*82. S.V.Fomin, *Sistemy schisleniya ("Populyarnye lektsii po matematike", 40)*, Nauka, 1968. (Russian)
83. I.M.Yaglom, *Sistemy schisleniya*, Kvant **6** (1970). (Russian)
84. I.M.Yaglom, *Dve igry so spichkami*, Kvant **2** (1971). (Russian)

See also [**5**], chapter 35; [**10**], chapter 14; [**25, 58**].

14. For "Inequalities"

*85. E.Beckenbach, R.Bellman, *An introduction to inequalities*, Berlin, Springer-Verlag, 1961.
86. E.Beckenbach, R.Bellman, *Inequalities*, New York, Random House, 1961.
*87. P.P.Korovkin, *Vvedenie v neravenstva ("Populyarnye lektsii po matematike", 5)*, Nauka, 1983. (Russian)
88. V.A.Krechmar, *Zadachi po algebre*, Nauka, 1964. (Russian)
89. G.L.Nevyazhskii, *Neravenstva*, Kvant **12** (1985). (Russian)

Added for the American edition

90. K.Ireland, M.Rosen, *A classical introduction to modern number theory*, Springer-Verlag, 1982.